Oct 89

PROGRESS IN CLINICAL AND BIOLOGICAL RESEARCH

Series Editors

Nathan Back
George J. Brewer

Vincent P. Eijsvoogel
Robert Grover

Kurt Hirschhorn
Seymour S. Kety

Sidney Udenfriend
Jonathan W. Uhr

RECENT TITLES

Vol 240: **Pathophysiological Aspects of Sickle Cell Vaso-Occlusion,** Ronald L. Nagel, *Editor*

Vol 241: **Genetics and Alcoholism,** H. Werner Goedde, Dharam P. Agarwal, *Editors*

Vol 242: **Prostaglandins in Clinical Research,** Helmut Sinzinger, Karsten Schrör, *Editors*

Vol 243: **Prostate Cancer,** Gerald P. Murphy, Saad Khoury, Réne Küss, Christian Chatelain, Louis Denis, *Editors*. Published in two volumes: Part A: *Research, Endocrine Treatment, and Histopathology*. Part B: *Imaging Techniques, Radiotherapy, Chemotherapy, and Management Issues*

Vol 244: **Cellular Immunotherapy of Cancer,** Robert L. Truitt, Robert P. Gale, Mortimer M. Bortin, *Editors*

Vol 245: **Regulation and Contraction of Smooth Muscle,** Marion J. Siegman, Andrew P. Somlyo, Newman L. Stephens, *Editors*

Vol 246: **Oncology and Immunology of Down Syndrome,** Ernest E. McCoy, Charles J. Epstein, *Editors*

Vol 247: **Degenerative Retinal Disorders: Clinical and Laboratory Investigations,** Joe G. Hollyfield, Robert E. Anderson, Matthew M. LaVail, *Editors*

Vol 248: **Advances in Cancer Control: The War on Cancer—15 Years of Progress,** Paul F. Engstrom, Lee E. Mortenson, Paul N. Anderson, *Editors*

Vol 249: **Mechanisms of Signal Transduction by Hormones and Growth Factors,** Myles C. Cabot, Wallace L. McKeehan, *Editors*

Vol 250: **Kawasaki Disease,** Stanford T. Shulman, *Editor*

Vol 251: **Developmental Control of Globin Gene Expression,** George Stamatoyannopoulos, Arthur W. Nienhuis, *Editors*

Vol 252: **Cellular Calcium and Phosphate Transport in Health and Disease,** Felix Bronner, Meinrad Peterlik, *Editors*

Vol 253: **Model Systems in Neurotoxicology: Alternative Approaches to Animal Testing,** Abraham Shahar, Alan M. Goldberg, *Editors*

Vol 254: **Genetics and Epithelial Cell Dysfunction in Cystic Fibrosis,** John R. Riordan, Manuel Buchwald, *Editors*

Vol 255: **Recent Aspects of Diagnosis and Treatment of Lipoprotein Disorders: Impact on Prevention of Atherosclerotic Diseases,** Kurt Widhalm, Herbert K. Naito, *Editors*

Vol 256: **Advances in Pigment Cell Research,** Joseph T. Bagnara, *Editor*

Vol 257: **Electromagnetic Fields and Neurobehavioral Function,** Mary Ellen O'Connor, Richard H. Lovely, *Editors*

Vol 258: **Membrane Biophysics III: Biological Transport,** Mumtaz A. Dinno, William McD. Armstrong, *Editors*

Vol 259: **Nutrition, Growth, and Cancer,** George P. Tryfiates, Kedar N. Prasad, *Editors*

Vol 260: **EORTC Genitourinary Group Monograph 4: Management of Advanced Cancer of Prostate and Bladder,** Philip H. Smith, Michele Pavone-Macaluso, *Editors*

Vol 261: **Nicotine Replacement: A Critical Evaluation,** Ovide F. Pomerleau, Cynthia S. Pomerleau, *Editors*

Vol 262: **Hormones, Cell Biology, and Cancer: Perspectives and Potentials,** W. David Hankins, David Puett, *Editors*

Vol 263: **Mechanisms in Asthma: Pharmacology, Physiology, and Management,** Carol L. Armour, Judith L. Black, *Editors*

Vol 264: **Perspectives in Shock Research,** Robert F. Bond, *Editor*

Vol 265: **Pathogenesis and New Approaches to the Study of Noninsulin-Dependent Diabetes Mellitus,** Albert Y. Chang, Arthur R. Diani, *Editors*

Vol 266: **Growth Factors and Other Aspects of Wound Healing: Biological and Clinical Implications,** Adrian Barbul, Eli Pines, Michael Caldwell, Thomas K. Hunt, *Editors*

Vol 267: **Meiotic Inhibition: Molecular Control of Meiosis,** Florence P. Haseltine, Neal L. First, *Editors*

Vol 268: **The Na^+,K^+-Pump,** Jens C. Skou, Jens G. Nørby, Arvid B. Maunsbach, Mikael Esmann, *Editors*. Published in two volumes: Part A: *Molecular Aspects*. Part B: *Cellular Aspects*.

Vol 269: **EORTC Genitourinary Group Monograph 5: Progress and Controversies in Oncological Urology II,** Fritz H. Schröder, Jan G.M. Klijn, Karl H. Kurth, Herbert M. Pinedo, Ted A.W. Splinter, Herman J. de Voogt, *Editors*

Vol 270: **Cell-Free Analysis of Membrane Traffic,** D. James Morré, Kathryn E. Howell, Geoffrey M.W. Cook, W. Howard Evans, *Editors*

Vol 271: **Advances in Neuroblastoma Research 2,** Audrey E. Evans, Giulio J. D'Angio, Alfred G. Knudson, Robert C. Seeger, *Editors*

Vol 272: **Bacterial Endotoxins: Pathophysiological Effects, Clinical Significance, and Pharmacological Control,** Jack Levin, Harry R. Büller, Jan W. ten Cate, Sander J.H. van Deventer, Augueste Sturk, *Editors*

Vol 273: **The Ion Pumps: Structure, Function, and Regulation,** Wilfred D. Stein, *Editor*

Vol 274: **Oxidases and Related Redox Systems,** Tsoo E. King, Howard S. Mason, Martin Morrison, *Editors*

Vol 275: **Electrophysiology of the Sinoatrial and Atrioventricular Nodes,** Todor N. Mazgalev, Leonard S. Dreifus, Eric L. Michelson, *Editors*

Vol 276: **Prediction of Response to Cancer Therapy,** Thomas C. Hall, *Editor*

Vol 277: **Advances in Urologic Oncology,** Nasser Javadpour, Gerald P. Murphy, *Editors*

Vol 278: **Advances in Cancer Control: Cancer Control Research and the Emergence of the Oncology Product Line,** Paul F. Engstrom, Paul N. Anderson, Lee E. Mortenson, *Editors*

Vol 279: **Basic and Clinical Perspectives of Colorectal Polyps and Cancer,** Glenn Steele, Jr., Randall W. Burt, Sidney J. Winawer, James P. Karr, *Editors*

Vol 280: **Plant Flavonoids in Biology and Medicine II: Biochemical, Cellular, and Medicinal Properties,** Vivian Cody, Elliott Middleton, Jr., Jeffrey B. Harborne, Alain Beretz, *Editors*

Vol 281: **Transplacental Effects on Fetal Health,** Dante G. Scarpelli, George Migaki, *Editors*

Vol 282: **Biological Membranes: Aberrations in Membrane Structure and Function,** Manfred L. Karnovsky, Alexander Leaf, Liana C. Bolis, *Editors*

Vol 283: **Platelet Membrane Receptors: Molecular Biology, Immunology, Biochemistry, and Pathology,** G.A. Jamieson, *Editor*

Vol 284: **Cellular Factors in Development and Differentiation: Embryos, Teratocarcinomas, and Differentiated Tissues,** Stephen E. Harris, Per-Erik Mansson, *Editors*

Vol 285: **Non-Radiometric Assays: Technology and Application in Polypeptide and Steroid Hormone Detection,** Barry D. Albertson, Florence P. Haseltine, *Editors*

Vol 286: **Molecular and Cellular Mechanisms of Septic Shock,** Bryan L. Roth, Thor B. Nielsen, Adam E. McKee, *Editors*

Vol 287: **Dietary Restriction and Aging,** David L. Snyder, *Editor*

Vol 288: **Immunity to Cancer. II,** Malcolm S. Mitchell, *Editor*

Vol 289: **Computer-Assisted Modeling of Receptor–Ligand Interactions: Theoretical Aspects and Applications to Drug Design,** Robert Rein, Amram Golombek, *Editors*

Vol 290: **Enzymology and Molecular Biology of Carbonyl Metabolism 2: Aldehyde Dehydrogenase, Alcohol Dehydrogenase, and Carbonyl Reductase,** Henry Weiner, T. Geoffrey Flynn, *Editors*

Please contact the publisher for information about previous titles in this series.

COMPUTER-ASSISTED MODELING OF RECEPTOR-LIGAND INTERACTIONS
Theoretical Aspects and Applications to Drug Design

COMPUTER-ASSISTED MODELING OF RECEPTOR-LIGAND INTERACTIONS
Theoretical Aspects and Applications to Drug Design

Proceedings of the 1988 OHOLO Conference
Held in Eilat, Israel, April 24–28, 1988

Editors

Robert Rein
Department of Biophysics
Roswell Park Memorial Institute
Buffalo, New York

Amram Golombek
Israel Institute for Biological Research
Ness-Ziona, Israel

ALAN R. LISS, INC. • NEW YORK

Address all Inquiries to the Publisher
Alan R. Liss, Inc., 41 East 11th Street, New York, NY 10003

Copyright © 1989 Alan R. Liss, Inc.

Printed in the United States of America

Under the conditions stated below the owner of copyright for this book hereby grants permission to users to make photocopy reproductions of any part or all of its contents for personal or internal organizational use, or for personal or internal use of specific clients. This consent is given on the condition that the copier pay the stated per-copy fee through the Copyright Clearance Center, Incorporated, 27 Congress Street, Salem, MA 01970, as listed in the most current issue of "Permissions to Photocopy" (Publisher's Fee List, distributed by CCC, Inc.), for copying beyond that permitted by sections 107 or 108 of the US Copyright Law. This consent does not extend to other kinds of copying, such as copying for general distribution, for advertising or promotional purposes, for creating new collective works, or for resale.

Library of Congress Cataloging-in-Publication Data

OHOLO Conference (33rd : 1988 : Eilat, Israel)
   Computer-assisted modeling of receptor-ligand interactions.
     (Progress in clinical and biological research ; v. 289)
     Includes bibliographies and index.
     1. Drug receptors--Computer simulation--Congresses.
2. Drugs--Design--Computer simulation--Congresses.
3. Drugs--Structure-activity relationships--Congresses.
I. Rein, Robert. II. Golombek, Amram. III. Title.
IV. Title: Receptor-ligand interactions. V. Series.
[DNLM: 1. Computer Simulation--congresses. 2. Models, Chemical--congresses. 3. Protein Binding--congresses.
4. Receptors, Endogenous Substances--physiology--congresses. 5. Structure-Activity Relationship--congresses. W1 PR668E v.289 / QU 55 O38c 1988]
RM301.41.306 1988   615'.7   88-32589
ISBN 0-8451-5139-8

Contents

Contributors and Participants xi
Preface
Robert Rein and Amram Golombek xix
Overview
Robert Rein and Amram Golombek xxi
Permanent Committee of the OHOLO Conference
1988 Conference Organizing Committee xxv
OHOLO Conferences Since 1956 xxvi
Acknowledgments xxvii

SECTION I. COMPUTATIONAL METHODS OF PROTEIN MODELING AND ANALYSIS OF PROTEIN-LIGAND RECOGNITION

Some Computational Problems in the Conformational Analysis of Polypeptides and Proteins
Harold A. Scheraga 3
Modelling the Structure of Highly Constrained Proteins
Cyrus Levinthal 19
Assorted Modules in Protein Architecture
Mitiko Gō and Michiko Nosaka 27
A Computer Modeling Study of Acetylcholine Receptor-Ligand Interactions
Masayuki Shibata and Robert Rein 39
Molecular Dynamics as a Tool for Structural and Functional Predictions: The Retinol Binding Protein and Chloroplast C-Terminal Fragment of the L12 Ribosomal Protein Cases
O. Tapia and J. Åqvist 55
Electrostatic Interactions in Proteins
Barry Honig, Kim Sharp, and Michael Gilson 65
Protein Structure Predictions: New Theoretical Approaches
Fred E. Cohen, Lydia Gregoret, Scott R. Presnell, and I.D. Kuntz 75
Normal-Mode Dynamics as a Tool for Predicting Antigenic Sites on Proteins
Peter S. Stern 87
A Theoretical Model of Metal Binding Sites in Proteins
Morris Krauss and Walter J. Stevens 95

Computer Modeling of Membrane-Anchored Cellular and Viral Proteins: Organization and Function
Yechiel Becker 109

C-H. . .X Hydrogen-Bonded Pseudo-Watson-Crick Base Pairing With 7-Deazanebularin and Canonical Bases in DNA and RNA
Rick L. Ornstein 131

SECTION II. PHYSICAL METHODS IN DRUG DESIGN

The Structure of Proteins and Their Binding Sites: NMR and Artificial Intelligence
Oleg Jardetzky, Olivier Lichtarge, James Brinkley, and Marcela Madrid 145

Folding and Dynamics of Globular Proteins Studies by Time Resolved Fluorescence Spectroscopy
Elisha Haas 157

Protein Adaptation to Extreme Salinity: The Crystal Structure of 2Fe-2S Ferredoxin From *Halobacterium marismortui*
J.L. Sussman, M. Shoham, and M. Harel 171

Partial Molal Volume and Pharmacodynamic Activity
Sasson Cohen, Zipora Brif, Frank Haberman, and Zvi Liron 189

Polymorphism in Drug Design and Delivery
Joel Bernstein 203

The Solution Conformation of the Antibiotic Anticancer Chromomycin A_3 by Two-Dimensional NMR Spectroscopy
Elisha Berman and Michal Kam 217

SECTION III. RECEPTORS AND TRANSMEMBRANE SIGNALING

Structural Models of Alpha Helical Membrane Peptides and the GABA Receptor Channel
H. Robert Guy and G. Raghunathan 231

Structure of Nicotinic Acetylcholine Receptors From Muscle and Neurons
Jon Lindstrom, Paul Whiting, Ralf Schoepfer, Michael Luther, and Manoj Das . 245

Voronoi Receptor Site Models
Laurent Boulu and G.M. Crippen 267

A Structural and Dynamic Model for the Nicotinic Acetylcholine Receptor
Edward M. Kosower 279

Conformational Differences Between Aged and Non-Aged Organophosphoryl Conjugates of Chymotrypsin
N. Steinberg, J. Grunwald, E. Roth, R. August, E. Haas, Y. Ashani, and I. Silman 293

Structural and Immunochemical Properties of Fetal Bovine Serum Acetylcholinesterase
B.P. Doctor, K.K. Smyth, M.K. Gentry, Y. Ashani, C.E. Christner, D.M. De La Hoz, R.A. Ogert, and S.W. Smith 305

Receptor-Effector Coupling Processes Probed by Monoclonal Antibodies
Enrique Ortega, R. Schweitzer-Stenner, and Israel Pecht 317

Structural and Functional Diversity of Muscarinic Acetylcholine Receptor Subtypes
J. Ramachandran, E.G. Peralta, A. Ashkenazi, J.W. Winslow, D.H. Smith, and D.J. Capon 327

Modelling the Cholinergic Binding Site: Considerations
Jonathan M. Gershoni 341

Sequence Similarities Between Human Acetylcholinesterase and Related Proteins: Putative Implications for Therapy of Anticholinesterase Intoxication
Hermona Soreq and Catherine A. Prody 347

Comparison of the Interaction of the Histamine H_2-Antagonists Histamine and Dimaprit
E.E.J. Haaksma, G.M. Donné-Op den Kelder, H. Timmerman, P. Vernooijs, and W. Ravenek 361

SECTION IV. RECEPTOR-DIRECTED DRUG DESIGN

Quantum Mechanical SCF/CI Studies as Probes of Macromolecular Structure: Methodological Aspects of Spectral Comparisons
James D. Petke, Gerald M. Maggiora, and Ralph E. Christoffersen 373

Comparative Modeling of Proteins in the Complement Pathway
Jonathan Greer, Karl W. Mollison, George W. Carter, and Erik R.P. Zuiderweg 385

A Molecular Theoretical Model of Recognition and Activation at a 5-HT Receptor
Gustavo A. Mercier, Roman Osman, and Harel Weinstein 399

Opiate Receptor Heterogeneity: Relative Ligand Affinities and Molecular Determinants of High Affinity Binding at Different Opiate Receptors
Gilda Loew, Lawrence Toll, John Lawson, Gernot Frenking, and Wilma Polgar 411

The Role of Hydrogen-Bonds in Drug Binding
Rebecca C. Wade and Peter J. Goodford 433

Hydrogen Bonding in Protein Ligand Interactions: A Theoretical Dimension of Aspartic Proteinase Crystallography
Amiram Goldblum 445

Computational Studies of Ligand/Receptor Interactions
Sid Topiol and Michael Sabio 455

Design and Synthesis of Antimuscarinics Based on Physical and Mechanical Attributes
Richard K. Gordon, Ruthann M. Smejkal, Eli Breuer, and Peter K. Chiang . . . 465

SECTION V. ROUND TABLE DISCUSSION

Round Table Discussion
Harold A. Scheraga, Chairman 473

Index 499

Contributors and Participants

K. Adams,* Warburton Court, Middlesex HA4 OAN, United Kingdom

Y. Amieal,* Israel Institute for Biological Research, 70450 Ness-Ziona, Israel

A. Amir,* Israel Institute for Biological Research, 70450 Ness-Ziona, Israel

G. Amitai,* Israel Institute for Biological Research, 70450 Ness-Ziona, Israel

J. Åqvist, Department of Molecular Biology, Swedish University of Agricultural Sciences, UPPSALA Biomedical Center, S-751 24 UPPSALA, Sweden [55]

N. Ariel,* Israel Institute for Biological Research, 70450 Ness-Ziona, Israel

Y. Ashani, Division of Biochemistry, Walter Reed Army Institute of Research, Washington, DC 20307-5100; present address: Israel Institute for Biological Research, 70450 Ness-Ziona, Israel [293,305]

A. Ashkenazi, Department of Developmental Biology and Molecular Biology, Genentech, Inc., South San Francisco, CA 94080 [327]

R. August, Department of Chemical Physics, Weizmann Institute of Science, Rehovot 76100, Israel [293]

D. Barak,* Israel Institute for Biological Research, 70450 Ness-Ziona, Israel

R. Barak,* Israel Institute for Biological Research, 70450 Ness-Ziona, Israel

Y. Beatus, Israel Institute for Biological Research, 70450 Ness-Ziona, Israel

Yechiel Becker, Department of Molecular Virology, Faculty of Medicine, Hebrew University, Jerusalem, Israel [109]

N. Ben-David,* Israel Institute for Biological Research, 70450 Ness-Ziona, Israel

Z. Berkowitch-Yellin,* Weizmann Institute of Science, 76110 Rehovot, Israel

Elisha Berman, Organic Chemistry Department, Weizmann Institute of Science, Rehovot 76100, Israel [217]

Joel Bernstein, Department of Chemistry, Ben-Gurion University of the Negev, Beer Sheva 84105, Israel [203]

I.R. Berry,* Pharmacaps, Inc., Elizabeth, NJ 07207

A.J. Beveridge,* Institute of Cancer Research, Biomedical Structure Unit, Surrey, United Kingdom

The numbers in brackets are the opening page numbers of the contributors' articles.
*Participants are indicated by an asterisk.

Contributors and Participants

N. Borkakoti,* Roche Products, Welwyn Garden City, Herts AL7 3AY, United Kingdom

Laurent Boulu, College of Pharmacy, University of Michigan, Ann Arbor, MI 48109 **[267]**

C. Brender,* Department of Chemistry, Bar-Ilan University, 52100 Ramat Gan, Israel

Eli Breuer, Department of Applied Biochemistry, Walter Reed Army Institute of Research, Washington, DC 20307-5100; present address: Department of Pharmaceutical Chemistry, Hebrew University, School of Pharmacy, Jerusalem, Israel **[465]**

Zipora Brif, Department of Physiology and Pharmacology, Tel Aviv University, Tel Aviv 69978, Israel **[189]**

S. Briggs,* Chemical Design, Unit 12, Oxford OX2 OJB, United Kingdom

James Brinkley, Knowledge Systems Laboratory, Stanford University, Stanford, CA 94305-5055 **[145]**

H. Bruns,* E. Merck, Postfach 4119, D-6100 Darmstadt, West Germany

D.J. Capon, Department of Developmental Biology and Molecular Biology, Genentech, Inc., South San Francisco, CA 94080 **[327]**

George W. Carter, Pharmaceutical Products Division, Abbott Laboratories, Abbott Park, IL 60064 **[385]**

Peter K. Chiang, Department of Applied Biochemistry, Walter Reed Army Institute of Research, Washington, DC 20307-5100 **[465]**

D.M. Chipman,* Department of Biology, Ben-Gurion University, Beer Sheva 84105, Israel

C.E. Christner, Division of Biochemistry, Walter Reed Army Institute of Research, Washington, DC 20307-5100 **[305]**

Ralph E. Christoffersen, The Upjohn Company, Kalamazoo, MI 49001 **[373]**

G. Chukat,* Department of Life Sciences, Bar-Ilan University, 52100 Ramat Gan, Israel

Fred E. Cohen, Department of Pharmaceutical Chemistry and Medicine, University of California, San Francisco, CA 94143 **[75]**

N.C. Cohen, CIBA–GEIGY Pharmaceutical Division, CH–4002 Switzerland

Sasson Cohen, Department of Physiology and Pharmacology, Tel Aviv University, Tel Aviv 69978, Israel **[189]**

R. Corett,* Israel Institute for Biological Research, 70450 Ness-Ziona, Israel

G.M. Crippen, College of Pharmacy, University of Michigan, Ann Arbor MI 48109 **[267]**

Manoj Das, The Salk Institute for Biological Studies, San Diego, CA 92138 **[245]**

D.M. De La Hoz, Division of Biochemistry, Walter Reed Army Institute of Research, Washington, DC 20307-5100 **[305]**

R.M. Dittel,* Bahnhofstr. 18, D-8919 Utting, West Germany

B. Dixon,* The Scientist, European Editorial Office, Uxbridge, Middx. UB8 1DP, United Kingdom

B.P. Doctor, Division of Biochemistry, Walter Reed Army Institute of Research, Washington, DC 20307-5100 **[305]**

G.M. Donné-Op den Kelder, Department of Pharmacochemistry, Vrije Universiteit, 1081 HV Amsterdam, The Netherlands **[361]**

M. Dornay, * Israel Institute for Biological Research, 70450 Ness-Ziona, Israel

E. Elhanaty, * Israel Institute for Biological Research, 70450 Ness-Ziona, Israel

S. Feinstein, * Israel Institute for Biological Research, 70450 Ness-Ziona, Israel

Gernot Frenking, Molecular Theory Department, Biomedical Research Laboratory, SRI International, Menlo Park, CA 94025 **[411]**

G. Frishman, * Israel Institute for Biological Research, 70450 Ness-Ziona, Israel

G. Gabor, * Israel Institute for Biological Research, 70450 Ness-Ziona, Israel

M.K. Gentry, Division of Biochemistry, Walter Reed Army Institute of Research, Washington, DC 20307-5100 **[305]**

Jonathan M. Gershoni, Department of Biophysics, The Weizmann Institute of Science, Rehovot, Israel **[341]**

Michael Gilson, Department of Biochemistry and Molecular Biophysics, Columbia University, New York, NY 10032 **[65]**

R. Glaser, * Department of Chemistry, Ben-Gurion University, 84105 Beer Sheva, Israel

Amiram Goldblum, Department of Pharmaceutical Chemistry, School of Pharmacy, Hebrew University of Jerusalem, Jerusalem 91120, Israel **[445]**

S. Goldstein, * U.C.B.s.a. Pharmaceutical Sector, Chemin du Foriest, 1420 Braine l'Alleud, Belgium

Amram Golombek, * Israel Institute for Biological Research, 70450 Ness-Ziona, Israel **[xix,xxi]**

Mitiko Gō, Department of Biology, Faculty of Science, Kyushu University, Fukuoka 812, Japan **[27]**

Peter J. Goodford, University of Oxford, Laboratory of Molecular Biophysics, Oxford OX1 3QU, United Kingdom **[433]**

Richard K. Gordon, Department of Applied Biochemistry, Walter Reed Army Institute of Research, Washington, DC 20307-5100 **[465]**

Jonathan Greer, Pharmaceutical Products Division, Abbott Laboratories, Abbott Park, IL 60064 **[385]**

Lydia Gregoret, Department of Pharmaceutical Chemistry and Medicine, University of California, San Francisco, CA 94143 **[75]**

J. Grunwald, Israel Institute for Biological Research, 70450 Ness-Ziona, Israel **[293]**

H. Robert Guy, Laboratory of Mathematical Biology, National Cancer Institute, National Institutes of Health, Bethesda, MD 20892 **[231]**

E.E.J. Haaksma, Department of Pharmacochemistry, Vrije Universiteit, 1081 HV Amsterdam, The Netherlands **[361]**

Elisha Haas, Department of Life Sciences, Bar-Ilan University, Ramat Gan 52100; and Departments of Chemical Physics and Biophysics, Weizmann Institute of Science, Rehovot 76100, Israel **[157, 293]**

xiv / Contributors and Participants

Frank Haberman, Institute for Biological Research, 70450 Ness-Ziona, Israel **[189]**

B. Halperin,* Israel Institute for Biological Research, 70450 Ness-Ziona, Israel

G. Halperin,* Israel Institute for Biological Research, 70450 Ness-Ziona, Israel

M. Harel, Department of Structural Chemistry, Weizmann Institute of Science, Rehovot 76100, Israel **[171]**

S. Harel,* Ministry of Industry and Trade, 91021 Jerusalem, Israel

G. Heran,* Department of Physical Chemistry, Weizmann Institute of Research, 76100 Rehovot, Israel

O. Herzberg,* CARB, National Bureau of Standard, Gaithersburg, MD 20899

Barry Honig, Department of Biochemistry and Molecular Biophysics, Columbia University, New York, NY 10032 **[65]**

D. Hurwitz,* Department of Structural Chemistry, Weizmann Institute of Science, 76100 Rehovot, Israel

E. Israeli,* Israel Institute for Biological Research, 70450 Ness-Ziona, Israel

Oleg Jardetzky, Stanford Magnetic Resonance Laboratory, Stanford University, Stanford, CA 94305-5055 **[145]**

Michal Kam, Organic Chemistry Department, Weizmann Institute of Science, Rehovot 76100, Israel **[217]**

H. Karfunkel,* CIBA-GEIGY AG, R-1046.2.13, CH-4002 Basel, Switzerland

Y. Karton,* Israel Institute for Biological Research, 70450 Ness-Ziona, Israel

R. Katz,* National Institutes of Health, Bethesda, MD 20892

S. Kinamon, Israel Institute for Biological Research, 70450 Ness-Ziona, Israel

A. Kohn,* Israel Institute for Biological Research, 70450 Ness-Ziona, Israel

H. Kooijman,* Lab. voor Kristal en Structuurchemie, 3508 TB Utrecht, The Netherlands

Edward M. Kosower, Biophysical Organic Chemistry Unit, School of Chemistry, Sackler Faculty of Exact Sciences, Tel Aviv University, Ramat-Aviv, Tel Aviv 69978, Israel; and Department of Chemistry, State University of New York, Stony Brook, NY 11794-3400 **[279]**

Morris Krauss, Molecular Spectroscopy Division, National Bureau of Standards, Gaithersburg, MD 20899 **[95]**

I.D. Kuntz, Department of Pharmaceutical Chemistry and Medicine, University of California, San Francisco, CA 94143 **[75]**

C. Lachman,* Israel Institute for Biological Research, 70450 Ness-Ziona, Israel

P. Lagant,* Univ. du Droit et de la Sante, Faculte de Pharmacie, I.N.S.E.R.M. U.279, 59045 Lille Cedex, France

John Lawson, Molecular Theory Department, Biomedical Research Laboratory, SRI International, Menlo Park, CA 94025 **[411]**

H. Leader,* Israel Institute for Biological Research, 70450 Ness-Ziona, Israel

Cyrus Levinthal, Department of Biological Sciences, Columbia University, New York, NY 10027 **[19]**

M. Levitt,* Department of Cell Biology, Stanford University School of Medicine, Stanford, CA 94305

D. Levy,* Israel Institute for Biological Research, 70450 Ness-Ziona, Israel

R. Levy,* Teva Pharmaceutical Industries Ltd., 49131 Petach Tikva, Israel

Olivier Lichtarge, Stanford Magnetic Resonance Laboratory, Stanford University, Stanford, CA 94305-5055 **[145]**

S. Lifson,* Department of Chemical Physiology, Weizmann Institute of Science, 76100 Rehovot, Israel

Jon Lindstrom, The Salk Institute for Biological Studies, San Diego, CA 92138 **[245]**

M. Lion,* Israel Institute for Biological Research, 70450 Ness-Ziona, Israel

Zvi Liron, Institute for Biological Research, 70450 Ness-Ziona, Israel **[189]**

Gilda Loew, Molecular Theory Department, Biomedical Research Laboratory, SRI International, Menlo Park, CA 94025 **[411]**

Michael Luther, The Salk Institute for Biological Studies, San Diego, CA 92138 **[245]**

Marcela Madrid, Stanford Magnetic Resonance Laboratory, Stanford University, Stanford, CA 94305-5055; present address: Department of Biological Sciences, Carnegie Mellon University, Pittsburgh, PA 15213 **[145]**

Gerald M. Maggiora, The Upjohn Company, Kalamazoo, MI 49001 **[373]**

G.R. Marshall,* Department of Pharmacology, Washington University School of Medicine, St. Louis, MO 63105

A. Mizrahi,* Israel Institute for Biological Research, 70450 Ness-Ziona, Israel

Gustavo A. Mercier, Departments of Physiology and Biophysics and Pharmacology, Mount Sinai School of Medicine, CUNY, New York, NY 10029 **[399]**

Karl W. Mollison, Pharmaceutical Products Division, Abbott Laboratories, Abbott Park, IL 60064 **[385]**

J. Moult,* CARB, National Bureau of Standards, Gaithersburg, MD 20899

Michiko Nosaka, Department of Biology, Faculty of Science, Kyushu University, Fukuoka 812, Japan **[27]**

R.A. Ogert, Division of Biochemistry, Walter Reed Army Institute of Research, Washington, DC 20307-5100 **[305]**

Rick L. Ornstein, Department of Biochemistry, Princeton University, Princeton, NJ 08544 **[131]**

Enrique Ortega, Department of Chemical Immunology, The Weizmann Institute of Science, Rehovot 76100, Israel **[317]**

Roman Osman, Departments of Physiology and Biophysics and Pharmacology, Mount Sinai School of Medicine, CUNY, New York, NY 10029 **[399]**

M. Ovaska,* Orion Pharmaceutica, SF-02101 Espoo, Finland

Israel Pecht, Department of Chemical Immunology, The Weizmann Institute of Science, Rehovot 76100, Israel **[317]**

M. Peled,* Israel Institute for Biological Research, 70450 Ness-Ziona, Israel

E.G. Peralta, Department of Developmental Biology and Molecular Biology, Genentech, Inc., South San Francisco, CA 94080 [327]

A. Perlman,* Israel Institute for Biological Research, 70450 Ness-Ziona, Israel

N. Perry,* Chemical Design, Unit 12, Oxford OX2 OJB, United Kingdom

James D. Petke, The Upjohn Company, Kalamazoo, MI 49001 [373]

R. Pniel,* Israel Institute for Biological Research, 70450 Ness-Ziona, Israel

Wilma Polgar, Molecular Theory Department, Biomedical Research Laboratory, SRI International, Menlo Park, CA 94025 [411]

Scott R. Presnell, Department of Pharmaceutical Chemistry and Medicine, University of California, San Francisco, CA 94143 [75]

Catherine A. Prody, Department of Biological Chemistry, The Life Sciences Institute, The Hebrew University, Jerusalem 91904, Israel [347]

G. Raghunathan, Laboratory of Mathematical Biology, National Cancer Institute, National Institutes of Health, Bethesda, MD 20892 [231]

J. Ramachandran, Department of Developmental Biology and Molecular Biology, Genentech, Inc., South San Francisco, CA 94980; present address: Neurex Corporation, Menlo Park, CA 94025 [327]

W. Ravenek, Department of Quantum Chemistry, Vrije Universiteit, 1081 HV Amsterdam, The Netherlands [361]

Robert Rein, Department of Biophysics, Roswell Park Memorial Institute, Buffalo, NY 14263 [xix,xxi,39]

R. Rigler,* Department of Medical Physics, Karolinska Institute, S-104 01 Stockholm, Sweden

M.J. Ross,* Genentech Inc., South San Francisco, CA 94080

E. Roth, Department of Neurobiology, Weizmann Institute of Science, Rehovot 76100, Israel [293]

Michael Sabio, Department of Medicinal Chemistry, Berlex Laboratories, Cedar Knolls, NJ 07927 [455]

Harold A. Scheraga, Baker Laboratory of Chemistry, Cornell University, Ithaca, NY 14853-1301 [3]

Ralf Schoepfer, The Salk Institute for Biological Studies, San Diego, CA 92138 [245]

R. Schweitzer-Stenner, Department of Chemical Immunology, The Weizmann Institute of Science, Rehovot 76100, Israel [317]

M. Seiffe,* Teva Pharmaceutical Industries Ltd., Kfar Sava 44102, Israel

A. Shafferman,* Israel Institute for Biological Research, 70450 Ness-Ziona, Israel

Kim Sharp, Department of Biochemistry and Molecular Biophysics, Columbia University, New York, NY 10032 [65]

Masayuki Shibata, Department of Biophysics, Roswell Park Memorial Institute, Buffalo, NY 14263 [39]

M. Shoham, Department of Structural Chemistry, Weizmann Institute of Science, Rehovot 76100, Israel [171]

Contributors and Participants / xvii

I. Silman, Department of Neurobiology, Weizmann Institute of Science, Rehovot 76100, Israel [293]

Ruthann M. Smejkal, Department of Applied Biochemistry, Walter Reed Army Institute of Research, Washington, DC 20307-5100 [465]

D.H. Smith, Department of Developmental Biology and Molecular Biology, Genentech, Inc., South San Francisco, CA 94080 [327]

S.W. Smith, Division of Biochemistry, Walter Reed Army Institute of Research, Washington, DC 20307-5100 [305]

K.K. Smyth, Division of Biochemistry, Walter Reed Army Institute of Research, Washington, DC 20307-5100 [305]

Hermona Soreq, Department of Biological Chemistry, The Life Sciences Institute, The Hebrew University, Jerusalem 91904, Israel [347]

M. Spiegelstein,* Israel Institute for Biological Research, 70450 Ness-Ziona, Israel

N. Steinberg, Department of Neurobiology, Weizmann Institute of Science, 76100 Rehovot, Israel [293]

Peter S. Stern, Chemical Physics Department, Weizmann Institute of Science, 76100 Rehovot, Israel [87]

Walter J. Stevens, Molecular Spectroscopy Division, National Bureau of Standards, Gaithersburg, MD 20899 [95]

C.-T. Su,* Department of Neurobiology, Weizmann Institute of Science, 76100 Rehovot, Israel

J.L. Sussman, Department of Structural Chemistry, Weizmann Institute of Science, Rehovot 76100, Israel [171]

O. Tapia, Department of Molecular Biology, Swedish University of Agricultural Sciences, UPPSALA Biomedical Center, S-751 24 Uppsala, Sweden [55]

H. Timmerman, Department of Pharmacochemistry, Vrije Universiteit, 1081 HV Amsterdam, The Netherlands [361]

Lawrence Toll, Molecular Theory Department, Biomedical Research Laboratory, SRI International, Menlo Park, CA 94025 [411]

Sid Topiol, Department of Medicinal Chemistry, Berlex Laboratories, Inc., Cedar Knolls, NJ 07927 [455]

R. Unger,* Department of Structural Chemistry, Weizmann Institute of Science, 76100 Rehovot, Israel

B. Velan,* Israel Institute for Biological Research, 70450 Ness-Ziona, Israel

M. Vered,* Teva Pharmaceutical Ind. Ltd., 49131 Petach Tikva, Israel

P. Vernooijs, Department of Quantum Chemistry, Vrije Universiteit, 1081 HV Amsterdam, The Netherlands [361]

J. Vidgren,* Orion Corporation, Orion Pharmaceutica, SF-02101 Espoo, Finland

G. Vriend,* Bekemaheerd 19, 9737 PP Groningen, The Netherlands

H. Vyplel,*Sandoz Forschungsinstitut, A-1235 Wein, Austria

Rebecca C. Wade, University of Oxford, Laboratory of Molecular Biophysics, Oxford OX1 3QU, United Kingdom [433]

U.G. Wagner,* Department of Structural Chemistry, Weizmann Institute of Science, 76100 Rehovot, Israel

M. Walkinshaw,* Sandoz Ltd., 4002 Basel, Switzerland

A. Warshel,* Department of Chemistry, University of Southern California, Los Angeles, CA 90089-0482

S. Weinberger,* Department of Organic Chemistry, Weizmann Institute of Science, 76100 Rehovot, Israel

Harel Weinstein, Departments of Physiology and Biophysics and Pharmacology, Mount Sinai School of Medicine, CUNY, New York, NY 10029 **[399]**

Paul Whiting, The Salk Institute for Biological Studies, San Diego, CA 92138 **[245]**

J.W. Winslow, Department of Developmental Biology and Molecular Biology, Genentech, Inc., South San Francisco, CA 94080 **[327]**

P.B. Youkharibache,* Polygen (Europe) Ltd., 94666 Rungis Cedex, France

Erik R.P. Zuiderweg, Pharmaceutical Products Division, Abbott Laboratories, Abbott Park, IL 60064 **[385]**

A. Zywiez,* Bayer AG., ZF-FWI, Q18, D-5090 Leverkusen 1, West Germany

Preface

The Israel Institute for Biological Research (IIBR), founded in 1952, is primarily oriented towards applied research, development, and production, mainly in the fields of biology, chemistry, ecology, and public health. Basic research studies related to and evolving from the applied projects are also part of the IIBR program. Expertise gained throughout more than 30 years of cooperative research has enabled IIBR to develop scores of advanced products and processes with applications in biotechnology and drug design.

Since 1956, IIBR has been organizing international meetings on selected topics in biology, chemistry, and the environmental sciences. The original purpose of these gatherings was to encourage the dialogue between Israeli scientists—especially the Institute's scientists—and their colleagues from abroad; perhaps even more important was the goal to introduce Israeli students to the world of living science and its controversies. The conferences were first held at OHOLO, a seminar center of the Labour Movement located on the Sea of Galilee. However, the OHOLO Conferences, as they came to be called, soon outgrew their first home, and in recent years they have been held at more luxurious recreation centers at Ma'alot, Zichron Yaacov, and Eilat.

These articles published here represent the content of most of the invited talks and a few selected posters from the 1988 OHOLO Conference, which celebrated the 35th Anniversary of the Israel Institute for Biological Research. This volume presents interdisciplinary studies of molecular aspects of receptors, transmembrane signaling, and drug design. The conference on which it is based, **Computer-Assisted Modeling of Receptor–Ligand Interactions: Theoretical Aspects and Applications to Drug Design**, was held at the King Solomon Hotel in Eilat, Israel, from April 24 to April 28, 1988.

These proceedings provide a broad overview of protein modeling and analysis of protein–ligand interactions, as well as a comprehensive discussion of the role of these interactions in processing transmembrane signals. There is a strong emphasis on interdisciplinary research methodology, with particular attention to integration of physical and computational methods with those of biochemistry, cellular biology, and pharmacology. The central objective of the 1988 OHOLO conference was the strengthening of the dialogue between experimental and theoretical scientists in the receptor field and drug design. The main effect of this volume will be its contribution to this interdisciplinary dialogue.

Robert Rein
Amram Golombek

Overview

Computer-Assisted Modeling of Receptor–Ligand Interactions: Theoretical Aspects and Applications to Drug Design is organized into five sections, just as the OHOLO Conference had five sections.

I. Computational Methods of Protein Modeling and Analysis of Protein–Ligand Recognition
II. Physical Methods in Drug Design
III. Receptors and Transmembrane Signaling
IV. Receptor-Directed Drug Design
V. Round Table Discussion

The first section contains papers on various methods of structural modeling of proteins and on analysis of the interactions responsible for ligand binding to active sites of proteins. Specific topics in this section include Scheraga's article on computational aspects of conformational analysis of proteins; Levinthal's paper discussing the modeling of the structure of highly constrained proteins; and Gō and Nosaka's report concerned with the make-up of globular domains of proteins from modules and the correlation of these modules with intron boundaries. According to Gō, modules play an important role in domain conformation and protein function and evolution. Shibata and Rein's contribution is one in the series of papers in this volume that deals with modeling of cholinergic receptor sites. The authors also discuss agonist and antagonist binding to the receptor active site. Tapia and Åqvist use molecular dynamics and fractal analysis to study complexes of apo-retinol binding protein with prealbumin. Honig et al. focus on electrostatic interactions in proteins and nucleic acids, using macromolecular charge distribution and explicit inclusion of environmental effects. Honig and coworkers present a number of interesting applications such as effects of site-directed mutagenesis on interaction energies and the effect of surface charges on pKs. Cohen et al. presented an integrated approach, comprised of homology secondary structure prediction and packing of structural elements, to the study of protein structure. Stern's paper uses normal mode dynamics to study atomic and segmental motion in proteins. The aim of this study is prediction of antigenic regions in proteins. Krauss and Stevens's study widens the scope of methodology to include quantum mechanics. This study focuses on metal ion binding to proteins. Becker discusses the computer modeling of membrane-anchored cellular and viral protein from the organizational and functional aspects. Ornstein shows results of computations aimed at establishing the structure of hydrogen-bonded pseudo-Watson-Crick base pairing.

The section on Physical Methods in Drug Design features a series of articles devoted to application of various spectroscopic techniques to the study of structure and dynamics of proteins and protein-ligand complexes. Jardetzky et al., for example, deal with the methodology of extracting structural information from NOE data. Bernstein's paper illustrates the use of crystallographic information, in particular polymorphis in drug design. Sasson Cohen et al. deal with partial molal volumes and their application in the calculation of pharmacodynamic activity. The paper by Haas features fluorescence energy transfer measurements for characterization of intermediate structure of BPTI derivatives. Berman and Kam show how they used two-dimensional NMR spectroscopy to solve the conformation of antibiotic anticancer drugs. This series of articles further demonstrate the diversity of techniques applicable for studying protein complexes.

The next section on Receptors and Transmembrane Signaling evolves around two major topics. The first deals with receptor binding, in particular of the structure and binding to cholinergic receptors. This series of papers include articles by Guy and Raghunathan on helix packing and structural models of membrane channels. The paper by Ortega et al. also deals with channel-forming proteins. Lindstrom et al. deal with nicotinic acetylcholine receptor; models of the acetylcholine receptor are also addressed in papers by Kosower, Gershoni, and Ramachandran et al. and Crippen describe a novel receptor site-modeling method that utilizes ligand structures and binding free energies. Haaksma and coworkers discuss the interaction between the histamine H_2 receptor and some of its antagonists.

The second series of papers, in this section, such as that of Doctor et al. and of Soreq and Prody, are concerned with acetylcholinesterase structure and characteristic properties. Ashani and Silman discuss conformational differences between aged and nonaged organophosphoryl conjugates of Chymotrypsin.

The fourth section, Receptor-Directed Drug Design, features several articles based on quantum mechanical studies. The paper by Petke et al. uses *ab initio* SCF and CI calculations to probe spectorscopic transitions of DNA bases and analogs that can be employed as new DNA structural probes. Loew et al. present an interdisciplinary study of opiate narcotics that involves synthetic and theoretical chemistry. Mercier et al. present a theoretical model of recognition and activation of the 5-HT receptor. Greer et al. discuss comparative modeling of the anaphylotoxin family. Wade and Goodford describe a method for searching for a receptor binding site, with emphasis on the role of the hydrogen bonding in ligand binding. Gordon et al. show how they designed and synthesized antimuscarinics based on physical and mechanical attributes.

Goldblum's semiempirical quantum mechanical study also emphasizes the importance of the hydrogen bonding term in protein-ligand interactions, and deals specifically with the analysis of the active site of aspartic proteinases. Topiol and Sabio describe computational studies of ligand-receptor interactions, based on a general model for receptor activity.

The fifth section of this volume presents the transcript of the Round Table Discussion, which revolves around such questions as these:

1. What can the present state of the art of computation and modeling contribute to ligand-receptor recognition and to understanding of the mechanism of biological function? In particular, how can our understanding the molecular basis of ligand-induced activation of transmembrane channels and signal transmission be enhanced?

2. What and how can molecular modeling and computation methods contribute to the design of more effective drugs?

3. An important point in the discussion dealt with the question of what we can expect in the future. In what directions should we go? Are bigger computers the answer? Are new ideas required? How important is the role of the solvent and ionic effects, and how much effort should be invested in pursuing more accurate treatment of their effects?

In summary, this volume is significant to those who are interested in obtaining an overview of the current state of ideas, methodologies, and results in molecular modeling of cellular receptors and their complexes. It also shows how such studies can be employed in drug design. The biologically oriented reader can benefit from the introduction to computer-based methodologies. By the same token, theoretical scientists can gain a comprehensive overview of biological and pharmacological aspects of receptor studies.

Robert Rein
Amram Golombek

Permanent Committee of the OHOLO Conference

S. Cohen,	Tel-Aviv University, Sackler School of Medicine, Tel-Aviv
M. Feldman,	Weizmann Institute of Science, Rehovot
A. Golombek,	Israel Institute for Biological Research, Ness-Ziona
N. Grossowicz,	Hadassah Medical School, Jerusalem
I. Hertman,	Israel Institute for Biological Research, Ness-Ziona
A. Keynan,	Hebrew University of Jerusalem, Jerusalem
A. Kohn,	Orgenics Ltd., Yavne
M. Sela,	Weizmann Institute of Science, Rehovot

1988 Conference Organizing Committee

Co-Chairmen

R. Rein,	Roswell Park Memorial Institute, Buffalo, New York
A. Golombek,	Israel Institute for Biological Research, Ness-Ziona

Members

D. Barak,	Israel Institute for Biological Research, Ness-Ziona
M. Dornay,	Israel Institute for Biological Research, Ness-Ziona
A. Goldblum,	Hebrew University of Jerusalem, Jerusalem
P. Stern,	Weizmann Institute of Science, Rehovot

Technical and Administrative Assistances

N. Ben-David,	Israel Institute for Biological Research, Ness-Ziona
R. Pniel,	Israel Institute for Biological Research, Ness-Ziona
M. Peled,	Israel Institute for Biological Research, Ness-Ziona
A. Perlman,	Israel Institute for Biological Research, Ness-Ziona
S. Kinamon,	Israel Institute for Biological Research, Ness-Ziona

OHOLO Conferences Since 1956

- 1956 Bacterial Genetics
- 1956 Tissue Cultures in Virological Research
- 1958 Inborn and Acquired Resistance to Infection in Animals
- 1959 Experimental Approach to Mental Diseases
- 1960 Cryptobiotic Stages in Biological Systems
- 1961 Virus—Cell Relationships
- 1962 Biological Synthesis and Function of Nucleic Acids
- 1963 Cellular Control Mechanism of Macromolecular Synthesis
- 1964 Molecular Aspects of Immunology
- 1965 Cell Surfaces
- 1966 Chemistry and Biology of Psychotropic Agents
- 1967 Structure and Mode of Action of Enzymes
- 1968 Growth and Differentiation of Cells In Vitro
- 1969 Behavior of Animal cells in Culture
- 1970 Microbial Toxins
- 1971 Interaction of Chemical Agents With Cholinergic Mechanisms
- 1972 New Concepts in Immunity in Viral and Rickettsial Diseases
- 1973 Strategies for the Control of Gene Expression
- 1974 Sensory Physiology and Behavior
- 1975 Air Pollution and the Lung
- 1976 Host-Parasite Relationships in Systemic Mycoses
- 1977 Skin: Drug Application and Evaluation of Environmental Hazards
- 1978 Extrachromosomal Inheritance in Bacteria
- 1979 Neuroactive Compounds and Their Cell Receptors
- 1980 New Developments With Human and Veterinary Vaccines
- 1981 Biomimetic Chemistry and Transition-State Analogs: Approaches to Understanding Enzyme Catalysis
- 1982 Behavioral Models and the Analysis of Drug Action
- 1983 Mechanisms of Viral Pathogenesis (From Gene to Pathogen)
- 1984 Boundary Layer Structure: Modelling and Application to Air Pollution and Wind Energy
- 1985 Basic and Therapeutic Strategies in Alzheimer's and Other Age-Related Neuropsychiatric Disorders
- 1986 Model Systems in Neurotoxicology: Alternative Approaches to Animal Testing
- 1987 Modern Approaches to Animal Cell Technology (A Joint Meeting With ESACT)
- 1988 Computer-Assisted Modeling of Receptor–Ligand Interactions: Theoretical Aspects and Applications to Drug Design
- 1989 Calcium Channel Modulators in Ischemic Heart Disease and Other Cardiovascular Disorders: Basic and Therapeutic Approach

Acknowledgments

The Organizing Committee of the 1988 OHOLO Conference gratefully acknowledges the generous support of the following organizations (alphabetical order):

Abott Laboratories, North Chicago, Illinois
American Cyanamid Co., Pearl River, New York
Bayer AG, Wuppertal, Federal Republic of Germany
British Council–London, United Kingdom
Fidia Research Laboratories, Abano Terme, Italy
Genentech, Inc., South San Francisco, California
ICI Americas Inc., Wilmington, Delaware
Israel Institute for Biological Research, Ness-Ziona, Israel
Merck Sharp & Dohme Co., Inc., Rahway, New Jersey
Ministry of Science and Development, Jerusalem, Israel
National Council for Research and Development, Jerusalem, Israel
Rhône Poulenc Santé, Antony Cedex, France
Sandoz AG, Basel, Switzerland
Snow Brand Milk Products Co. Ltd., Tokyo, Japan
U.S. Air Force EOARD, London, United Kingdom

Editor R. Rein also wishes to thank both the National Aeronautics and Space Administration and the National Foundation for Cancer Research for their generous support.

SECTION I. COMPUTATIONAL METHODS OF PROTEIN MODELING AND ANALYSIS OF PROTEIN-LIGAND RECOGNITION

SOME COMPUTATIONAL PROBLEMS IN THE CONFORMATIONAL ANALYSIS OF POLYPEPTIDES AND PROTEINS

Harold A. Scheraga

Baker Laboratory of Chemistry,
Cornell University, Ithaca, New York 14853-1301,
U.S.A.

INTRODUCTION

A number of problems of interest in the field of biological macromolecules have been solved by means of conformational energy calculations. We thus have an understanding of how interatomic interactions lead to (a) formation of the fundamental structures from which proteins are built (i.e. the α-helix, β-sheet, and β-turn), (b) interactions among these fundamental structures (i.e. α...α, α...β, β...β, and β...α...β), and (c) transformations between these structures (e.g. the helix-coil transition) (Scheraga, 1984, 1986; Chou et al., 1988a,b). The methodology has also been extended to small naturally-occurring linear peptides (e.g. enkephalin), with cognizance taken of the fact that the conformations of such flexible molecules are influenced by both intra- and intermolecular interactions (Glasser and Scheraga, 1988). The conformations of small cyclic oligopeptides, e.g. gramicidin S, are less susceptible to the influence of intermolecular interactions, and the calculated structure of this isolated decapeptide has been confirmed by X-ray diffraction, as shown in Figure 1.

Similar success has been achieved with regular-repeating collagen-like poly(tripeptides), i.e. fibrous proteins. For example, the non-hydrogen-(i.e. heavy)-atom coordinates of the computed structure of poly(Gly-Pro-Pro) (Miller and Scheraga, 1976) agree with those obtained subsequently from a single crystal X-ray diffraction study of (Gly-Pro-Pro)$_{10}$ (Okuyama et al., 1976) within an r.m.s.

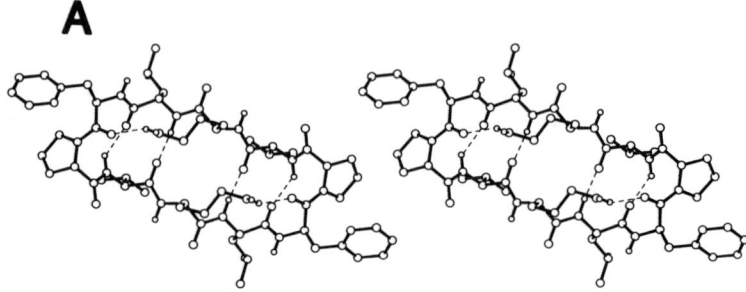

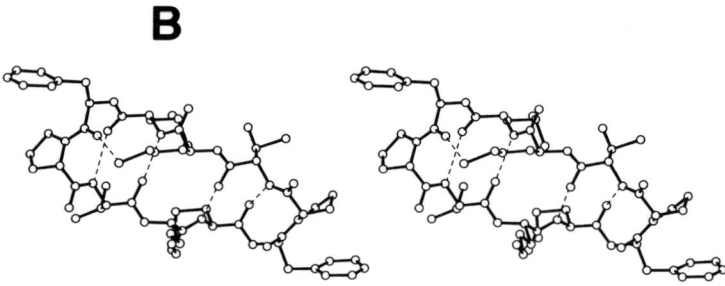

Figure 1. Computed (A) (Dygert et al., 1975; Némethy and Scheraga, 1984) and X-ray (B) (Hull et al., 1978) structures of gramicidin S showing (among other things) a hydrogen bond between the ornithine side chain and the phenylalanine backbone carbonyl group.

deviation of 0.3 Å.

The application of this methodology to globular proteins is more involved. In searching for the global minimum of the potential energy of a globular protein, with proper account taken of hydration (Kang et al., 1987; Gibson and Scheraga, 1987) and conformational fluctuational entropy (Gō and Scheraga, 1976), the more difficult multiple-minima problem is encountered. On the other hand, some progress has been made in developing methods to surmount this problem for small polypeptides (up to about 20 residues in length), and these are described here. Current efforts are in progress to scale them up to larger polypeptides -- of the order of 100-200 residues.

METHODS TO SURMOUNT THE MULTIPLE-MINIMA PROBLEM

Numerous procedures are available to surmount the multiple-minima problem (Gibson and Scheraga, 1988), the most promising of these being:

a. Build-up procedure
b. Optimization of electrostatics
c. Monte Carlo-plus-minimization
d. Electrostatically driven Monte Carlo
e. Inclusion of distance restraints
f. Adaptive importance sampling Monte Carlo
g. Increase of dimensionality.
h. Pattern recognition-plus-minimization

The characteristics of these procedures, and some applications, are described in the following sections. The procedures are used separately, and in various combinations with each other, to locate the approximate native conformation of a globular protein. They are all intended as the _initial_ approaches in the computations. In the _final_ stages, the results from all of these procedures are collated into an approximate three-dimensional structure whose energy should lie in the potential well containing the global minimum (i.e. this structure should be a good approximation of the native structure). Then, the conformational energy of this structure is minimized, taking _all_ pairwise interactions (over the whole molecule) into account.

BUILD-UP PROCEDURE

In a renormalization group type approach, one starts with the low-energy structures of single residues, and uses these to build up low-energy structures of dipeptides, tripeptides, etc., carrying out energy minimization at each stage. This requires storage of _many_ (backbone and side-chain) low-energy conformations. For example, at the single-residue level, the numbers of such conformations are in the tens, whereas they are already in the hundreds at the dipeptide level. It might, at first sight, seem to be an impossible task to store so many minimum-energy conformations, as the peptides get longer and longer. However, various strategies permit a legitimate elimination of many of these as longer peptides are built up. For example, in the pentapeptide of sequence A-B-C-D-E, built from the tetrapeptides A-B-C-D and B-C-D-E, we can eliminate all tetrapeptide conformations in which residues B-C-D (being common to both tetrapeptides) are not in the same conformational state. The conformational states that are stored are ordered according to their energies, taking hydration into account with a solvent-shell model. As the peptides become longer and longer, long-range interactions alter the (energetic) order of their conformations. The variety of structures being stored at each stage have conformations that allow such long-range interactions to come into play, as residues are added to the growing chain. From a practical point of view, this procedure and others have been applied to polypeptides as large as 25-30 residues. These 25-30-residue segments are then "stitched" together to build up even larger structures. Thus far, about half of the 156-residue interferon molecule has been built up (Gibson et al, 1986). In combination with distance restraints (3.3 restraints per residue), the structure of the complete BPTI (bovine pancreatic trypsin inhibitor) molecule has been computed (Vasquez and Scheraga, 1988b); the backbone atoms of the resulting structure (shown in Figure 2) have an r.m.s. deviation of 1.19 Å from an idealized model of the X-ray structure.

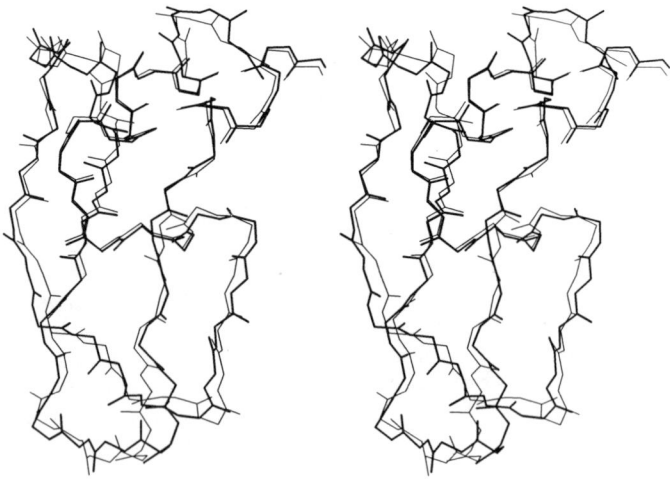

Figure 2. Stereo diagram illustrating superposition of backbone (N, C^{α}, C', O atoms) of computed structure of BPTI (thick line) and backbone of native structure (thin line) (Vasquez and Scheraga, 1988b).

Attempts are currently being made to use empirical prediction schemes to reduce the number of conformations being stored at each stage of the build-up procedure, and thereby eliminate the need to introduce distance restraints. For example modification of the nearest-neighbor Ising model (Vasquez and Scheraga, 1988a) can be used, together with experimental data on the helix-coil transition in host-guest random copolymers, to predict the locations of α-helical sequences. Likewise, experimental data on the influence of inter-residue interactions on the formation of an intramolecular disulfide bond can be used to predict the locations of β-bend and anti-parallel β-sheet structures (Milburn et al., 1988).

OPTIMIZATION OF ELECTROSTATICS

In this SCEF (Self-Consistent Electrostatic Field) procedure, we make an initial approximation by neglecting all components of the total energy except the electrostatic, and assume that each residue must have optimal electrostatic energy; i.e. the dipole moment of each residue must be optimally aligned in the electrostatic field created by the _whole_ molecule. If it is not, we change the orientation of the dipole moment (of each residue, successively) to improve its electrostatic energy. Since this involves a _local_ movement (in the field of the whole molecule), it is computationally very fast. Then the energy of the whole molecule (taking all interactions, not only electrostatic, into account) is minimized, and the whole procedure is repeated iteratively.

Thus far, we have tested this procedure on a 19-residue poly-(L-alanine) chain with acetyl- and N-methyl amide terminal blocking groups. The global minimum of this structure is presumably an α-helix. We started with conformations _very far_ from the helical conformation, and in trivially short computation time achieved the global minimum (Piela and Scheraga, 1987). The top stereo diagram of Figure 3 illustrates one of the starting conformations (optimized by conventional energy minimization), which is very far from a helical one. After application of the electrostatic-optimization procedure (and subsequent energy minimization with the complete energy function), the global-minimum (α-helix) structure at the bottom of Figure 3 was obtained. Unlike the usual minimization procedures, which make _small_ changes in the dihedral angles, this new procedure can make very large changes (even 100°-200°) in these independent variables. Further improvements of this method are described in a later section.

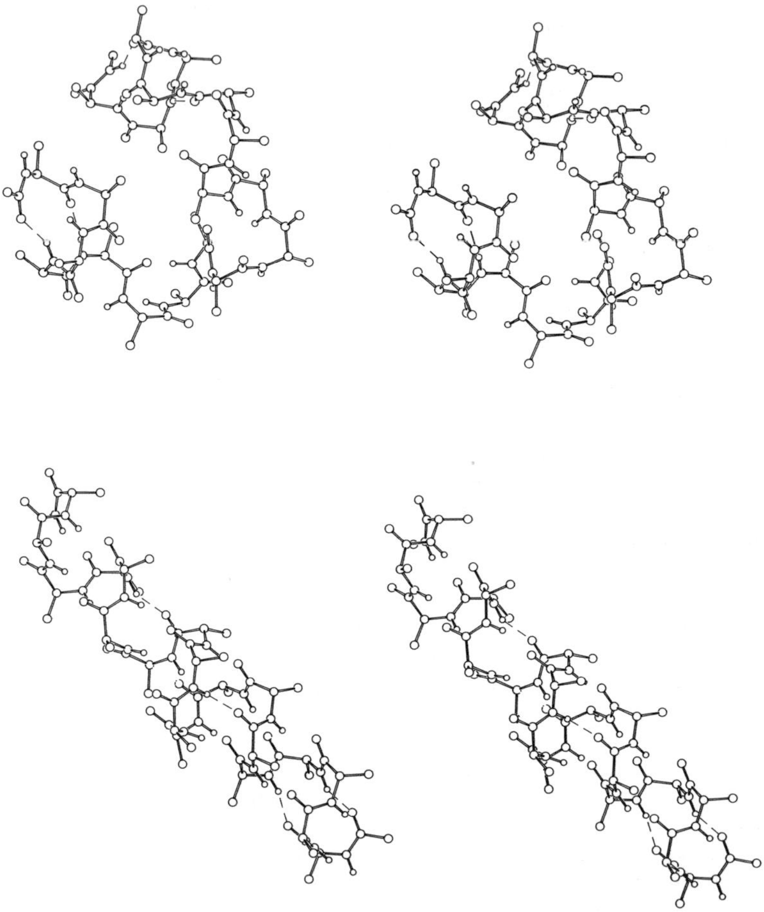

Figure 3. Stereo diagrams illustrating (top) a compact conformation of N-acetyl-(Ala)$_{19}$-NHCH$_3$ after complete energy minimization to reach this particular local minimum, and (bottom) the global-minimum (α-helix) structure attained by first optimizing the local electrostatic interactions and then carrying out a complete energy minimization (Piela and Scheraga, 1987).

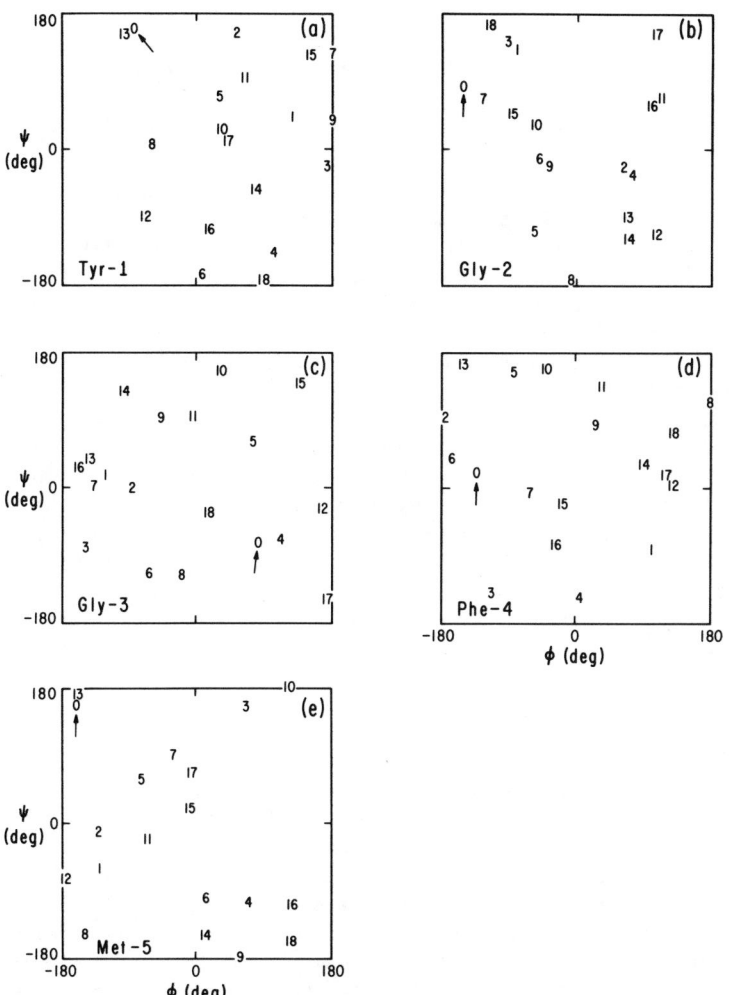

Figure 4. Illustration of random starting conformations of enkephalin (indicated by the numbers 1-18), and the global minimum (indicated by 0) which was reached from 13 of the 18 starting conformations (Li and Scheraga, 1987). In a subsequent variation of the procedure (Li and Scheraga, 1988), <u>all 18</u> starting conformations converged to the same global minimum.

MONTE CARLO-PLUS-MINIMIZATION

To overcome the inefficiency of Metropolis Monte Carlo, which searches all of conformational space very slowly, we have devised a procedure to move rapidly through the space of local minima (Li and Scheraga, 1987). The energy of a random starting conformation is minimized. Then a random change (0 to 2π) is made in a randomly-chosen dihedral angle, and the energy of this new conformation is minimized. The Metropolis criterion is used to decide whether to accept this new minimum, and the procedure is then iterated. Figure 4 shows how 13 of 18 random starting conformations of the pentapeptide enkephalin converged to the same global minimum; the remaining 5 runs were trapped in (recognizable) higher-energy local minima. This MCM (Monte Carlo-plus-minimization) procedure was subsequently improved (Li and Scheraga, 1988) by allowing random changes in more than one dihedral angle with successively lower probabilities and by sampling fluctuations in backbone dihedral angles more frequently then side-chain dihedral angles; with this modified sampling strategy, the 5 initial conformations, which previously did not converge, all achieved convergence. We are now trying to scale up this procedure to larger molecules, and include the effect of hydration.

ELECTROSTATICALLY DRIVEN MONTE CARLO

The electrostatically driven Monte Carlo (EDMC) procedure incorporates the best features of the SCEF and MCM methods, combined with random conformational changes to simulate the effect of thermal motion (Ripoll and Scheraga, 1988). This technique analyzes a given conformation (the current one), producing an electrostatic diagnosis based on the orientations of the dipole moments of the protein with respect to the local electric field. This diagnosis is used in combination with a random sampling technique to generate new conformations each of which is subjected to conventional energy minimization to reach a local energy minimum. This local minimum is compared with that corresponding to the current conformation with the aid of the Metropolis criterion. Each time that a conformation is accepted, it replaces the current one and is subjected to an identical analysis. If all the electrostatic diagnoses fail to produce an acceptable conformation, and this situation remains unalterable after generating a significant number of random conformations, the process is forced to choose one of the conformations generated previously (but rejected) and to accept that one as the subsequent current conformation. This procedure is equivalent to a perturbation due to thermal effects. Figures 5 and 6 illustrate how the right-handed α-helix [the presumed global-minimum structure of a terminally-blocked 19-residue chain of poly(L-alanine)] was achieved in successive stages of this

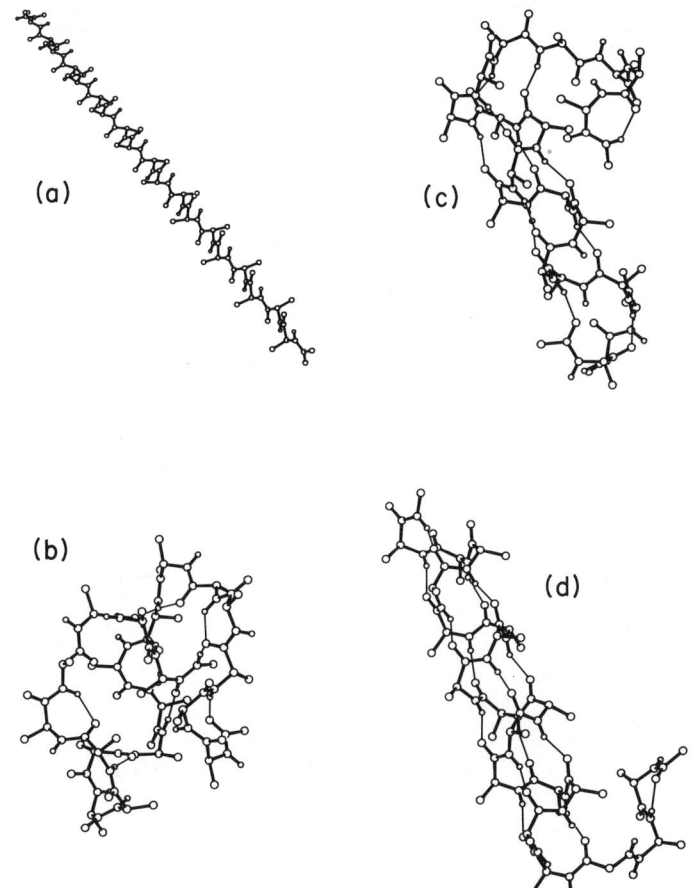

Figure 5. Diagrams of a set of conformations of poly(L-alanine) encountered at various stages on a conformational pathway during a folding simulation, starting from a fully-extended conformation (Ripoll and Scheraga, 1988).

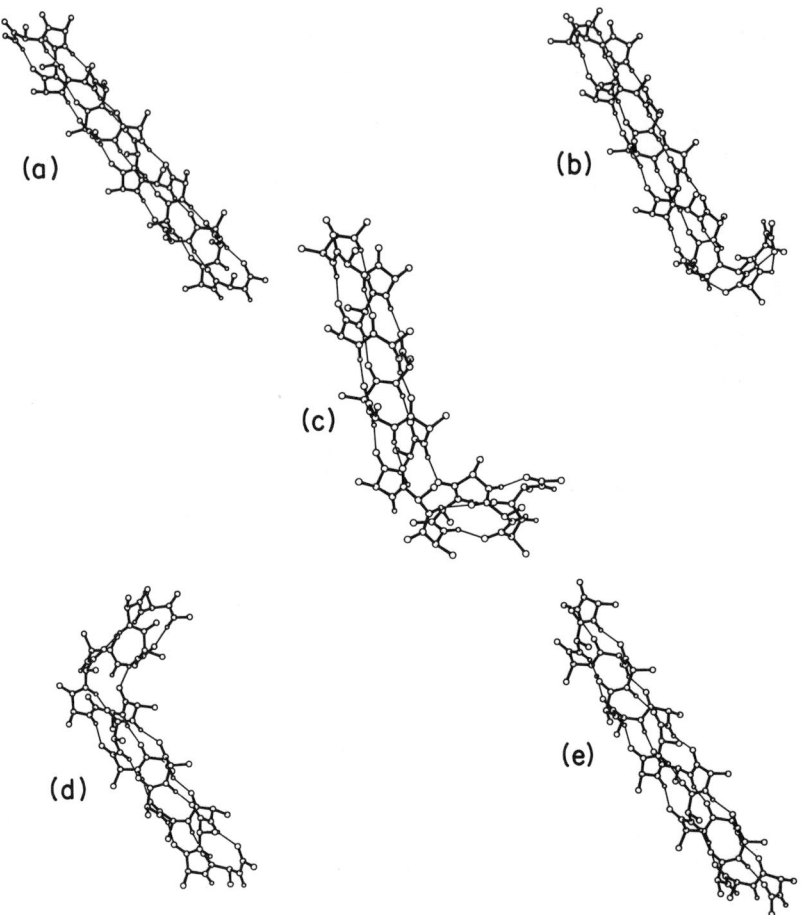

Figure 6. Diagrams of a set of conformations of poly(L-alanine) encountered at various stages on a conformational pathway during a folding simulation, starting from a <u>left</u>-handed α-helix (Ripoll and Scheraga, 1988).

---

procedure, by starting from the fully-extended structure and from the left-handed α-helix, respectively. Application of this method to BPTI has resulted in the location of a new set of local minima ranging from 35 to 60 kcal/mol lower than the best one found previously (-442 kcal/mol) (D.R. Ripoll, L. Piela and H.A. Scheraga, in preparation).

INCLUSION OF DISTANCE RESTRAINTS

As already mentioned above, the structure of BPTI has been computed by energy minimization, with distance restraints included. Such distance information is obtainable, for example, from two-dimensional NMR experiments or from non-radiative energy transfer experiments.

Another use of distance restraints, together with energy minimization, is in the computation of the structures of homologous proteins. The known structure of one such protein provides enough distance restraints which, together with energy minimization, can lead to the computed structure of the other. Several years ago, we computed the structure of α-lactalbumin, based on its homology to lysozyme (Warme et al, 1974). Recently, the X-ray structure of α-lactalbumin has been determined (Smith et al, 1987), and it agrees very well with the computed structure (D.C. Phillips and H.A. Scheraga, in preparation). Figure 7 provides a superposition of the two structures.

Figure 7. Stereo diagram illustrating superposition of computed structure of α-lactalbumin (Warme et al, 1974), in light lines, on the X-ray structure (Smith et al, 1987), in dark lines (D.C. Phillips and H.A. Scheraga, in preparation).

ADAPTIVE IMPORTANCE SAMPLING MONTE CARLO

Another procedure to overcome the inefficiency of Metropolis Monte Carlo is adaptive importance sampling (Paine and Scheraga, 1987; Scheraga and Paine, 1986). In this technique, the partition function (and quantities derived from it, such as the probability of a given conformation) is evaluated by continually upgrading the distribution function (ultimately, to the Boltzmann distribution) to concentrate the sampling in the region(s) where the probabilities are highest. Tests with enkephalin led to a low-energy structure.

INCREASE OF DIMENSIONALITY

Vanderbilt and Louie (1984) developed an annealing approach in which the temperature of the system is raised (when the minimization becomes trapped in a local minimum) and a Monte Carlo procedure is carried out to allow the system to escape from the local potential well. We have recently developed a method for relaxing a system, not by raising the temperature but by raising the dimensionality of the space (Purisima and Scheraga, 1986, 1987). In higher dimensional space, there are many more degrees of freedom in which the atoms can move about, making it easier to adjust to a low-energy conformation. Many potential barriers in three dimensions do not exist in higher dimensions. Our method starts from a very low-energy high-dimensional conformation and obtains a low-energy three-dimensional structure from it by gradual contraction of the dimensionality. The contraction in dimensionality is achieved by use of Cayley-Menger determinants, of which we have derived a simplified form (Sippl and Scheraga, 1986; Purisima and Scheraga, 1986, 1987).

In using this procedure, the energy-minimization problem is recast in terms of distances as the primary variables. In this respect, there is a similarity to the distance-geometry approach of Crippen (1981) and Braun et al (1981); however, our method differs from these in the manner in which the distances are used. Each distance variable is initially set to its global- energy minimum subject to whatever geometric constraints may be imposed on it. This starting set of distances is then at a global-energy minimum if the dimensionality is not a consideration; i.e. this is a lower bound on the energy. To obtain a realizable three-dimensional structure, a penalty function is added to the objective function to be minimized. This penalty function consists of the Cayley-Menger five- and six-point determinants. The purpose of these is to force the four- and five-dimensional volumes of the structure to zero (to satisfy the necessary and sufficient condition that a structure be embeddable in a three-dimensional space). A penalty function consisting of upper and lower bounds on the distance is also added. These bounds are obtained from covalent geometric constraints but may optionally contain bounds from other theoretical or experimental considerations. By steadily increasing the weight of the determinants, the distances are forced to three-dimensionality, and the three-dimensional global energy minimum is approached from below rather than from above. The expectation is that, as the dimensionality is gradually reduced, the structure, having started from a high-dimensional global minimum, will evolve into the three-dimensional global energy minimum. Preliminary results (Purisima and Scheraga, 1986) for a virtual-bond pentapeptide and for full-atom representations of several terminally blocked amino acids are encouraging, and application of the method to enkephalin (Purisima and Scheraga, 1987) led to the same global minimum described in Figure 4.

PATTERN RECOGNITION PLUS MINIMIZATION

Pattern recognition techniques are used to predict a series of probable backbone structures, whose energies are then minimized to locate the global minimum (Lambert and Scheraga, 1988). The (ϕ,ψ) map of each residue is divided into four regions (α, ε, $\alpha*$ and $\varepsilon*$), and all possible tripeptides from a properly selected set of X-ray structures from the Brookhaven Protein Data Bank (Bernstein et al, 1977) are collected and grouped according to conformation (e.g. $\alpha\alpha\alpha$, $\alpha\varepsilon\varepsilon$, $\alpha\varepsilon\varepsilon*$, etc). These data are then used to calculate the probability that each tripeptide in a protein under study has a given conformation.

The polypeptide chain is built up from the N-terminus, fitting the most probable tripeptide conformations together, one tripeptide at a time, allowing for proper overlap of the tripeptides. As the build-up proceeds, the probabilities of the growing chain (conformation) are calculated, and only the 1000 most probable are retained. Thus, when the C-terminus is reached, there are 1000 different predictions of the backbone structure of the protein, sorted in order of decreasing probability.

The symbolic representation (in terms of the regions α, ε, $\alpha*$, $\varepsilon*$) of the conformation of a protein is converted to a dihedral-angle representation by randomly generating values of ϕ and ψ in each of the assigned regions from appropriate probability distributions. A bivariate (two-dimensional) Gaussian distribution parameterized on (ϕ,ψ) values from the known X-ray structures (Bernstein et al, 1977) is used, together with standard techniques for generating random numbers from Gaussian distributions. Several such random structures are generated for each backbone prediction, and the energy of each of them is minimized. The lowest-energy structure is taken to represent the backbone prediction. The aforementioned probabilities serve to reduce, to a manageable size, the set of conformations whose energies have to be minimized. This procedure is currently being tested on avian pancreatic polypeptide (Lambert and Scheraga, 1988).

CONCLUSIONS

Progress has been made in surmounting the multiple-minima problem for the small polypeptides illustrated in the foregoing sections. With the availability of super-computers, the possibility exists to scale up these procedures to treat a protein of 100-200 residues. Efforts in this direction are in progress.

ACKNOWLEDGMENTS

This work was supported by research grants from the National Science Foundation (DMB84-01811) and from the National Institutes of Health (GM-14312).

REFERENCES

Bernstein FC, Koetzle TF, Williams GJB, Meyer EF, Jr, Brice MD, Rodgers JR, Kennard O, Shimanouchi T, Tasumi M (1977). J Mol Biol 112: 535-542.
Braun W, Bösch C, Brown LR, Gō N, Wüthrich K (1981). Biochim Biophys Acta 667:377-396.
Chou KC, Maggiora GM, Némethy G, Scheraga HA (1988a). Proc Natl Acad Sci, USA, in press.
Chou KC, Némethy G, Pottle M, Scheraga HA (1988b). J Mol Biol, submitted.
Crippen GM (1981). "Distance Geometry and Conformational Calculations." Chichester: Research Studies Press.
Dygert M, Gō N, Scheraga HA (1975). Macromolecules 8:750-761.

Gibson KD, Scheraga HA (1987). J Comput Chem 8:826-834.
Gibson KD, Scheraga HA (1988). In Sarma MH, Sarma RH (eds): "Structure and Expression: Vol. I, From Proteins to Ribosomes", Guilderland, N.Y.: Adenine Press, pp. 67-94.
Gibson KD, Chin S, Pincus MR, Clementi E, Scheraga HA (1986). In Dupuis M (ed.): "Supercomputer Simulations in Chemistry", Berlin: Springer-Verlag, pp. 198-213.
Glasser L, Scheraga HA (1988). J Mol Biol 199:513-524.
Gō N, Scheraga HA (1976). Macromolecules 9:535-542.
Hull SE, Karlsson R, Main P, Woolfson MM, Dodson EJ (1978). Nature 275:206-207.
Kang YK, Némethy G, Scheraga HA (1987). J Phys Chem 91: 4105-4120.
Lambert MH, Scheraga HA (1988). Biopolymers, to be submitted.
Li Z, Scheraga HA (1987). Proc Natl Acad Sci, USA 84:6611-6615.
Li Z, Scheraga HA (1988). Theochem, in press.
Milburn PJ, Meinwald Y-C, Takahashi S, Ooi T, Scheraga HA (1988). Intntl J Peptide and Protein Res 31:311-321.
Miller MH, Scheraga HA (1976). J Polymer Sci: Polymer Symposia, No. 54, pp. 171-200.
Némethy G, Scheraga HA (1984). Biochem Biophys Res Commun 118:643-647.
Okuyama K, Tanaka N, Ashida T, Kakudo M (1976). Bull Chem Soc Japan 49:1805-1810.
Paine GH, Scheraga HA (1987). Biopolymers 26:1125-1162.
Piela L, Scheraga HA (1987). Biopolymers 26:S33-S58.
Purisima EO, Scheraga HA (1986). Proc Natl Acad Sci, USA 83:2782-2786.
Purisima EO, Scheraga HA (1987). J Mol Biol 196:697-709.
Ripoll DR, Scheraga HA (1988). Biopolymers, in press.
Scheraga HA (1984). Carlsberg Research Commun 49:1-55.
Scheraga HA (1986). Israel J Chem 27:144-155.
Scheraga HA, Paine GH (1986). Annals NY Acad Sci 482: 60-68.
Sippl MJ, Scheraga HA (1986). Proc Natl Acad Sci, USA 83: 2283-2287.
Smith SG, Lewis M, Aschaffenburg R, Fenna RE, Wilson IA, Sundaralingam M, Stuart DI, Phillips DC (1987). Biochem J 242:353-360.
Vanderbilt D, Louie SG (1984). J Comput Phys 56:259-271.
Vasquez M, Scheraga HA (1988a). Biopolymers 27:41-58.
Vasquez M, Scheraga HA (1988b). J Biomolec Structure & Dynamics 5:705-755, 757-784.

Warme PK, Momany FA, Rumball SV, Tuttle RW, Scheraga HA (1974). Biochemistry 13:768-782.

Modelling the Structure of Highly Constrained Proteins

Cyrus Levinthal

Department of Biological Sciences
Columbia University
New York, New York 10027

During the past several decades an enormous amount has been learned concerning the general principles of protein structure from the large number of proteins whose coordinates had been determined crystallographically. However, even the improvements in X-ray sources and detectors which are now taking place and the development of new procedures such as NMR spectroscopy, the rate of experimental structure determination will continue to be at least two orders of magnitude slower than the rate of protein sequence determination deduced from DNA sequencing. New automatic techniques of DNA sequencing, automation of DNA preparation and purification, and the intensified effort to sequence the human genome will obviously increase by a large factor the rate at which amino acid sequences will be deduced from the genes encoding both known and unknown proteins.

However, as the rate which DNA sequences has increased, it has become more and more clear that proteins are found in families having strong homologies both in sequence and structure and this information presents a new and very approachable challenge for theoreticians. Can we deduce by computational means the 3-D conformation of members of such families if the conformation of at least one member has been determined crystallographically?

In addition to having problems which seem to be approachable, those of us who would like to predict conformations have the prospect of being able to use site-directed mutagenesis to evaluate the predictions when they have been made.

There are three activities in our laboratory which deal with these predictions and verification aspects of protein conformation. The first has to do with studies of a small protein, colicin E1, a bacterial toxin which kills other E. coli by forming an ion channel releasing the transmembrane potential and sodium and potassium ions from within the target cells(1). The portion of the colicin molecule which we study is the carboxyl terminal end which rapidly attaches to a phospholipid bilayer in vitro and when a transmembrane potential is applied will produce a ion channel. We first attempted to model the channel-forming region with six alpha helices stacked to form a barrel. These models seemed computationally reasonable although we were not able to produce an effectively minimized version. However, a critical test was provided by the production of short peptides which would not provide enough material to produce these 6 helices. The short peptides were made by using site-directed mutagenesis to insert new Methionine codons into the gene which encodes the colicine. So we can then treat the protein with CNBr and cleave protein at the location of the new mets. As a result of these manipulations, it was possible to show that a peptide of only 88 amino acids could still produce an ion channel(2) in a black lipid membrane. Of course this experiment would only require that the short peptide could make an ion switchable channel by itself if it could be shown that a single molecule is used to form the channel. There was good kinetic evidence to this effect from many investigators ever since the original experiments of Jacob and his collaborators(3) 35 years ago. However in order to establish the conclusion by stoichiometry Wayne Hubbell and Paul Todd of UCLA and Francoise Levinthal and I did an experiment to establish the molecularity of the channel by direct measurement. This was done using phospholipid vesicles in which a voltage gradient was established by loading the inside of the vesicles with potassium while the bathing medium was equimolar in sodium and then treating the suspension with valinomycin which allow passage of the potassium but not sodium. In order to determine the fraction of the vesicles which retained the transmembrane potential formed by valinomycin we used an electron spin labeled phosphonium ion which Hubbell and his collaborators had previously shown to partitions into the membrane if there is a transmembrane potential but not otherwise(4). They had furthermore shown that the spin label when it is in the membrane has a very much broadened EPR spectrum, a result which can be readily detected. Our experiment provided a clear demonstration that one molecule of colicin could discharge the potential of one vesicle and the experimental results were inconsistent with the hypothesis that more than one molecule was needed to form the channel.

Thus, we were faced with the question of how one could form an ion channel with a peptide of only 88 amino acids when 18 to 20 are needed to form an alpha helix which can span the thickness of the lipid region of a bilayer. In order to model an ion channel with a lumen large enough to match the experimentally determined conductance six helices would be needed and there is clearly not a sufficient stretch of peptide to form such a structure. However, when a series of beta helices were studied, it seemed clear that plausible models could be formed with a structure similar to that of gramicidin but with all amino acids of the normal L conformation. The proposed gramicidin structure(5) is energetically possible only because the sequence alternates between L and D amino acids and it is this alteration which avoids the excessively close contacts of the carbon-beta atoms which would otherwise be found on the inside of the helix. However, for the much larger channnel we propose for colicin this difficulty does not arise. Mr. Huajun Wang and Dr. Philippe Youkharibache has shown by extensive model building, molecular dynamics and energy minimization that at least two different models of this general type are energetically possible and we are now using a large set of new mutants to determine whether this type of model is in fact correct and if so which residues point into the lumen and which point out into the lipid. Mutants which we are now using introduced new unique cysteine residues into the peptide so that they can be used as attachment sites for either electron spin labels or groups which can block the passage of ions through the channel.

There are two other projects being pursued in our laboratory which are more directly related to the topic of this meeting. In order to make a serious attempt to determine the conformation of a protein which has a high degree of homology to others which have been determined crystallographically, it is necessary to find the conformation of lowest free energy for those regions of the protein which are different from the known structure. Frequently the difference region is composed of one or more loops which have their ends connected to the rest of the protein by bonds at known positions. For example, to find the structure of antibody combining sites when the amino acids sequence is known we must determine the lowest free energy state which would be accessible to the set of complementarity determining (hyper variable) loops. Our group (6) and (7) has been carrying out this effort for some time using many hours of computer time on a STAR-100 array processor which for these problems has somewhat more speed than a CYBER-205. Our approach to finding the global energy minima for these loops is to deal with them first separately and then together using randomly generated starting conformations with each followed by molecular dynamics and energy minimization.

The first test we made of this procedure was to see if we could reach the crystallographically determined conformations of the four smaller loops of the antibody MPCP-603 by selecting the lowest energy states reached from an ensemble of starting structures. For these short loops ranging in size from five to nine amino acids the results were quite good in that the lowest energy state showed either one or very small number of conformations which were essentially identical to the crystal structure(8) of the loops.

However, this procedure which can be termed "the brute force approach" is very computationally demanding and its full exploitation waits for more computing power than we have on the STAR alone. In addition new programs are being written so that all aspects of the free energy estimates can be made. To supplement the interaction energy which gives the enthalpy we can use a rapid program for obtaining the accessible surface area of the protein to make an estimation of the entropy contribution due to the immobilization of water molecules around the protein. This surface area program has been written by Mr. Huajun Wang to run on the STAR-100 and is about 100 times faster than the corresponding program currently used on a VAX.

The primary question which must be addressed in considering an approach of the type proposed here is how can one determine the number of different starting conformations which must be used to have a high probability of finding the global minimum for a loop. The answer to this question depends on the answers to two other questions: 1) How well can the dynamics and minimization routines used get over small bumps in the energy surface and still reach a deep minimum; that is, how far from a deep minimum can the starting conformation be and still reach it andd 2) how broad is the energy minimum which we must find. There are several ways in which these questions can be approached theoretically although they are not as yet very satisfactory. However, the initial results we obtained for the antibody loops leads to some optimism that the final answer will be that with the methods now being developed the problem for loops of the size necessary to deal with the homology problem can be handled successfully.

However, it is clear that in order to pursue this approach successfully one needs a large amount of computer power. It was for this reason that Dr. Richard Fine and I drew the design for a special purpose attached processor several years ago. We believe that such a processor could be made which would calculate in a synchronouslong pipe-line cyncrinous system the two body terms required for the force and energy calculations used in the molecular mechanics calculations, either dynamics or energy minimization. This part of the calculation requires over 90% of the total time for a large molecule on any of the existing computers--super or otherwise. We have used quadratic look-up tables for all force and energy function evaluations in which the function, the first derivative and the second derivative are stored for each of many bins as a function of distance and as a function of the type of atoms in the atom pair being considered. Since these tables can be easily changed from the host computer we do not believe that this type of attached processor will become obsolete as the algorithms develop since any force field can be fed into these tables and regardless of the change in the methods for determining free energy one will certainly need to use the two body interactions even if other terms must be added.

This system called FASTRUN(9) has had its engineering design and construction done at the Brookhaven National Laboratory under the direction of Mr. Gurd Dimmler with Dr. Fine directeing the testing of the system on a virtually continuous basis. FASTRUN should be ready for operation at Columbia in July of this year although not uet at its design speed but at a sufficient speed so that the entire system of STAR plus FASTRUN will provide computational speed several times greater than a CRAY-XMP. This Should be a sufficiently powerful system so that a great deal of science can be done and within a few months the change in two boards of the system will allow the speed to increase to its original design specifications and the overall speed of the system for large molecule-molecular mechanics calculations will be about 8-10 times that of a CRAY-XMP.

In conclusion it seems to be a reasonable and not overly optimistic statement to suggest that homology model building either by the methods we propose or others being developed elsewhere will allow theoreticians to make reasonable predictions as to the conformation of protein molecules which are members of homologous families as long as one member of the family has been solved crystallographically and the use of site-directed mutagenesis should allow one to valuate these predictions and modify them if wrong. However, it does not seem likely that there will be any way in which this type of computation can be used to predict the structure of a protein from the amino acid sequence alone if we do not have the crystal structure for a very similar molecule.

BIBLIOGRAPHY

1. Schein, S.J., Kagan, B.L., and Finkelstein, A.; Colicin K acts by forming voltage-dependent channels in phospholipid bilayers membranes; **Nature** 267: 159-163, 1978.

2. Lin, Q.R., Crozel, V., Levinthal, F., Slatin, S., Finkelstein, A. and Levinthal, C.; A very short peptide makes a voltage-dependent ion channel: The critical length of the channel domain of colicin E1. **Proteins: Structure, Function and Genetics** 2: 218-229; 1986.

3. Jacob, F., Siminovitch, L. and Wollman, E.; Sur le biosynthese d'une colicin et son mode d'artion; **Ann. Inst. Pasteur** 83: 295-315; 1952.

4. Cafiso, D. and Hubbell, W.L.; Estimation of transmembrane potentials from phase equilibria of hydrophobic paramagnetic ions; **Biochemistry** 17: 187-195; 1978.

5. Urry, D.W.; The gramacidin A transmembrane channel: A proposed $\pi_{(L, D)}$ Helix; **Proc. Natl. Acad. Sci. USA** 68: 672-676; 1971

6. Shenkin, P., Yarmush, D.L., Fine, R.M., Wang, H. and Levinthal, C; Predicting antibody hypervariable loop conformation. I. Ensembles of random conformations for ringlike structures; **Biopolymers** 26: 2053-2085; 1987.

7. Fine, R.M., Wang, H. Shenkin, P.S., yarmush, D.L. and Levinthal, C; Predicting antibody hypervariable loop conformations II: Minimization and molecular dynamics studies of MCPC603 from many randomly generated loop conformations; **Proteins: Structure, Function, and Genetics** 1: 342-362; 1986.

8. Segal, D.M., Padlan, E.A., Cohen, G.H., Rudikoff, S., Potter, M. and Davies, D.R; The three dimensional structure of a phosphocholine-binding mouse immunoglobulin Fab and the nature of the antigen binding site; **Proc. Natl. Acad. Sci. USA** 71: 4298-4302; 1974.

9. Levinthal, C., Fine, R. and Dimmler, G.; FASTRUN: A special purpose hard-wired computing device fro molecular mechanics; In **"Proc. Mol. Dyn. Prot. Structure;"** Hermans, J. ed. pp 126-129; 1985.

ASSORTED MODULES IN PROTEIN ARCHITECTURE

Mitiko Gō and Michiko Nosaka

Department of Biology, Faculty of Science,
Kyushu University, Fukuoka 812, Japan

INTRODUCTION

It is an attractive idea to assume that proteins have been designed by combining small pieces of polypeptides in early stage of evolution. Globular domains which are well defined by X-ray crystallographic studies, are actually one of such parts. Similar globular domains are sometimes observed in a variety of proteins (Forthergill-Gilmore, 1986). They appear to have been integrated into various proteins by gene duplication, shuffling and fusion. These mechanisms are strongly supported by recent accumulation of DNA sequences encoding the proteins, particularly by the presence of introns close to the boundaries of the globular domains (Blake, 1985; Gō, 1985).

Introns are located at some boundaries of the domains in the genes encoding various multi-domain proteins. However, no introns are often located at such boundaries. Globular domains are functional as well as structural units in several known proteins. Domains of immunoglobulin are typical of such domains and the introns are located at sites corresponding to the boundaries of the domains in the genes of differentiated B-cells (Tonegawa, et al., 1978). However, the intron positions in many eukaryotic genes are not restricted only at the boundaries of the domains. Many introns were found within the coding sequences of the globular domains. Why are there such introns in the coding regions of globular domains? Were they inserted randomly in the eukaryotic genes? Or, are there any relationships between their positions and protein structures?

It has been found that globular domains or proteins are decomposable into compact units called modules and the boundaries of the modules are closely correlated with the intron positions of the genes encoding the proteins (Gō, 1981, 1983, 1985). We will summarize briefly such correlations in various proteins and describe new results of research on modules; i.e., on their conformations, functions and evolution.

MODULES AND INTRONS IN SMALL DOMAINS OR SUBUNITS

Modules were introduced as relatively compact structural segments in globular domains (Gō, 1981). In the monolayer domains or subunits, having no core modules buried inside, the boundaries of the modules are characterized by their locations not being close to the surface of the domains or subunits. The modules were identified in hemoglobin (Gō, 1981), lysozyme (Gō, 1983) and other small proteins or domains (Gō, 1985). Those modules were shown to be closely correlated with the exons of their genes, i.e., the introns are positioned close to the module boundaries (Gō, 1985). In small subunits or domains, modules have been identified by using distance map and by focusing on the amino acid residue pairs separated more than a certain distance in the three dimensional space (Gō, 1981).

The genes of hemoglobin α- and β-chains are split by two introns at the homologous positions (Konkel, Tilghman and Leder, 1978; Nishioka and Leder, 1979) and those positions are closely correlated with a junction between modules F1 and F2 and another between F3 and F4, respectively. By expecting one-to-one correspondence between the modules and exons, the existence of one more intron was predicted at the junction between modules F2 and F3 in related genes to hemoglobin. The expected intron was in fact found in leghemoglobin gene of soy bean (Jensen et al., 1981). Recently it was reported that also non-nodulating plants have hemoglobins (Bogusz, et al., 1988) and their genes are split by three introns at exactly the same positions to those of leghemoglobins. A simple and reasonable explanation is that the ancestral gene of hemoglobins were split into four exons by the three introns. And the contemporary plant globin genes conserved the ancestral intron positions. Since the divergence time of leghemoglobin from other globins was estimated more than one billion years ago (Dayhoff, et al., 1972), the origin of the introns should then be traced back more than one billion years ago.

Domain 3 of ovomucoid is decomposable into four modules M1-M4 (Gō, 1985). M1 corresponds to the short connecting region between domains 2 and 3, and the introns of its gene are found located close to the junctions M1-M2 and M3-M4 but not at the junction M2-M3. Interestingly, the gene of pancreatic secretory trypsin inhibitor (PSTI), a protein homologous to ovomucoid, is also split at the same positions to ovomucoid (Horii et al., 1987). It is suggested that an intron existed at the junction M2-M3 in an ancestral or related genes to ovomucoid and PSTI. The positions of introns in the families of serine protease inhibitors and globins are generally well conserved during long period of evolution, although some introns appear to have been lost in a meantime. The origin of the introns is probably older than the divergence of the eukaryotes and prokaryotes. A supposed role of the modules and introns in prebiological evolution has been discussed (Gō, 1985).

MODULES AND INTRONS IN LARGE PROTEINS

We have developed and applied an extended and improved method to identify modules in large or multi-layer proteins having core modules (Gō and Nosaka, 1987). By using this algorithm it is also possible to identify the modules with more accuracy than by the use of the distance map.

In monolayer domains or subunits having no core modules, the junctions between modules are characterized by their locations not being close to the surface of the domains or subunits (Gō, 1981). Utilization of this character as well as the compactness itself makes easier the identification of modules in small domains or subunits. In larger domains or subunits, however, junctions between modules are not necessarily located close to its center. Even in such cases, they are located close to local centers, i.e., to centers of long contiguous segments. This property allows us to detect module boundaries by considering not a whole chain, but only contiguous segments of a certain length and by locating their center (Gō and Nosaka, 1987).

Such centers were identified as local minima of centripetal profile (Gō and Nosaka, 1987) and their locations were found stable for the change of the lengths of the segments from 81 to 181 (Fig.1(a)). The compactness of the modules identified by the centripetal profile were confirmed by extension profile. Extension of a backbone segment, which has a certain length, was defined and calculated (Gō and Nosaka, 1987). Local maxima of the extensity profile were identified (Fig.1(b)) and modules were identified as the segments bordered by two consecutive local maxima of the profile.

Triose phosphate isomerase (TIM; EC 5.3.1.1) is one enzyme belonging to the glycolytic pathway. Chicken TIM functions as a dimer of identical subunits, each of which is composed of 247 amino acid residues. One subunit of chicken TIM consists of a single domain having a typical barrel structure where (α/β) repeats eight times (Banner et al., 1975).

All of the identified local minima in the centripetal profile, except one at residue 10 (Fig.1(a)), have correspondents in the local maxima of the extensity profile (Fig.1(b)). Boundaries of the modules identified by the two profiles sometimes differ by a few residues. Module boundaries can be identified with this resolution. Local minima or maxima within the window length from the N- or C-terminals are less reliable. Thus at least 13 modules, possibly 14 modules, are identified in TIM by using a new algorithm (Gō and Nosaka, 1987).

The human (Brown et al., 1985) and chicken (Straus and Gilbert, 1985) TIM genes are split into seven exons; the positions of their introns are the same. The genes of maize (Marchionni and Gilbert, 1986) and *Aspergillus nidulans* (McKnight et al., 1986) are interrupted by eight and five introns, respectively.

Since the tertiary structure is conserved better than the primary structure, in the evolutionary time scale, the tertiary structures of human, maize, and *A.*

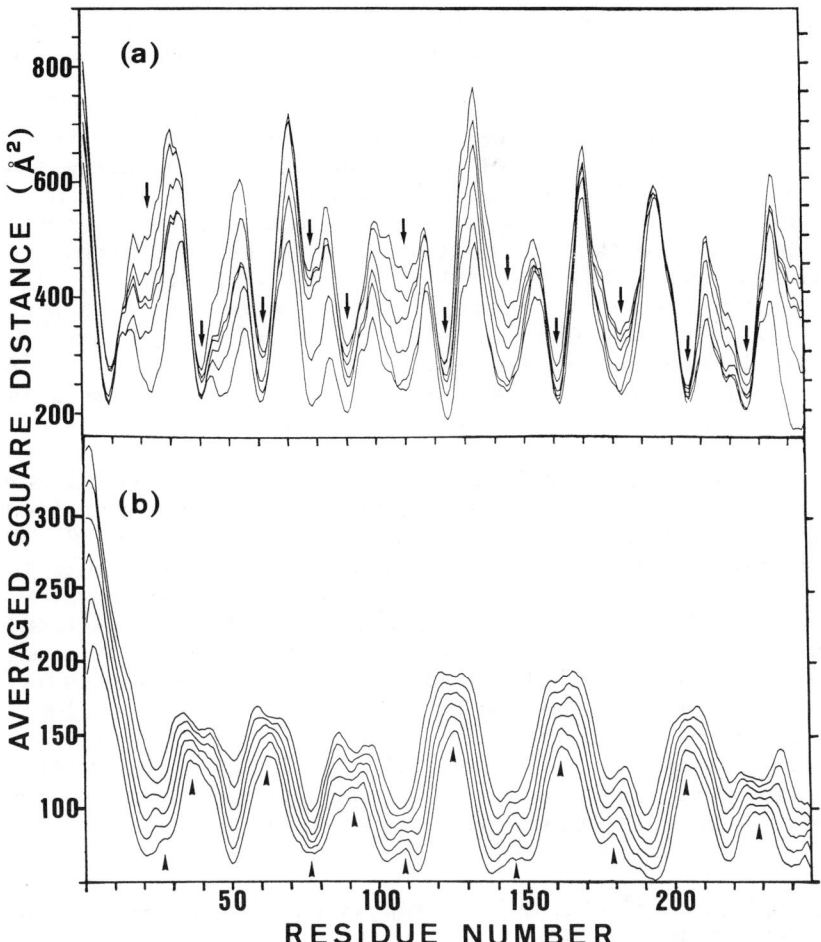

Fig.1 Centripetal (a) and extension profiles (b) of chicken TIM are shown. Centripetal profile is shown with window lengths of 81, 101, 121, 141, 161 and 181 residues and extension profile with those of 21, 23, 25, 27, 29 and 31 residues. The local minima in (a) and the local maxima in (b) marked with the arrows or arrow heads were identified as module junctions. Reproduced from Gō and Nosaka (1987) with permission.

nidulans TIMs are expected to be similar to that of chicken. Therefore, we can discuss the intron positions of these genes based on the module organization of chicken TIM (Gō and Nosaka, 1987).

All of the introns in chicken (Straus and Gilbert, 1985) and human (Brown et al., 1985) TIM genes correspond to the boundaries of the identified modules within a deviation of up to five residues and most of the introns except those close to the terminals in maize (Marchionni and Gilbert, 1986) and *A. nidulans* (McKnight et al., 1986) genes also correspond to these boundaries within up to six residues. One intron of *A. nidulans* TIM gene is located close to the N-terminal minimum at residue 10 in the centripetal profile, however, no close maximum is observed in the extensity profile (Fig.1(b)).

Such close correlation of modules with introns may not happen only by chance. The most probable explanation for the correspondence is that an ancestral gene of TIM was interrupted by at least 12 introns corresponding to the module boundaries identified; some of these, however, were lost during evolution. Then at least two more introns should have been located in an ancestral TIM gene close to the junctions of M3-M4 and M5-M6 (Gō and Nosaka, 1987). Close correlation between modules and intron positions, although there is an exceptional case seen in one extra intron in mouse renin gene (Miyazaki et al., 1984), implies that proteins have evolved by combining various modules into globular domains (Gō, 1985).

ARE MODULE JUNCTIONS OR INTRONS LOCATED ON THE SURFACE OR IN THE INTERIOR OF PROTEINS ?

Majority of the module junctions in small domains or proteins consisting of less than 200 residues are located in the interior of the domains (Nosaka and Gō, unpublished results). The module junctions as well as intron positions of hemoglobin α and β chains, lysozyme and ovomucoid third domain are buried in the interior of the proteins (Fig.2). This feature is expected from the definition of the modules in monolayer proteins (Gō, 1981), i. e. the junctions are not far from all the other residues. Relative accessibility normalized according to the size of the side chains (Gō and Miyazawa, 1980) is a useful

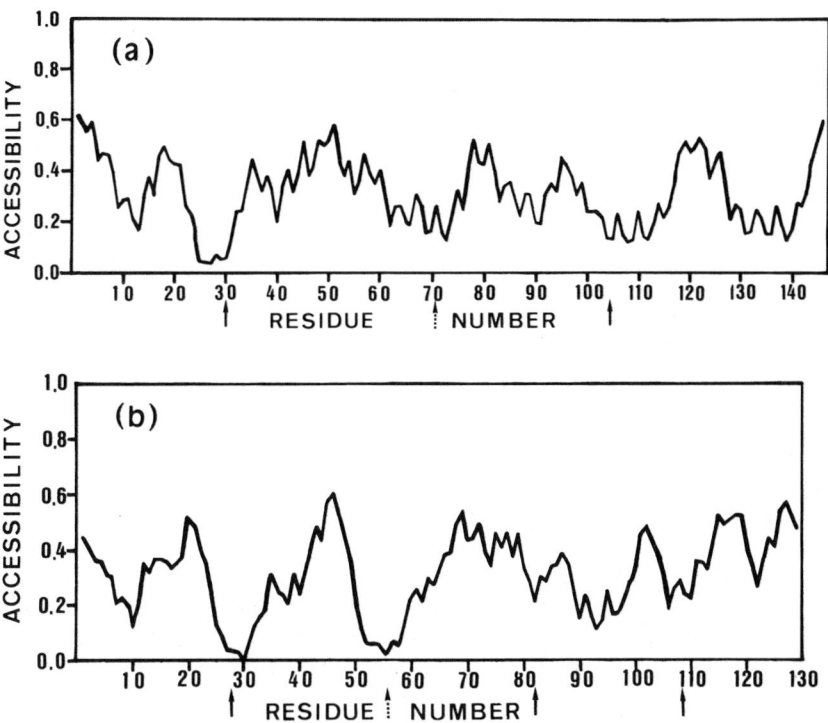

Fig.2 (a) Relative accessibility of hemoglobin β chain. The arrows at residue 30 and beween 104-105 show the positions of the introns in its gene. The arrow between 70-71 corresponds to the intron position found in leghemoglobin gene. (b) Relative Accessibility of hen egg white lysozyme. The arrows at residues 28, 82 and 108 show the intron positions in its gene. The arrow between 55-56 shows the boundary of modules M2 and M3. It was suggested that an intron was lost here in the gene.

measure to see whether the junctions are located on the surface of the proteins or not. Accessible surface area (ASA) of a residue itself is not a good measure to know the location of the residues, because the difference of ASA's of small and large sidechains can't be neglected.

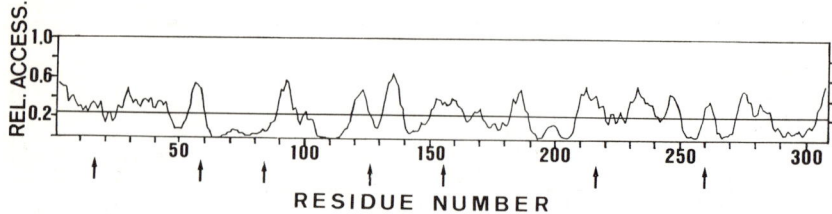

Fig.3 Relative accessibility of carboxypeptidase A. The arrows show the intron positions in its gene. Reproduced from Gō and Nosaka (1987) with permission.

This point was not taken into consideration properly before (Craik et al., 1983).

However, the boundaries of modules in large domains or proteins, where the module organization is complex, are not always found in the interior of the proteins. The compactness of modules does not always guarantee the locations of the module junctions in the inteior. The modules in large proteins are identified by searching local centers of the segments having certain length of the proteins and they are confirmed by their compactness. As a result, in large proteins some junctions are located on the surface and the others in the interior of the proteins. An example of such cases is carboxypeptidase A; the intron positions vary from the interior to the surface (Fig.3). A similar fact is seen in many other large proteins.

MODULES AND FUNCTIONS

Modules are structural elements in the three dimension, i.e., they are defined as contiguous polypeptide segments having least extended conformation. Various protein functions are closely related with the protein conformation. How catalytic sites are distributed on modules? Are modules functional units themselves or not? Catalytic residues glu-35 and asp-52 in chicken lysozyme are located together on module M2 and there is no sites involved in the interaction with the substrate on module M1 (Gō, 1983). ATP binding region in adenylate kinase corresponds close to one module (Nosaka and Gō, to be submitted).

However, the residues in contact with hem group in hemoglobin α and β chains and in other globins are equally distributed on modules M2 and M3 (Gō, 1981). It seems that the both modules M2 and M3 are necessary for function of hemoglobin. The catalytic sites of serine proteases and aspartyl proteases, nucleotide binding sites of alcohol dehydrogenase are distributed on different modules. CRO repressor of λ phage interacts with DNA through two α helices. This region is divided into two modules (Gō, unpublished results). Possible catalytic site glu-164 and phosphate binding site gly-232 of TIM are located on modules M10 and M13, respectively. It is noted that the modules having even numbers except M2 are involved in TIM function, i.e., the active sites (Banner et al., 1975) are distributed mainly on M4, M6, M8, M10 and M12, and on the terminal modules M1 and M13.

MODULES AND THEIR SECONDARY STRUCTURES

Since introns are good mediators of exon shuffling by making crossing over on themselves (Gilbert, 1978), the intron-module correspondence implies that proteins are designed by putting modules together; primitive proteins might have evolved by assembling compact modules, such consisting of about 20 amino acid residues. It is suggested that exon shuffling can bring out new functional proteins by various combinations of the modules. Most of the module junctions are located on the secondary structures (Gō and Nosaka, 1987). Secondary structures seem to hold flexible loops, on which active sites are often located, by facilitating structural foundations of protein architecture. The fact that many of the module junctions, not only of TIM but also of other proteins, are found on the secondary structures implies that proteins are designed in nature by making module boundaries rigid by use of the secondary structures as the joining regions of modules. This may be used as an idea in design of new artificial proteins by connecting short segments into stable globular proteins.

ACKNOWLEDGMENT

A part of this work was supported by Grants-in-Aid for Special Project Research Grant and for Scientific Research from the Ministry of Education, Science and Culture of Japan and by the Kurata Research Grant.

REFERENCES

Banner DW, Bloomer AC, Petsko GA, Phillips DC, Pogson CI, Wilson IA (1975). Structure of Chicken Muscle Triose Phosphate Isomerase Determined Crystallographically at 2.5 Å Resolution. Nature 255:609-614.
Blake CCF (1985). Exons and the Evolution of Proteins. Int Rev Cytology 93:149-185.
Bogusz D, Appleby CA, Landsmann J, Dennis ES, Trinick MJ, Peacock WJ (1988). Functioning Haemoglobin Genes in Non-nodulating Plants. Nature 331:178-180.
Brown JR, Daar IO, Krug JS, Maguart LE (1985). Characterization of the Functional Gene and Several Processed Pseudogenes in the Human Triose-phosphate Isomerase Gene Family. Mol Cell Biol 5: 1694-1706.
Craik SC, Sprang S, Fletterick R, Rutter WJ (1982). Intron-exon Splice Junctions Map at Protein Surfaces. Nature 299:180-182.
Dayhoff MO, Hunt LT, McLaughlin PJ, Jones DD (1972). Gene Duplications in Evolution: The Globins. In Dayhoff MO (ed): "Atlas of Protein Sequence and Structure," Vol.5 Washington: National Biomedical Research Foundation, pp 17-30.
Horii A, Kobayashi T, Tomita N, Yamamoto T, Fukushige S, Murotsu T, Ogawa M, Mori T, Matsubara K (1987). Primary Structure of Human Pancreatic Secretory Trypsin Inhibitor (PSTI) Gene. Biochem Biophys Res Commun 149: 635-641.
Forthergill-Gilmore LA (1986). The Evolution of Glycolytic Pathway. Trends Biochem Sci 11:47-51.
Gilbert W (1978). Why genes in pieces? Nature 271:501.
Gō M, Miyazawa S (1980). Relationship between Mutability, Polarity and Exteriority of Amino Acid Residues in Protien Evolution. Int J Peptide Protein Res 15:211-224.
Gō M (1981). Correlation of DNA Exonic Regions with Protein Structural Units in Haemoglobin. Nature 291:90-92.
Gō M (1983). Modular Structural Units, Exons, and Function

in Chicken Lysozyme. Proc Natl Acad Sci USA 80:1964-1968.
Gō M (1985). Protein Stuructures and Split Genes. Adv Biophys 19:91-131.
Gō M, Nosaka M (1987). Protein Architecture and the Origin of Introns. Cold Spring Harbor Symposia on Quantitative Biology, 52:915-924.
Jensen EO, Pauldan JK, Hyldig-Nielsen JJ, Jorgensen P, Marcker KA (1981). The Structure of a Chromosomal Leghaemoglobin Gene from Soybean. Nature 291:677-679.
Konkel DA, Tilghman S, Leder P (1978). The Sequence of the Chromosomal Mouse β-Globin Major Gene: Homologies in Capping, Splicing and poly(A) Sites. Cell 15:1125-1132.
Marchionni M, Gilbert W (1986). The Triosephosphate Isomerase Gene from Maize: Introns Antedate the Plant-Animal Divergence. Cell 46:133-141.
McKnight GL, O'Hara PJ, Parker ML (1986). Nucleotide Sequence of the Triosephosphate Isomerase Gene from Aspergillus nidulans: Implications for a Differential Loss of Introns. Cell 46:143-147.
Miyazaki H, Fukumizu A, Hirose S, Hatashi T, Hori H, Ohkubo H, Nakanishi S, Murakami K (1984). Structure of the Human Renin Gene. Proc Natl Acad Sci USA 81: 5999-6003.
Nishioka Y, Leder P (1979). The Complete Sequence of Chromosomal Mouse α Globin Gene Reveals Elememts Conserved throughout Vertebrate Evolution. Cell 18:875-882.
Straus D, Gilbert W (1985). Genetic Engineering in the Precambrian: Structure of the Chicken Triosephosphate Isomerase Gene. Mol Cell Biol 5:3497-3506.
Tonegawa S, Maxam AM, Tizard R, Bernard O, Gilbert W (1978). Sequence of a Mouse germ-line for a Variable Region of an Immunoglobulin Light Chain. Proc Natl Acad Sci USA 75:1485-1489.

A COMPUTER MODELING STUDY OF ACETYLCHOLINE RECEPTOR-LIGAND INTERACTIONS

Masayuki Shibata and Robert Rein

Department of Biophysics
Roswell Park Memorial Institute
666 Elm Street
Buffalo, New York 14263

INTRODUCTION

The nicotinic acetylcholine receptor (AChR) is composed of four different polypeptide chains assembled into a transmembrane pentamer with a stoichiometry of $\alpha_2\beta\delta\gamma$ (for recent reviews, see Changeux et al., 1984, McCarthy et al., 1986, Stroud and Finer-Moore, 1985, Kosower, 1987). The cholinergic binding site is in the extra cellular domain of the α-subunit. Based on affinity labeling, it was suggested that a sulfhydryl group is located in the vicinity of the binding site (Karlin, 1980). Cysteine residues 128, 142, 192, 193 are suggested to be involved in the formation of disulfide bridges at the binding site (Noda et al., 1982). The importance of these disulfide bridges in cholinorgic binding was confirmed by site directed mutagenesis which demonstrated that a mutation of any of the above mentioned cysteins to a serine eliminates the responsiveness of AChR to acetylcholine(ACh) (Mishina et al., 1985).

A primary ACh binding site in the region of Cys 128-Cys 142 is supported by the fact that the region contains the glycosylation site Asn 141 which is known to exist at the ACh binding site (Noda et al., 1982). More recent evidence from various studies including affinity labeling, toxin binding and antibody recognition (Criado et al., 1986, Dennis et al., 1986, Kao et al., 1984, Kao and Karlin, 1986, Lennon et al., 1985, Mulac-Jericevic and Atassi, 1986, Wilson et al., 1985) now suggest that the region containing Cys 192 and Cys 193 is involved in the primary Ach binding site. It was clearly demonstrated, by Neumann et al. (1986) with synthetic dodecapeptides corresponding to residues 185 through 196 of AChR, that the ACh and neurotoxins bind to this later region of AChR. Molecular model building studies of an ACh derivative and/or toxin binding to the peptide corresponding to AChR residues 136 to 142 (Smart et al. 1984 and Srinivasan et al. 1987) and of the toxin binding to the AChR models in the region of Cys 192 and Cys 193 (Garduno-Juarez et al., 1988) have also been reported.

The objective of the present work is to obtain better structural information about the cholinergic binding site of AChR by comparing and further refining the two different previously proposed models. Distance information obtained from AChR/CTX models served to guide the placement of ACh onto the AChR in the ACh/AChR complex models. The comparison of these receptor agonist/antagonist binding models may present clues for further characterization of AChR structure/function relationship.

METHOD

Our two previous α-cobratoxin - acetylcholine receptor (CTX/AChR) binding models (Srinivasan et al., 1987 and Garduno-Juarez et al., 1988) were further refined by the use of molecular model building programs developed in our laboratory (Rein et al., 1978 and Srinivasan et al., 1986). Since detailed modeling procedures were given previously, only a brief description of the models is provided here. Also, throughout the study, special attention was paid to maintaining correct chirality during the energy optimization and modeling process.

The first AChR model (Model 1), originally proposed by Smart et al., (1984), consists of a β pleated sheet spanning the residues 136-142 (PFDQQNC) and having a Type II' β bend (Rose et al., 1985) at Asp 138 and Gln 139. The earlier form for this CTX/AChR model (Srinavasan et al., 1987) was further refined by the molecular mechanics program written by us using a potential function described by Weiner et al., (1984).

The second model (Model 2) of AChR is a dodecapeptide corresponding to the residues 185-196 of the α-subunit of the Torpedo AChR. This synthetic dodecapeptide (KHWVYYTCCPDT) was shown to contain essential elements of the neurotoxin binding site (Neumann et al., 1986). This peptide also contains an unusual disulfide bond formed by adjacent cysteine residues, which together with the proline residue generate a turn in the otherwise β strand structure. This gives the receptor the overall shape of a hook attached to the concave surface of the CTX. Although the available crystallographic data on the synthetic L-cysteinyl-L-cysteine and its derivatives (Capasso et al., 1977 and Hata et al., 1977) suggests the existence of the unusual cis peptide bond linkage, the possibility of adapting a trans peptide bond conformation in the dodecapeptide model was examined. Both previous models, cis (model 2C) and trans (model 2T) conformers, were further refined in this study using the same CTX/AChR residues as before (Garduno-Jaurez et al., 1988).

After the CTX/AChR complex models were generated, the placement of ACh onto the receptor models was initiated by searching for potential binding sites utilizing distance information obtained from the CTX/AChR model complexes. Least square juxtaposition of ACh onto the interacting CTX residues and/or distance constrained (between AChR and ACh) molecular mechanics optimization followed the selection of the binding sites. The ACh charges obtained by Margheritis and Corongiu (1988) with ab initio minimal basis set (7s/3p) were used in the molecular mechanics optimization. A few new force field parameters were added to the original parameter set (Weiner et al., 1984) in order to reproduce the crystal structures of the ACh molecule. No attempt was made to optimize these parameters in this study.

After an initial placement of the ACh onto the AChR model, the distance constraints were removed. The model was then further refined, first, retaining the fixed AChR structure found for the CTX/AChR model and, second, allowing the AChR molecule to move freely.

RESULTS

The optimized structure of the model 1 CTX/AChR complex is shown in Figure 1. The hydrogen bonding and ionic contact distances between CTX and AChR (Table 1)

Table 1. Hydrogen bonding and ionic contact distances between cobratoxin (CTX) and the acetylcholine receptor (AChR) as found in the Model 1 structure.[a]

AChR Residues		CTX Residues		Distance (Å)
Asp 138	O	$N_\zeta(+)$	Lys 49	2.69
Asp 138	$O_{\delta 1}(-)$	$N_\zeta(+)$	Lys 49	2.53
Asp 138	$O_{\delta 2}(-)$	$N_\zeta(+)$	Lys 23	2.64
Gln 139	$O_{\varepsilon 1}$	$N_\zeta(+)$	Lys 23	2.70
Gln 139	$N_{\varepsilon 2}$	O_γ	Thr 47	2.85
Gln 140	$O_{\varepsilon 1}$	$N_\zeta(+)$	Lys 23	2.72
Gln 140	$N_{\varepsilon 2}$	$O_\delta(-)$	Asp 38	2.65

a) Throughout this paper, the IUPAC (1969) symbols for atoms were used. Atoms without any subscripts correspond to those in the peptide backbone.

Figure 1. Optimized structure of AChR/CTX model 1. Thick line indicates the receptor model corresponding to the residues 136-142 and the thin line describes the CTX residues used in the calculation. Only hydrogen bonding hydrogens are treated explicitly and others are treated as the united atoms. The thin broken line indicates the CTX C_α trace.

indicate the strong ionic nature of the intermolecular interaction found in this complex. In addition, a stacking interaction formed by the aromatic rings of Phe 137 of AChR and Trp 25 of CTX is shown. The Asp 138 of AChR interacts with both Lys 23 and Lys 49 of CTX while Gln 140 interacts with both Lys 23 and Asp 38 of CTX as originally suggested by Smart et al. (1984). In the present model, Gln 139 of AChR has one extra interaction with Lys 23 of CTX in addition to the interaction with Thr 47.

The optimized structures of AChRs in models 2T and 2C are shown in Figure 2. Although the orientations of the disulfide bond with respect to the CTX are quite different between the two models, the conformations of the AChR models in the region of residues 185-191 are practically identical, making the majority of CTX/AChR interactions similar for both models. This was expected since the formation of the cis or trans peptides occures in the region far from those residues interacting with CTX. The optimized structures and the interaction distances of CTX/AChR complex (model 2T) are shown in Figure 3 and Table 2. In addition to hydrogen bonding and ionic interactions shown in Table 2, two tryptophan residues, Trp 187 of AChR and Trp 25 of CTX, exhibit a stacking interaction.

The search for the ACh binding sites was carried out by carefully examining the AChR/CTX complex models. In the family of cholinergic molecules, four functional groups, namely, trimethylammonium, ester oxygen, carbonyl group and

Table 2. Hydrogen bonding and ionic contact distances between cobratoxin (CTX) and the acetylcholine receptor (AChR) as found in Model 2T.

AChR	Residues	CTX	Residues	Distance (Å)
Lys 185	$N_\xi(+)$	$O_{\delta 1},_{\delta 2}(-)$	Asp 27	2.55
Val 188	O	$N_\xi(+)$	Lys 23	2.65
Tyr 190	N	O_η	Tyr 21	2.93
Tyr 190	O	O_η	Tyr 21	2.67
Tyr 190	O_η	O	Thr 44	2.74
Thr 191	$O_{\gamma 1}$	N	Cys 41	2.84
Thr 191	$O_{\gamma 1}$	O	Cys 41	2.74
Asp 195	O_γ	O_γ	Ser 11	2.68
Thr 196	$O_{\gamma 1}$	$O_{\delta 2}(-)$	Asp 8	2.64

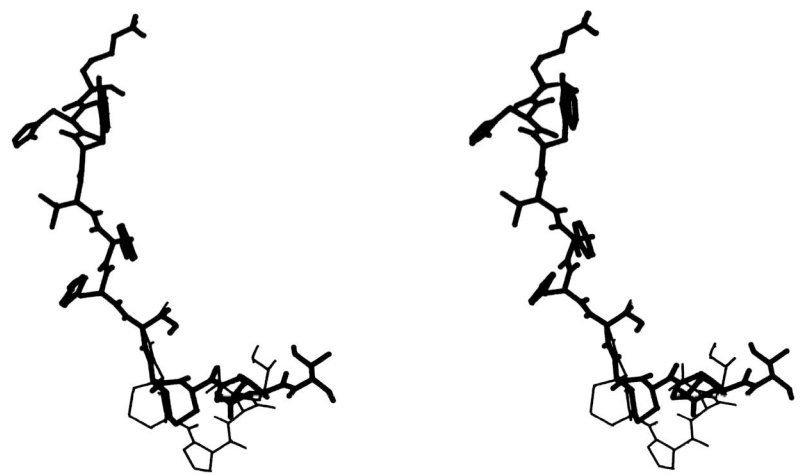

Figure 2. Two AChR models of the dodecapeptide corresponding to residues 185-196. Thick line for model 2T (trans) and thin line for model 2C (cis).

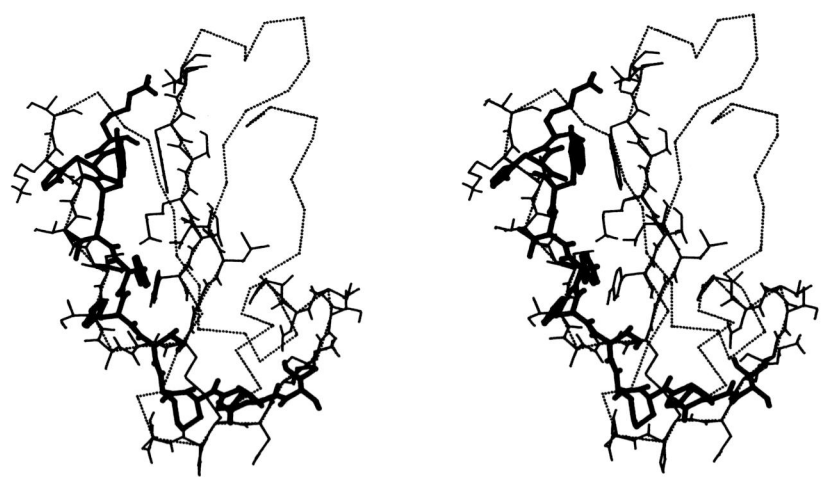

Figure 3. Optimized structure of AChR/CTX model 2T. Thick line indicates the receptor model corresponding to the residues 185-196. Other lines are the same as in Figure 1.

Table 3. Distances (Å) between the plus charge and the proton acceptor in ACh Crystals.[a]

Plus Charge	Proton Acceptor	Crystals[b]		
		ACh-Cl	ACh-Br	ACh-I
N	O_1	3.25	3.20	3.29
C_1	O_1	3.17	2.99	3.06
C_2	O_1	4.53	4.44	4.45
C_3	O_1	3.95	4.00	4.18

a) The atom numbers of ACh used throughout this paper are those used by Herdklotz and Sass (1970). The C_1, C_2, C_3, are the trimethyl group of quarternary ammonium ion while O_1 is an ester oxygen: b) References: ACh-Cl Herdklotz and Sass (1970), ACh-Br Svinnining and Sorum (1975) and ACh-I Jagner and Jensen (1977).

Table 4. Distances (Å) between the plus charge and the proton acceptor in CTX models.

Plus Charge	Proton Acceptor		Model 1	Model 2 T
Lys 49 N_ζ	$O_{\gamma 1}$	Thr 47	2.69	6.40
Lys 49 N_ζ	O	Val 48	3.43	5.84
Lys 49 N_ζ	O	Thr 47	3.53	8.11
Lys 49 N_ζ	O	Lys 49	4.88	5.24
Lys 23 N_ζ	$O_{\delta 1}$	Asp 38	2.80	2.59
Lys 23 N_ζ	O	Asp 38	5.08	4.71
Lys 23 N_ζ	O	Thr 49	5.81	6.16
Lys 23 N_ζ	O	Lys 23	6.11	5.53
Lys 12 N_ζ	O	Ser 11	--	4.27

methyl group have been identified (Weinstein et al., 1974). The orientation of the trimethylammonium group and the ester oxygen is the critical factor (Schulman et al., 1983). Although ACh exists in two different conformations in three crystal structures, the orientation of a plus charge and a proton acceptor is very similar as shown in Figure 4 and Table 3. If the ACh binds to its receptor in the same way as CTX interacts with AChR, the CTX residues from CTX/AChR models must also contain a plus charge and a proton acceptor interacting with the AChR separated by the distance matching that found in the ACh molecule. The distances between a plus charge and a proton acceptor of CTX residues found in the CTX/AChR

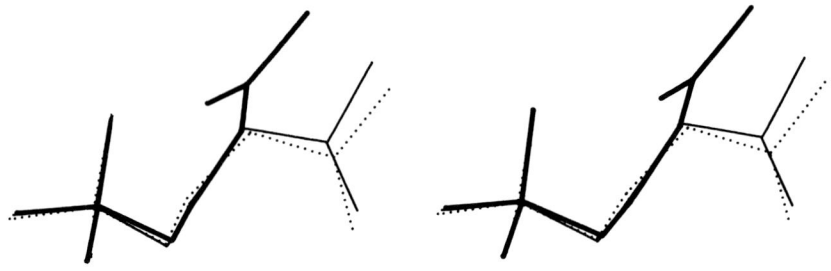

Figure 4. Crystal structures of ACh molecule using a thick line for ACh chloride, a thin line for ACh bromide and a thin broken line for ACh iodide. All hydrogens are treated as a part of united atom carbons and are not shown here.

Table 5. Distances between the minus charge and the proton donor in AChR Model 1 structure.[a]

Minus Charge	Proton Donor	Distance (Å)
Asp 138 O_δ	$N_{\varepsilon 2}$ Gln 139	2.90
Asp 138 O_δ	N Gln 139	4.09
Asp 138 O_δ	N Asp 138	4.37
Asp 138 O_δ	$N_{\varepsilon 2}$ Gln 140	7.14
Asp 138 O_δ	N Gln 140	6.59
Asp 138 O_δ	N Phe 137	7.65

a) The distances shown in this table are obtained by averaging the two O_δs of Asp 138.

complexes are shown in the Table 4. The corresponding groups on the AChR which interact with the above groups, complements to a plus charge and a proton acceptor, are a minus charge and a proton donor. The separation distances between these groups obtained from the model 1 are shown in Table 5.

Only two (Site A and B) of the potential binding sites in the Model 1 system were examined in this study. The resulting ACh/AChR complexes with the fixed AChR structures are shown in Figures 5 and 6. The characteristic distances between a plus charge and an ester oxygen and the torsional angles of ACh molecules in the complexes are given Tables 6 and 7. The results indicated that the ACh in the site A model is similar to the ACh bromide crystal structure while the one found in the site B model is similar to the ACh chloride crystal structure. The geometrical parameters of ACh/AChR interactions are given in Table 8 which indicate the presence of ionic interaction and hydrogen bonding also seen from Figures 5 and 6. In the site A model, the trimethylammonium replacing Lys 49 of CTX interacts with Asp 138 while the ester oxygen interacts with amino group of Gln 139 as the Thr 49 $O_{\gamma 1}$ does in the

Table 6. Distances (Å) between the plus charge and the ester oxygen of ACh found in the ACh/AChR model.

Atoms	Crystal[a]	Site A	Site B
$N\text{--}O_1$	3.25	3.15	3.12
$C_1\text{--}O_1$	3.07	2.96	2.91
$C_2\text{--}O_1$	4.47	4.47	4.42
$C_2\text{--}O_1$	4.04	3.75	3.76

a) The average distances obtained from three crystal structures shown in Table 3.

Table 7. ACh conformations (deg) found in the ACh/AChR model.

Dihedral Angle[a]		Crystal[b]	Site A	Site B
τ_1	$C_2\text{-}N\text{-}C_4\text{-}C_5$	-179.3	176.8	179.0
τ_2	$N\text{-}C_4\text{-}C_5\text{-}O_1$	83.9	66.3	67.3
τ_3	$C_4\text{-}C_5\text{-}O_1\text{-}C_6$	-166.9:81.1[c]	72.8	-176.9
τ_4	$C_5\text{-}O_1\text{-}C_6\text{-}C_7$	-176.1	-179.9	173.2

a) See Table 3 for atom numbering scheme; b.) The average values obtained from the three crystal structures shown in Table 3; c) The trans value is from ACh chloride and gauche value is an average from ACh bromide and ACh iodine structures.

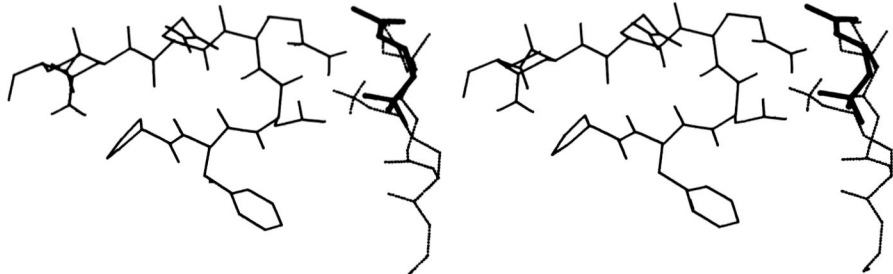

Figure 5. Structure of the ACh/AChR complex site A model. The thin line is for the AChR, the thick line for ACh. The corresponding CTX residues are shown with a thin broken line. Only the interacting side chains of the CTX residues are shown.

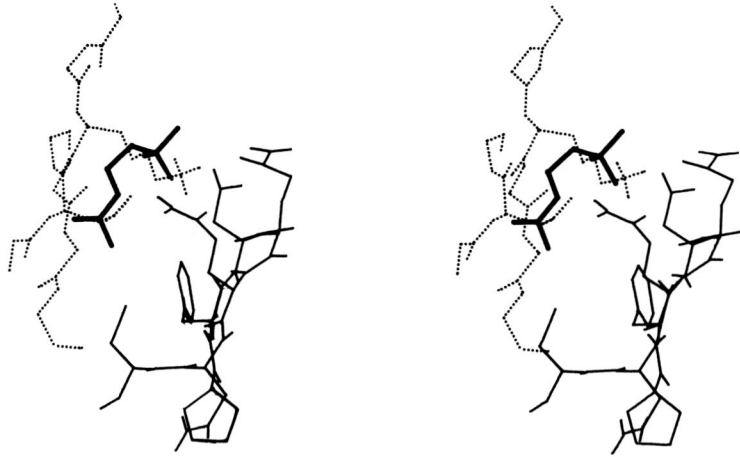

Figure 6. Structure of the ACh/AChR complex site B model; line definition are the same as for Figure 5.

CTX/A ChR complex. The corresponding interactions on site B model involve the Asp 138 and the amino group of Gln 140 which were interacting with Lys 23 and Asp 38 in the CTX/AChR model. In both models (sites A and B), it seems that relaxing the receptor model makes the ACh/AChR interaction weaker, especially for the site B case.

Table 8. Geometrical Parameters[a] of ACh/AChR Interactions.

AChR[d]	ACh[e]	Site A[b] Fixed[f]	Free[g]	Site B[c] Fixed[f]	Free[g]
O_δ	N	3.48	3.47	3.20	3.21
O_δ	C_1	4.19	3.92	2.83	2.98
O_δ	C_2	2.92	3.06	3.03	3.05
O_δ	C_3	2.91	2.92	3.18	3.03
$N_{\varepsilon 2}$	O_1	3.05	2.96	3.23	3.27
$HN_{\varepsilon 2}$	O_1	2.05	2.55	2.27	2.83
$N_{\varepsilon 2}$–$HN_{\varepsilon 2}$	O_1	168.9	103.9	154.4	106.7

a) Distances in Å and angles in degree; b) The ACh interacts with Asp 138 and Gln 139 of AChR; c) The ACh interacts with Asp 138 and Gln 140 of AChR; d) The O_δ represent either $O_{\delta 1}$ or $O_{\delta 2}$ of Asp 138; e) The $N_{\varepsilon 2}$ represent the amino nitrogen of Gln 139 and Gln 140; f) The model obtained with the fixed AChR at its CTX/AChR complex structure; g) The model obtained by allowing AChR to move freely.

DISCUSSION

Two different AChR/CTX complex models emerged from the experimental observation that there may be two different cholinergic agonist/antagonist binding sites on the α-subunit of AChR (Mular-Jerevic and Atassi, 1986). The primary objective of the present study is to gain an insight into AChR binding site structure by comparing these two different models.

The amino acid sequences involved in the AChR/CTX model complexes are shown in Figure 7. Since the CTX residues of model 2 cover all of the CTX residues used in model 1, the interacting residues of CTX are very similar in both models. Both AChR models contain an aromatic

CTX	10	20	30	40	50
	IRCFITPDIT	SKDCPNGHVC	YTKTWCDAFC	SIRGKRVDLG	CAATCPTVKT
			111111	111	11111
	2222	222222	2222222	222	2222222222

AChR Mdl 1	136	137	138	139	140	141	142				
	Pro	Phe	Asp	Gln	Gln	Asn	Cys				

AChR Mdl 2	185	186	187	188	189	190	191	192	193	194	195	196
	Lys	His	Trp	Val	Tyr	Tyr	Thr	Cys	Cys	Pro	Asp	Thr

Figure 7. The sequences of CTX and AChR used in the models 1 and 2. The numerals 1 and 2 indicate the CTX residues used in the model 1 and model 2 respectively.

residue, Phe 137 in model 1 and Trp 187 in model 2 interacting with Trp 25 of CTX which determines the basic orientation of CTX/AChR interaction so that the AChR interacts with CTX in its concave surface in both cases. The similarity between the two models ends there. A completely different picture emerges if one examines the shape of AChR models and the nature of CTX/AChR

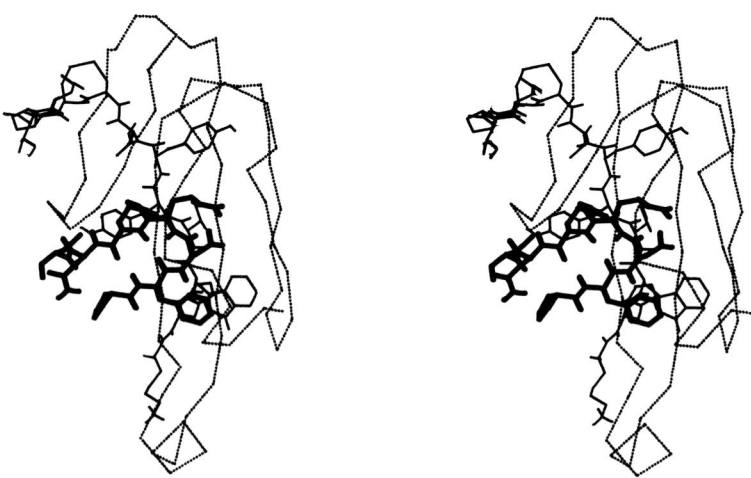

Figure 8. Comparison of AChR models, Model 1 (thick line) and Model 2T (thin line), with respect to the C_α trace of the CTX (thin broken line).

interactions. The model 1 AChR has the shape of a compact protrusion interacting with CTX perpendicular to its long axis direction (Figure 8). The model 2 AChR has the shape of a hook with a long tail interacting with CTX parallel to its long axis (Figure 8). Furthermore, the nature of AChR/CTX interaction in model 1 is quite different from that in model 2. In model 1, a small number of residues, three residues at the tip of the AChR protrusion are interacting with four CTX residues. Each of the AChR residues are interacting with two different residues of CTX resulting in an intertwined ionic network. In contrast, the AChR/CTX interactions in model 2 are spread over the different parts of the molecules covering regions of CTX which are not involved in model 1 and without the intensive hydrogen bonding and ionic interaction network found in model 1.

Another objective of this research was the placement of ACh on its receptor. Only two out of several possible sites on the AChR model 1 were used. The interaction scheme in these ACh/AChR complexes are the same as those

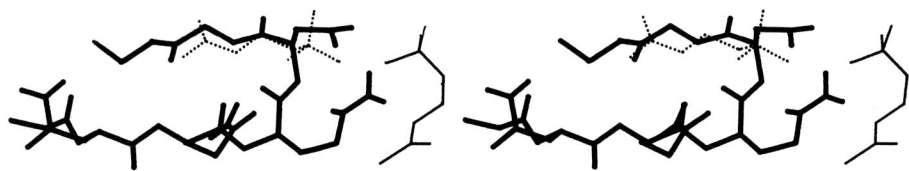

Figure 9. The Asp 138 (a negative charge), hydrogen bonding donor side chains, and amide backbone hydrogens (drawn in thick lines) provide the potential ACh binding sites of the AChR model 1 structures. The model AChs on site A (thin line) and site B (thin broken line) are also shown.

found in the CTX/AChR complexes. The two sites (A and B) represent two possibilities located near the tip of the turn on two different sides of the protrusion created by the AChR model 1. It is also possible that the ACh binds to its receptor in a totally different orientation than CTX while using the same binding residues. Alternatively, the ACh binds to an area where no corresponding CTX binding residues can be found. The NH group in AChR peptide backbone has potential as a proton donor (see Table 5). This possibility provides, as a third potential region for binding sites, the interior of the β sheet (see Figure 9). This third possibility is especially favored when the Asp 138 side chain flips over to the other side of the CTX/AChR model.

Although the ACh binding sites for the model 2 AChR were not fully examined in this study, no direct correspondence between CTX/AChR interaction and ACh/AChR interaction was found for this model. The Asp 194 of AChR is located near the tip of the hook where no positively charged CTX residues can be found. It is possible that the ACh might interact very differently than CTX for this receptor model or the dodecapeptide may interact differently from the model proposed. Examination of this conformation and CTX/AChR interaction of the region including Cys 192-Cys 193 is underway.

We have used the known crystal structure of CTX (Walkinshaw et al., 1980) as a mold on which to model the AChR cholinergic binding site structure. According to the models studied, it seems that the Asp 138 in the region containing the Cys 142 plays a key role in the ACh binding. It is also expected that Asp 194 plays a similar role for the second region due to its proximity to the disulfyde bond created by Cys 192 and Cys 193. The importance of these aspartic residues in biological functions may be tested by transforming them into non charged residues with site directed mutagenesis. This may also provide a clue as to which region plays a primary role in triggering the channel opening.

ACKNOWLEDGEMENT

This work was partially supported by a grant from NASA (NSG-7305) and pursurant, in part, to a contract with the

National Foundation for Cancer Research. The authors would like to thank Dr. Theresa Julia Zielinski for helpful discussions and a critical review of the manuscript and Miss Jennifer Brackett for her assistance in the preparation of this manuscript.

REFERENCES

Capasso S, Mattia C, Mazzarella L (1977). Acta Cryst B33:2080-2083.
Changeux J-P, Devillers-Thiery A, Chemovilli P (1984). Science 225:1335-1345.
Criado M, Sarin V, Fox JL, Lindstrom J (1986). Biochem 25:2839-2846.
Dennis M, Giraudat J, Kotzyba-Hibert F, Goeldner M, Hirth C, Chang J-Y, Changeux J-P (1986). FEBS Letters 207:243-249.
Garduno-Juarez R, Shibata M, Zielinski TJ, Rein R (in press). Acta Biochim Biophys Hungarica.
Hata Y, Matsuura Y, Tanaka N, Ashida T, Kakudo M (1977). Acta Cryst B33:3561-3564.
Herdklotz JK, Sass RL (1970). Biochem Biophys Res Comm 40:583-588.
Jagner S, Jensen B (1977). Acta Cryst B33:2757-2762.
Kao PN, Dwork AJ, Kaldany R-RJ, Silver ML, Wideman J, Stein S, Karlin A (1984). J Biol Chem 259:11662-11665.
Kao PN, Karlin A (1986). J Biol Chem 261:8085-8088.
Karlin A (1980). In Cotman, CW, Poste, G Nicolson, GL (eds.): 'Cell Surface and Neuronal Functions,' Amsterdam: Elsevier, pp 191-260.
Kosower EM (1987). Eur J Biochem 168:431-449.
Lennon VA, McCormick DJ, Lambert EH, Greismann GE, Atassi MZ (1985). Proc Natl Acad Sci USA 82:8805-8809.
Margheritis C, Corongiu G (1988). J Comput Chem 9:1-10.
McCarthy MP, Earnest JP, Young EF, Choe S, Stroud RM (1986). Ann Rev Neurosci 9:383-413.
Mishina M, Tobimatsu T, Imoto K, Tanaka K, Fujita Y, Fukuda K, Kurasaki M, Takahashi H, Morimoto Y, Hirose T, Inayama S, Takahashi T, Kuno M, Numa S (1985). Nature 313:364-369.
Mulac-Jericevic B, Atassi MZ (1986). FEBS Letters 199:68-74.
Neumann D, Barchan D, Fridkin M, Fuchs S (1986). Proc Natl Acad Sci USA 83:9250-9253.
Noda M, Takahashi H, Tanabe T, Toyosato M, Furutani Y,

Hirose T, Asai M, Inayama S, Miyata T, Numa S (1982). Nature 299:793- 797.
Rein R, Nir S, Haydock K, MacElroy RD (1978). Proc Indian Acad Sci 87A:95-113.
Rose G, Gierasch LM, Smith JA (1985). Advs Protein Chem 37:1-109.
Schulman JM, Sabio ML, Disch RL (1983). J Med Chem 26:817-823.
Smart L, Meyers HW, Hilgenfeld R, Saenger W, Maelicke A (1984). FEBS Letters 178:64-68.
Srinivasan S, McGroder D, Shibata M, Rein R (1986). In Sarma RH, Sarma MH (eds.): 'Biomolecular Stereodynamics III,' New York: Adenine Press, pp 299-304.
Srinivasan S, Shibata M, Rein R (1987). In Shahar A, Goldberg AM (eds.): 'Model Systems in Neurotoxicology. Alternative Approaches to Animal Testing,' New York: Alan R. Liss, pp 171-180.
Stroud RM, Finer-Moore J (1985). Ann Rev Cell Biol 1:317-351.
Svinning T, Sorum H (1975). Acta Cryst B31:1518-1586.
Walkinshaw MD, Saenger W, Maelicke A, (1980). Proc Natl Acad Sci USA 77:2400-2404.
Weiner SJ, Kollman PA, Case DA, Singh UC, Ghio C, Alagona G, Profetal S, Weiner P (1984). J Amer Chem Soc 106:765-784.
Weinstein H, Srebrenik S, Pauncz R, Maayanie S, Cohen S, Sokolvsky M (1974). In Bergman ED, Pullman B, (eds.): 'Chemical and Biochemical Reactivity,' Jerusalem: The Israel Academy of Sciences and Humanities, pp 493-512.
Wilson PT, Lentz TL, Hawrot E (1985). Proc Natl Acad Sci USA 82:8790-8794.

MOLECULAR DYNAMICS AS A TOOL FOR STRUCTURAL AND FUNCTIONAL PREDICTIONS: The retinol binding protein and Chloroplast C-Terminal Fragment of the L12 Ribosomal Protein Cases.

O. Tapia and J. Åqvist

Swedish Unversity of Agricultural Sciences, Uppsala Biomedical Center, Department of Molecular Biology, P.O.Box 590, S-751 24 UPPSALA, Sweden.

INTRODUCTION

Model built structures (MBSs) provide invaluable information for computer-assisted modeling of protein-ligand interactions (Bedarkar et al. 1977, Blundell et al. 1983, Eklund et al. 1985, Greer, 1981, Read et al. 1984, Strynatha and James, 1988). Graphics display programs, such as FRODO (Jones, 1978), energy minimisation (EM) techniques and X-ray structures of parent molecules (XPS) are the basic elements to help produce MBSs. EM eliminates bad contacts between side chains, while introducing small changes in the location of main chain atoms. The existence of an EM structure does not grant, however, its thermal stability. Insertions, deletions and/or substitutions of aminoacids can change the intramolecular packing as well as the pattern of atom dynamics. Thermal fluctuations can lead to significant changes in the folding, and, in patological cases, they may even lead to a partial or total unfold. Information of dynamical properties of the protein is required to complete the picture obtained from static MBSs. Such information may be obtained from molecular dynamics (MD) simulations.

MD studies of the XPS and MBS have been used to compare dynamical stability, conformational changes and patterns of atomic fluctuations. The simulation of the XPS serves also to check the quality of the force field used. In this paper two illustrative examples are reviewed:

1) the MBS of Chloroplast carboxy terminal fragment (Chl-CTF) of the L7 ribosomal protein by Leijonmark et al.(1984); 2) the apo retinol binding protein structure (Åqvist et al.1986). The force field and method are those described in Åqvist et al. (1985, 1986).

RIBOSOMAL PROTEIN

The sequences of Chloroplast and E.Coli (Leijonmark et al.1984) and the folding pattern are presented in Figure 1.

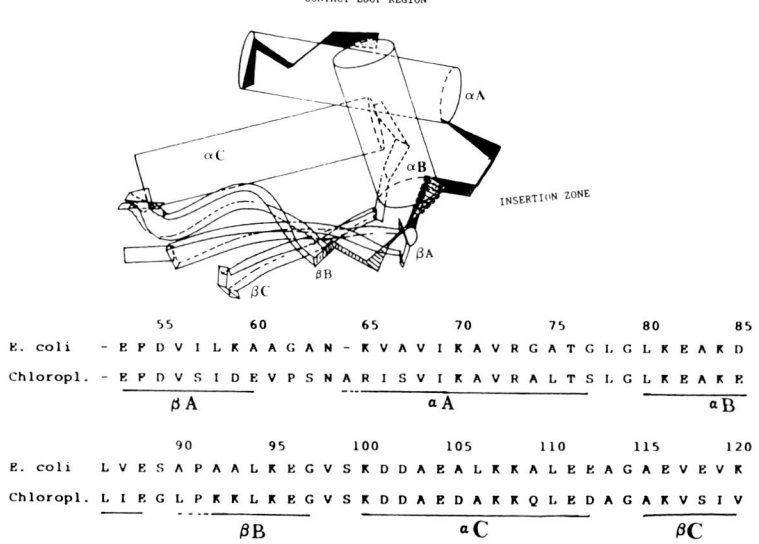

Figure 1. Schematic drawing of E. coli CTF. The α-helices are represented by cylinders and the β-strands by ribbons. The contact loop region designates the zone where two monomers of CTF form a dimer.

The structure of E.coli CTF has recently been described (Leijonmak and Liljas, 1987). The sequencial arrangement of α-helices and β-strands is βA-αA-αB-βB-αC-βC.
Two domains are distinguishable: 1)a right handed $\alpha\alpha$-corner (Efimov, 1984) formed by αA, a short extended connection (residues 75 to 79) and αB; 2)a twisted antiparallel β-sheet with the αC helix firmly attached by

two short loops. The $\alpha\alpha$-corner is articulated to the β-sheet by a reverse II' turn (residues 61 to 64) and the connection of αB to βB is mediated by cis-Pro 91.

Ch-CTF has 60% sequence identity with E.coli. There are 27 aminoacid substitutions. The most altered sequence is found in the region overlapping the reverse II' turn that connect the domains. For the 10 aminoacids from 57 to 67 there is only one residue identical to the E.coli sequence and one insertion: Ala is inserted between positions 64-65 of E.coli. Note that Gly 61 is replaced by a proline. This region will be referred to as β-α-turn in the following.

The MBS was obtained by Leijonmark et al.(1984) using FRODO software (Jones,1978). The bad contacts were relieved by manually changing the side-chain torsion angles. Horjales et al.(1987) produced an EM structure that could not be distinguished from the manually built one. The EM structure was used to perform MD simulations.

A 170ps trajectory was calculated. The first section of 40ps had the appearance of a stationary trajectory. A conformational change affecting the β-α-turn was found. The relative positions of the two domains were modified, the $\alpha\alpha$-corner moving away from the β-sheet domain. The trajectory was extended 40 ps more to build up statistics. Surprisingly, the conformational changes affecting the β-α-turn continued during this period. The reorientation of the domains was achieved and an equilibrated trajectory was finally obtained from 80ps to 170ps. The dynamical and structural properties of Chl-CTF were studied on this latter section.

The averaged <80-170> MD structure has the same gross spatial structure of the domains as in E.coli CTF. The domains appear to be displaced as rigid bodies as no sign of unfolding was detected. The nature of the surfaces of possible functional importance are not significantly altered.

A dimeric form of CTF has been proposed as functionally relevant in the ribosome. Leijonmark and Liljas (1987) suggested the crystallographic two-fold axis to be a real molecular twofold axis. The contact area involves residues 77 to 79 which are mostly conserved. An MD study of the E. coli dimer (unpublished) clearly established its thermal stability. A collective motion associated with the relative displacement of the monomers was found to have the same frequency (5 cm^{-1}) as the $\alpha\alpha$-corner motion in the isolated monomer. The dimer contacts involve both main-chain and side-chain atoms. In the $\alpha\alpha$-corner a short piece of the polypeptide chain, residues 77 to 79 in E. coli, forms a fourth β-strand antiparallel to βB on the neighbouring molecule (Leijonmarck and Liljas, 1987). There are also contacts between residues in the C-terminus of αA in one monomer and βB on the other. The collective motion of the $\alpha\alpha$-corner is communicated to the dimer via the interface contacts thereby producing a flexible dimer. This property may improve adaptive potentialities for interactions with other protein molecules in the ribosome.

Chl-CTF has retained the dynamical properties of E.coli CTF. A careful study of the rms atomic fluctuations shows that the dynamics have not been changed, except for the β-α-turn region. The librational mode of the αB helix axis has a frequency at about 6cm^{-1} and the spatial displacement is similar to the 5cm^{-1} in E. coli CTF. In view of the structural invariance in the interface regions leading to a dimer structure, and to the invariance in the dynamical pattern, it can be concluded that Chl-CTF will form dimers ressembling the one proposed by Leijonmark and Liljas(1987). The functionality of the Chl-CTF, in so far as it depends on adaptive flexibility, is not expected to differ much from the E. coli form.

RETINOL BINDING PROTEIN

The 3-D structure of holo-RBP (Newcomer et al.,1984) folds as a single domain, with an N-terminal coil, a β-sheet core, an α-helix and a C-terminal coil. The β-sheet core makes an eight stranded β-barrel that encapsulates the retinol molecule (cf. Figure 2).

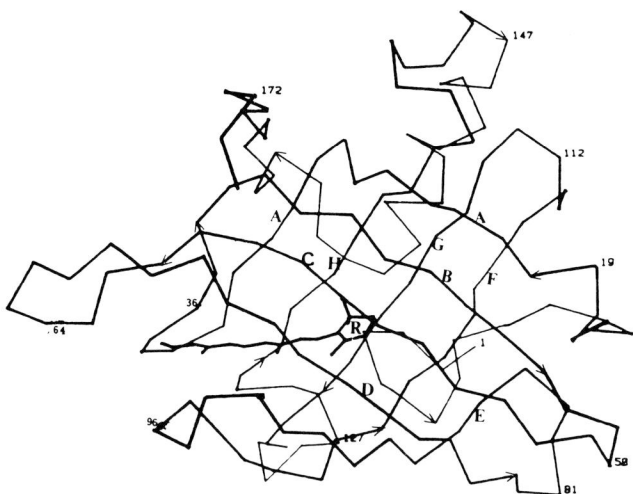

Figure 2. Schematic draw of the holo retinol binding protein.

Retinol binding protein is the transport protein for vitamin A. One molecule of vitamin A alcohol (retinol) in the liver hepatocytes is bound to RBP and secreted into the plasma; there, a complex is formed between holo-RBP and thyroxine binding prealbumin. RBP, free or in complex with prealbumin, is recognised by a surface receptor on the target cell which transfers the retinol molecule to the cell. The affinity of apo-RBP for prealbumin is virtually lost. It has also been shown that a-RBP has little or no capacity of displacing holo-RBP from the cell receptor (for original references see those quoted in Åqvist et al.,1986, Sandblom et al.,1986 and Åqvist and Tapia, 1987). It seems clear that upon delivery of retinol, RBP undergoes a conformational change affecting both its affinity for prealbumin and its interaction with the cell surface receptor.

MD trajectories were calculated for holo-RBP and apo-RBP, respectively. The starting structure for the apo simulation was obtained by energy minimisation after removal of retinol from the XMS (h-structure).

The average MD holo and apo structures (denoted by <h-MD> and <a-MD>) maintain the same β-barrel folding in the simulations as in the crystallographic holo structure. The coordinate deviations with respect to the X-ray structure are somewhat larger for <a-MD> than for <h-MD>.

The most relevant differences between <a-MD> and the two h-structures involve the three loops forming the entrance to the β-barrel; in the <a-MD> structure the β-barrel is effectively closed, mainly due to a movement of βE-βF loop, which in <a-MD> makes hydrogen bonds to the βA-βB loop. The region containing the α-helix also presents some features in the a-form simulation that differ from the h-form one. In the <a-MD> structure, a slight tilt of helix around its C-terminal is observed, while the X-ray and <h-MD> structures agree well in this region.

Several experimental results point towards the entrance loop region as being involved in the binding of prealbumin. Labelling experiments indicate that one Trp residue on RBP becomes buried upon the formation of the holo-RBP prealbumin complex. The best candidates are Trp61 in the βC-βD loop and Trp91 in the βE-βF loop. Furthermore, reconstitution experiments of apo-RBP with retinol analogues have shown different affinities for prealbumin depending on which analogue is bound to RBP. Conformational changes in this region can explain the difference in prealbumin affinity between the holo and apo-forms.

From the MD simulations two regions of RBP have been localised where conformational changes have taken place. The entrance loop and the α-helix zone. A MBS of the prealbumin-holo-RBP complex was obtained (Åqvist and Tapia, 1987). In Figure 3, the contact region is depicted. Besides the extraordinary surface complementarity, one of the Trp's becomes buried in agreement with available experimental information.

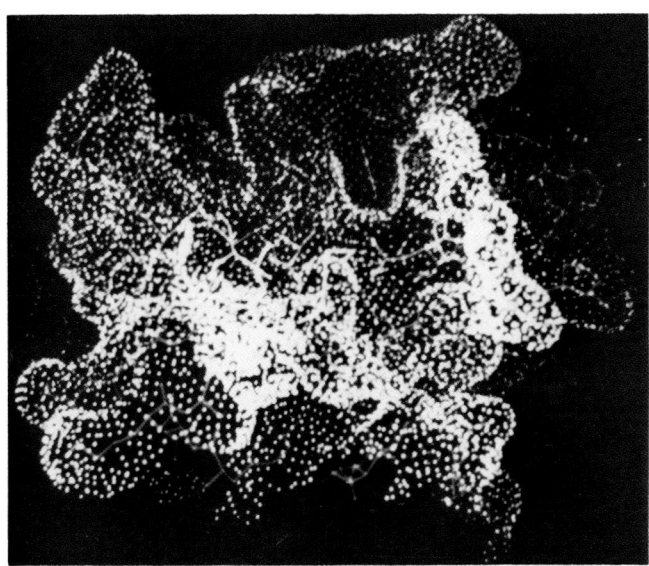

Figure 3. Contact surface of the complex between holo-RBP and prealbumin. RBP is on top of PAB.

The helix also shows different dynamical behaviour in the two simulations. While the longitudinal stretching of the helix in the apo-form displays a fairly small amplitude, which is dominated by a few frequencies in the region 14 to 33 cm^{-1}, the helix in the holo-form has more complex motions. These are reflected by a rather large bending of the helix, corresponding to variation in the curvature of its "axis". If this region of RBP is involved in receptor recognition, the structural and dynamical differences observed may be related to the differences in the interaction with the receptor between apo and holo-RBP.

DISCUSSION

As the number of protein sequences increases, basically due to efficient DNA sequencing methods, there is a high pressure put on model builders to produce reliable three-dimensional structures. Standard procedures based on graphics display programs have been fairly successful in predicting the structure of highly homologous molecules (for limitations and critical evaluation see for instance Greer 1981, Read et al. 1984). The prediction of thermal stability, conformational changes and fluctuation patterns of primary, secondary and tertiary structures is an important issue for model building. In this paper, two cases have been reviewed that illustrate the interest of producing this type of information.

Conformational changes have been elicited by MD simulations that, otherwise, would have been completely overlooked by EM techniques. MD helps sense changes in contact regions due to aminoacid replacements and insertions. Such contacts are essential for protein stability. A poly-alanine structure with the original glycines retained was built into the XPS of CTF. The contacts are essentially changed here. The EM structure had a rms coordinate shift of 0.1Å for the C_αs. The MD simulation showed this structure to be thermally unstable.

In general, the trajectories have to last sufficiently long to ensure statistical significance. A short MD shaking up would probably help moving the structure to regions not accessible when EM techniques are used. However, studies of this type might be useful as the work on antibody hypervariable loop conformations by Fine et al. (1988) has shown. When the properties of the global protein are at stake, it is necessary, although not sufficient, that the dynamical behaviour of highly refined parent structures be reproduced by MD simulations. The computer power required to accomplish such simulations will no longer be the bottle neck as progress in computer technology continues at present pace.

ACKNOWLEDGEMENTS

The authors wish to express their thanks to Marie Leijonmark, A. Liljas, T.A. Jones, M. Newcomer, E. Horjales and P. Sandblom for lively discussions and fruitful collaboration. This work has been supported by the Swedish Research Council (NFR).

REFERENCES

Aqvist J, van Gunsteren W.F, Leijonmark M, Tapia O (1985). A Molecular Dynamics Study of the C-terminal Fragment of the L7/L12 Ribosomal Protein. J.Mol.Biol 183:461-477.

Aqvist J, Tapia O (1987). Surface fractality as a guide for studying protein-protein interactions. J.Mol.Graph 5:30-34.

Aqvist J, Sandblom P, Jones T.A, Newcomer M.E, van Gunsteren W.F, Tapia O (1986). Molecular Dynamics Simulations of the Holo and Apo Forms of Retinol Binding Protein. J.Mol.Biol 192:593-604.

Bedarkar S, Turnell W.G., Blundell T.L. and Schwabe C. (1977). Relaxin has conformational homology with insulin, Nature (London) 270:449-451.

Blundell T., Sibanda B.L. and Pearl L. (1983). Three-dimensional structure, specificity and catalytic mechanism of renin. Nature (London) 304, 273-275.

Eklund H, Horjales E, Jörnvall H, Brändén C.-I, Jeffery J (1985). Molecular Aspects of Functional Differences between Alcohol and Sorbitol Dehydrogenases. Biochemistry 24: 8005-8012.

Efimov A.V (1984). A novel super-secondary structure of proteins and the relation between the structure and the amino acid sequence. FEBS Letters 166:33-38.

Fine R.M., Wang H, Shenkin P.S., Yarmush D.L., and Levinthal C., (1986). Predicting Antibody Hypervariable Loop Conformations II: Minimization and Molecular Dynamics Studies of MCPC603 From Many Randomly Generated Loop Conformations. Proteins 1:342-362

Greer, J. (1981). Comparative model-building of the mamalian serine proteases. J.Mol.Biol. 153:1027-1042.

Horjales E, Brändén C.-I (1985). Docking of Cyclohexanol into the Active Site of Liver Alcohol Dehydrogenase. J.Biol.Chem 260: 15445-15451.

Jones T.A (1978) A Graphics Model Building and Refinement System for Macromolecules. J.Appl.Cryst. 11:268-272.

Leijonmark M, Liljas A, Subramanian A (1984). Computed Spatial Homology Between the L12 Protein of Chloroplast Ribosome and 1.7E Structure of Escherichia Coli L12 Domain. Biochem.Int 8:69-76.

Leijonmark M, Liljas A (1987). Structure of the C-terminal Domain of the Ribosomal Protein L7/L12 from Escherichia coli at 1.7E. J.Mol.Biol. 195: 555-580.

Newcomer, M.E., Jones, T.A., Åqvist, J., Sundelin, J., Eriksson, U., Rask, L. and Peterson, P. (1984). The three-dimensional structure of retinol-binding protein. EMBO J. 3: 1451-1454.

Read, R.J., Brayer, G.D., Jurásek, L., James M.N.G. (1984). Critical evaluation of comparative model building of **Streptomyces grises** trypsin. Biochemistry 23: 6570-6575.

Sandblom P, Åqvist J, Jones T.A, Newcomer M.E, van Gunsteren W.F, Tapia O (1986). Structural Changes in Retinol Binding Protein Induced by Retinol Removal. A Molecular Dynamics Study. Biochem.Biophys. Res.Com. 139: 564-570

Strynadka N.C.J. and James M.N.G. (1988). Two Trifluoperazine-binding sites on Calmodulin Predicted From Comparative Molecular Modeling with Troponin-C. Proteins 3:1-17.

ELECTROSTATIC INTERACTIONS IN PROTEINS

Barry Honig, Kim Sharp and Michael Gilson

Department of Biochemistry and Molecular Biophysics, Columbia University, 630 West 168 St. New York, NY 10032, USA

INTRODUCTION

Electrostatic interactions play a central role in protein structure and function (for reviews see Matthew, 1985; Honig et al., 1986). However, despite their great importance, it has been extremely difficult to treat electrostatic interactions theoretically. The problem is not just one of obtaining accurate numbers. It has often been the case that even a qualitative estimate of the magnitude of electrostatic interactions in different systems has been difficult to achieve.

The underlying problem is that Coulomb's law is only valid for charges immersed in an infinite medium of a single dielectric constant. Since most biological phenomena take place near interfaces between water and a macromolecule or membrane, Coulomb's Law, which is a special case solution to the Poisson equation, is not valid. Our approach to the problem is to solve the Poisson equation (or the Poisson-Boltzmann (PB) equation when ionic strength effects are of interest) directly. The advantages of this approach is that the dielectric constant of the solvent can be set at any desired value, ionic strength-dependent phenomena can be treated, computational demands though significant are not intractable and the reliance on simple elctrostatic concepts provides intuitively pleasing explanations to a variety of phenomena.

In the following we first outline the major features of our model and methodology, describing both its strengths and weaknesses, and comparing its to other methods. We then describe the numerical methods and programs we have developed and illustrate how they may be used in a variety of different applications. Directions for future developemnt are presented in a concluding section.

MICROSCOPIC AND MACROSCOPIC MODELS

The basic model we have developed (thee FDPB method) has been described in a number of recent papers (Gilson et al., 1985; Gilson and Honig, 1986; Klapper et al., 1986; Gilson et al., 1988; Gilson and Honig, 1988). The protein is described in terms of its three dimensional structure with the location of all real and partial charges defined from the coordinates of the appropriate atoms. These charges are embedded in a low dielectric medium consisting of the volume enclosed by the solvent accessible surface of the protein. The surrounding solvent is treated as a continuum of dielectric constant 80 with an electrolyte behaving according to the Poisson-Boltzmann equation. In the following we consider the various assumptions implicit or explicit in the model. These fall into three categories: a) a PB treatment of the solvent ions, b) treatment of water as a continuous dielectric medium, c) the treatment of electronic polarizability in the macromolecule in terms of a single dielectric constant.

a) The use of the PB equation for polyelectrolytes has been criticized principally because it neglects the finite size of solvent ions and spatial correlations between them. However, quantitative comparisons have been made of cylindrical solutions to the PB equation with Monte-Carlo simulations (Murphy et al., 1985) and hypernetted chain calculations (Bacquet and Rossky, 1984) where these discrete ion effects can be included explicitly. These studies have shown that for univalent ions with radii that are small compared to the size of the macromolecule, the non-linear PB equation is a very good approximation. The PB equation generally underestimates the concentration of counterions very close to the surface by only about 12-18% and produces essentially identical results in other regions.

It should be emphasized that at this point the PB equation offers the only practical method to account for ionic strength effects on biological macromolecules. If one were to consider, say, Monte-Carlo or molecular dynamics simulation on a protein, the volume of the simulation space would have to be large enough to include ions at least two Debye lengths from each charged group in the system. At physiological ionic strength this would imply a layer of water and ions at least 16Å thick around the macromolecule.

b) A major issue regarding the solvent concerns the extent to which a dielectric constant of 80 is appropriate for boundary layer water. Of course, waters that are bound tightly enough to be seen crystallographically can be considered as part of the macromolecule. In fact, including bound waters significantly improved the agreement with experiment we obtained in a study of subtilisin discussed below (Gilson and Honig, 1987a). For waters that are loosely bound but still not as mobile as bulk water there is reason to believe that the dielectric constant is still quite high. For example, the remarkable success of the Born model of ion solvation (Rashin and Honig, 1985), even for charges as high as +3 or +4, suggests that water retains its dielectric properties at high fields and does not saturate easily. We have been carrying out free energy simulations of ion solvation and have found no significant saturation effects for ions as large as +2 (unpublished results). That is, the change in free energy of the system as the charge is increased varies as the square of the charge (as is observed experimentally) which is the expected dielectric response if no saturation is occurring.

Two other treatments of the solvent are currently in use. Warshel (see e.g. Warshell and Russell, 1984) treats water in terms of Langevin dipoles located on a cubic grid. It is not generally recognized that this is essentially a macroscopic model and it is not clear that it offers any advantages over simply treating the solvent as a medium of dielectric constant of 80. Its disadvantages include the absence of any treatment of ionic strength and uncertainties as to the dielectric properties of the model, particularly in its treatment of solvated ions.

The more common treatment of the solvent is in terms of specific potential functions for each water molecule. This approach, with suitable parametrization, can account for phenomena which are sensitive to the details of water structure and, as such, offers distinct advantages over a macroscopic treatment of the solvent. On the other hand, there are serious uncertainties associated with the ability of current potential functions to treat electrostatic phenomena. Problems include: a) inaccuracies in the calculated dielectric constants, b) large errors in calculating ion solvation energies (which depend on three body terms that are difficult to include in the potential functions), c) the difficulties discussed above in treating ionic strength dependent phenomena and d) the enormous computational demands associated with including a large number of water molecules in the simulations. Thus, it seems clear that a parallel development of microscopic and macroscopic models offers the best path to a comprehensive treatment of solvent effects in biological systems.

c) The dielectric constant assigned to the protein depends on the specific application. Consider, for example, the problem of calculating an electric field around a protein. In this case the dipolar groups on the protein can be taken to be fixed, or in a time-averaged position, and they will make no contribution to the dielectric constant. Under these circumstances we use a dielectric constant of 2 to account for electronic polarizability, which we do not treat explicitly. It would, however, be straightforward to incorporate variations in electronic polarizability into our calculations by specifying a different dielectric constant for each atom in the macromolecule. This would be completely equivalent to accounting for variations in polarizability by allowing each atomic center to contain an inducible dipole.

If the protein atoms are expected to move in a particular application, (as in their response to the titration of a surface group) a dielectric constant of about 3-5 (Gilson and Honig, 1986) will account for the response of the protein to a weak field. Finally, if regions of the protein are expected to significantly rearrange, for example upon ion binding, it would be inappropriate to use a dielectric constant to describe the response of the atoms in question.

A common misconception concerning the FDPB method is that it does not account for the microscopic nature of the protein. (See for example the headline, "What About Protein Polarity," that accompanied a recent News and Views in Nature; (Warshel, 1987)). As stated above the FDPB method treats the protein in microscopic detail, it is only the solvent that is treated as a continuum. Each atom in the protein is treated in an identical fashion to its treatment in standard molecular mechanics simulations. The limitation of the method is that it only deals with a single protein conformation. It does, however, provide a means of devoloping dielectric functions which will offer a more reliable treatment of solvent effects in molecular dynamics simulations than is currently available (see below).

NUMERICAL AND COMPUTATIONAL METHODS

We have recently developed a series of computer programs (called delPhi) which allow us to calculate a variety of phenomena associated with the electrostatic properties of proteins (Klapper et al., 1986; Gilson et al., 1988). These programs use Brookhaven data format as input and produce maps of electrostatic potential inside and outside the protein. Since the Poisson-Boltzmann equation is solved numerically, with a finite difference algorithm (the FDPB method), the maps are stored in the form of numerical values at each point in a cubic lattice. Recent enhancements of our numerical algorithm allow us to obtain solutions that are accurate to within 5% as compared to analytic solutions which are available for simple spherical models. Problems associated with discontinuities that arise for charges placed near interfaces are avoided by centering each charge at the atomic nucleus, thus guaranteeing that no charge is located closer than one atomic radius from the solvent.

A number of recent enhancements of the program include the solution of the non-linear PB equation and the introduction of periodic boundary conditions both necessary for the treatment of DNA (Jayaram et al., submitted) and a new method to calculate solvation energies and electrostatic contributions to ion binding energies (Gilson and Honig, 1988).

The maps of electrostatic potential are the starting point for the analysis of experimental data. In the following sections we outline approaches to the analysis of a variety of electrostatic phenomena.

APPLICATIONS

The Electrical Potential Around Proteins

It is possible to obtain the electrical potential around a protein with a single run of the program. All real and partial charges would be included in such a run. Potential maps, which can be displayed on 2D or 3D graphics devices, can be used to gain intuitive insights into the interactions of the protein with substrates or other macromolecules, or used as a basis for studying dynamical behavior. In recent work we have shown how the electrical potential of superoxide dismutase (Klapper et al., 1986) enhances substrate association rates by a factor of 30 (Sharp et al., 1987a). It was found that the PB equation made it possible to account for the ionic strength dependence of substrate diffusion while simple Coulombic models produced results which were contrary to experiment (Sharp et al., 1987b).

Effects of Single Site Mutations

The application of site directed mutagenesis makes it possible to directly determine the effect of changing a single charge on a protein. One of the numerous applications of such experiments is that they provide the opportunity of directly testing different electrostatic models. In a recent study, Thomas et al. (1985) have determined the pK changes of the active site histidine in subtilisin, brought about by a number of mutations located about 14 Å away. We have calculated these pK changes with a variety of electrostatic models (Gilson and Honig, 1987a, 1987b). The extent of agreement with experiment depends strongly on the method used to calculate the interaction. The pK shifts calculated with the Poisson-Boltzmann equation are close to the experimental values at all ionic strengths while simpler Coulombic models fail to reproduce the experimental data.

Conformational Energy Calculations

Potential energy functions used in energy mimimizations or molecular dynamics simulations have an effective dielectric constant in the denominator of the term used to calculate coulombic interactions. An effective dielectric constant is one which accounts, in an average way, for dielectric screening due to both the protein and solvent. For example, the distant dependent dielectric constant, commonly used in simulations, is an attempt to account for the fact that solvent screening will increase as the distance between two interacting groups increases. We have tested this model by comparing its predictions to those of the FDPB method for interactions between a large number of pairs of atoms in the protein, rhodanese (Gilson and Honig, 1987b). The distance dependent dielectric constant overestimates weak electrostatic interactions for groups near the protein surface but yields relatively good results for buried groups.

We have recently developed an ionic strength dependent dielectric function to describe phosphate-phosphate interactions in DNA (Jayaram et al., submitted). The basic idea is to fit a simple functional form to the potentials calculated from the FDPB method. This approach can be extended to the interactions between charged groups on proteins. That is, to calculate pairwise interactions between charged amino acids and find an appropriate effective dielectric function. This function can then be used in a dynamics simulation. Ultimately, with faster computers, it may be possible to calculate forces directly by incorporating a new FDPB run after a fixed number of steps in a simulation.

Solvation Contributions to Protein Stability and Ion Building

As the ability to predict protein conformation improves, for example through the use of homology model building, it will become increasingly important to devise methods for assessing the relative stability of different computer generated conformations. A particularly important parameter is the total electrostatic energy of the protein. This number depends not only on pairwise electrostatic interactions between charged and polar groups but also on the solvation energies of individual groups. Consider a conformation in which a charged lysine is surrounded by hydrophobic atoms and which is not in contact with water. There is no term in conventional force fields that indicates that this is an unstable configuration, nor is the extent of destabilization (relative to a configuration in which the lysine is exposed) easy to evaluate.

We have recently introduced a method for calculating the total electrostatic energy of molecules of arbitrary shape and charge distribution, accounting for both Coulombic and solvent polarization terms (Gilson and Honig, 1988). In addition to the solvation energies of individual molecules, the method can be used to calculate the electrostatic energy associated with conformational changes in proteins as well as changes in solvation energy that accompany the binding of charged substrates. Good agreement with experiemnt is obtained for the hydration energies of acetate, ammonium, methyl ammonium and methanol. The method was used to explore the relationship between the depth of a charge within a protein and its interaction with solvent. This charge-solvent term is the missing interaction in conventional force fields which do not account for energetics of removing a lysine from the solvent. We have shown that the same term can successfully distinguish between stable and misfolded protein structures (Gilson and Honig, 1988).

CONCLUSION

We now have available the theoretical and computational tools to treat a variety of phenomena associated with electrostatic interactions in proteins. The underlying model is conceptually straightforward and the computational demands, though not insignificant, are well within the resources of most labs working on protein design. This makes it possible to apply the method to different problems with only a moderate investment of initial effort. The available programs are interactive and require only Protein Data Base files as input.

One of the attractions of a continuum treatment of solvent is that there are essentially no adjustable parameters. However, the method is open to further refinements such as defining a separate dielectric constant to boundary layer water or combining a continuum treatment of bulk solvent with an explicit all-atom description of solvent molecules in certain regions. In any case, our initial studies indicate that, for many applications, satisfactory agreement with experiment is to be expected with the current model. This implies that, at the very least, it is now possible to obtain qualitatively reasonable estimates of the magnitude of electrostatic interactions, a feature which should prove useful in the design and interpretation of experiments.

ACKNOWLEDGEMENTS

Supported by the NIH (GM-30518) and ONR (N00014-86-K-04833).

REFERENCES

Bacquet R, Rossky P (1984). Ionic Atmosphere of Rodlike Polyelectrolytes. J Phys Chem 88:2260-2269.
Gilson M, Rashin A, Fine R, Honig B (1985). On the Calculation of Electrostatic Interactions in Proteins. J Mol Bio 183:503-516.
Gilson M, Honig B (1986). The Dielectric Constant of a Folded Protein. Biopolymers 25:2097-2119.
Gilson M, Honig B, (1987a). Calculations of Electrostatic Potentials in an Enzyme Active Site. Nature 330:844-86.
Gilson M, Honig B (1987b). Energetics of Charge-Charge Interactions in Proteins. Proteins 3:32-52.
Gilson M, Sharp K, Honig B (1988). Calculating the Electrostatic Potential of Molecules in Solution. Method and Error Assessment. J Comp Chem (in press).
Gilson M, Honig B (1988). The Energetics of Charge-Solvent Interactions in Proteins. Proteins (in press).
Jayaram B, Sharp K, Honig B. The Electrostatic Potential of B-DNA (submitted for publication).
Klapper I, Hagstrom R, Fine R, Sharp K, Honig B (1986). Focussing of Electric Fields in the Active Site of Cu-Zn Superoxide Dismutase. Proteins 1:47-59.
Matthew JB (1985). Electrostatic Effects in Proteins, Annu Rev Biophys Biophys Chem. 14:387-401.
Murthy C, Bacquet R, Rossky P (1985). Ionic Distribution Near Polyelectrolytes. J Phys Chem 89:701-710.
Rashin A, Honig B (1985). Reevaluation of the Born Model of Ion Hydration. J Phys Chem 89:5588-5593.
Sharp K, Fine R, Honig B (1987a). Computer Simulations of the Diffusion of a Substrate to an Active Site of an Enzyme. Science 236:1460-1463.
Sharp K, Fine R, Schulten K, Honig B (1987). Brownian Dynamics Simulation of Diffusion to Irregular Bodies. J Phys Chem 91:3624-3631.

Thomas PG, Russell AJ, Fersht AR (1985). Tailoring the pH Dependence of Enzyme Catalysis Using Protein Engineering. Nature 318:375.

Warshel A, Russell S (1984). Calculations of Electrostatic Interactions in Biological Systems and In Solutions. Q Rev Biophys 17:283-422.

Warshel A (1987). What About Protein Polarity. Nature 330:15-16.

PROTEIN STRUCTURE PREDICTIONS: NEW THEORETICAL APPROACHES

Fred E. Cohen, Lydia Gregoret, Scott R. Presnell and I.D. Kuntz

Department of Pharmaceutical Chemistry and Medicine,
University of California-San Francisco, California 94143

Early experiments by Anfinsen et al. (1961) proved than an amino acid sequence contains sufficient information to determine the precise three dimensional structure of a native globular protein. Since that time, theoreticians (Levitt & Warshel, 1975; Nemethy & Scheraga, 1977; Karplus & Weaver, 1976; Schulz & Schirmer, 1977; Sternberg, 1983) and experimentalists (Creighton, 1978; Shoemaker et al. 1987) have searched for constructive solutions to the chain folding problem. In spite of the large number of protein structures known to atomic resolution and the much larger data base of protein sequences, solutions to the folding problem remain elusive.

Theoretical efforts have focused along three lines: energetic, heuristic and statistical. Following the physically reasonable assumption that a protein folds so as to minimize the free energy of the system, many investigators (Levitt & Warshel, 1975; Nemethy & Scheraga, 1977; McCammon et al., 1977; Weiner et al., 1984) have developed potential functions to describe the energy surface of a polypeptide chain. Minimum energy conformations are sought computationally. Alternatively, conformational space is probed from a starting point by integrating the equations of motion over time. Energy minimization schemes have failed to predict chain folding accurately (Hagler & Honig, 1978; Cohen & Sternberg, 1980a). Presumably this stems from difficulties in modeling protein-solvent interactions, the use of analytic functions to approximate the chemical potential and the compounding of these errors in the computed gradient, and because the energy surface has multiple minima which makes it nearly impossible to locate the global minimum. Efforts to overcome these problems are underway. Molecular dynamics offers many attractive features but remains computationally limited as a technique for studying chain folding (McCammon et al, 1977; Karplus & McCammon, 1983; Hermans, 1985; Beveridge & Jorgenson, 1986). Presumably, chain folding will take place on the millisecond time scale (Baldwin, 1980). Extreme computing resources must be expended to sample 100 nanoseconds in the life of a small protein

(Post et al., 1986).

The success of statistical or data base methods rely heavily on the ability to recognize a known structural theme in a new protein sequence. While highly homologous sequences can be recognized through a variety of methods (Needleman & Wunsch, 1970; Dayhoff, 1978; Martinez et al., 1983; Doolittle et al., 1984) the significance of limited homology is much harder to evaluate. Certainly, some structures which have been proved similar by x-ray crystallography bear little or no sequence homology (Schultz & Schirmer, 1977). In principle, local homology could be used to identify structural features in a piecemeal fashion. However, several investigators have observed a five or six amino acid sequence adopting α-helical structure in one protein and β-structure in another (Kabsch & Sander, 1984). Clearly, this poses great difficulty for data base methods. In spite of these caveats, it is likely that these methods will offer significant structural information for a limited set of proteins (Blundell et al., 1987). To the extent that there is a very finite set of possible protein folds that can be easily catalogued, this approach would be increasingly useful.

Heuristic methods attempt to exploit structural inferences gained from the study of known protein structures. A large number of analyses of the topological, sequential, packing, and surface characteristics of proteins have been done (Lee & Richards, 1971; Chou & Fasman, 1974, 1977; Richards, 1977; Sternberg and Thornton, 1978; Richardson, 1981; Sheriden et al., 1985). These conclusions have been incorporated into a hierarchic condensation model.

To facilitate the accurate prediction of core secondary structure, the protein class (α/α, β/β, α/β, misc) is determined (Levitt & Chothia, 1976; Sheriden et al., 1985). Such an approach provides information on the number and distribution of secondary structure elements, however, it is important to consider whether this approach is valid. Since protein class can often be determined from primary sequence information, it can be considered simply another form of secondary structure prediction.

The actual definition of secondary structure has some bearing on the success of structure prediction by effecting the calculated residue conformational propensities and serving as the final arbiter of prediction accuracy. Presently, several criteria can be used to determine the location of secondary structure elements. Some of these include, 1) the distribution of backbone dihedral angles, 2) interatomic distances of the alpha carbons of the polypeptide chain, and 3) the hydrogen bonding patterns found in the local sequence (Kabsch & Sander, 1984; Farrah et al., 1988). Even though the entire length of a secondary structure element is necessary for the prediction of final tertiary structure, we would postulate that the identification of the central region of each element is sufficient to allow for tertiary structure modeling.

Secondary structure can be located along the protein chain through statistical (Chou & Fasman, 1974, 1977; Taylor & Thornton, 1984) or pattern based (Lim, 1974; Cohen et al., 1983, 1986a) algorithms. To predict secondary structure using pattern based algorithms, a pattern-matching language, the Pattern Language for Amino and Nucleic Acid Sequences (PLANS) was developed (Abarbanel, 1984).

Primary patterns described using the PLANS language are generally limited to short range interactions (from 2 to 5 residues). To describe patterns that span a longer distance of residues, PLANS allows the inclusion of primary patterns and logical operators such as "or" and "and" to synthesize higher order patterns. Using these techniques, patterns written in the PLANS language can describe short and medium range interactions that determine secondary structure. Turn prediction accuracy is enhanced by searching for segments of chain well suited for a protein surface location which are spaced through the chain at intervals compatible with secondary structure pitch and domain diameter. For each class of proteins a different set of patterns describing secondary structure elements was determined. α-helices and β-strands can be more easily identified between these turns because of the restrictions imposed by protein class.

As noted above, an isolated section of five to six residues in a polypeptide may adopt either a α-helical or a β-strand structure. This implies that short to medium range interactions are not sufficient to determine final secondary structure in a folded protein: some aspect of long range interaction (greater than 15 amino acids) must be included to fully describe secondary structure. While it is possible to describe long range patterns in the PLANS language, the types of interactions necessary for the determination of secondary structure are not well characterized. Furthermore, patterns for short and medium range interactions are based on a preconception of what secondary structure should (or will) look like in three dimensional space. For the most part, these assumptions are well founded, and indeed hold true (Cohen et al., 1986a). For α/α proteins, turns could be located with 95% accuracy on the test deck of sequences used to develop the turn patterns. For α/α proteins excluded from the test deck, 90% of the turns were identified correctly. Similar conclusions apply for α/β and β/β proteins. However, there exists no set of commonly observed, specific, long range interactions that could be used to develop patterns in the PLANS language.

Two of the many approaches that could be employed are empirical determination of long range interactions, and computer learning of long range interactions. In the first case, crystal structures are inspected for interactions such as salt bridges and hydrogen bonds that span secondary structural elements. Postulated interactions could then be used as prototype patterns in PLANS. In the second case, secondary structure prediction could be left at the present level of accuracy. This would allow for the formation of many structures using the ambiguities in the predicted secondary structure. Assuming that one or more of the generated tertiary structures was correct, this would provide a training set of structures which contained both positive and negative examples of individual long range interactions. Such training sets could then be examined either by hand, or automatically to construct patterns in the PLANS language.

Once identified, all possible juxtapositions of these secondary structures are generated using combinatorial algorithms (Richmond & Richards, 1978; Cohen et al., 1979, 1980, 1982; Cohen and Sternberg, 1980b). Proposed structures which are inconsistent with the known chain connectivity and steric properties of the secondary structure segments are rejected. Additional structures which violate topological properties, surface-volume relationships, or experimental data are discarded.

When applied to myoglobin, 13 docking sites were identified on 6 of the 8 helices (Richmond & Richards, 1978). 3.4×10^8 possible tertiary structures were generated which matched helix docking sites in geometries consistent with known packing constraints (Chothia et al., 1977; Cohen et al., 1979). Of these, only 20 were connected and sterically reasonable (Cohen et al., 1979), and two could bind a heme group between the proximal and distal histidine (Cohen & Sternberg, 1980b). These two structures were very similar (r.m.s. deviation 0.3Å) and resembled the myoglobin x-ray structure (r.m.s. deviation 3.6Å). Subsequent studies on a number of proteins with known structures suggested that similar results could be obtained in approximating the x-ray structure. However, frequently many alternative conformations were produced which could not be distinguished from the native-like structure on theoretical grounds.

Recently, this approach has been applied to two proteins of known sequence but unknown structure: Interleukin-2 (Il-2) (Cohen et al., 1986b) and Human Growth Hormone (HGH) (Cohen & Kuntz, 1987). Since the completion of these calculations, preliminary x-ray crystallographic information has become available. This provides a unique opportunity to document the successes and failures of the heuristic approach and points to key problems for future study.

Il-2 is a lymphokine which promotes T-cell growth (Morgan et al., 1976; Gillis & Smith, 1977). Analysis of the amino acid sequence suggested a predominance of α-helical structure. This was confirmed by circular dichroism (Cohen et al., 1986b). Four helices were identified in the human sequence A(17-31), B(64-73), C(83-97) and D(116-132) by identifying turns (Cohen et al., 1986a) and subsequently locating helical patterns between the turns. Ten helix docking sites were located using the method of Richmond and Richards (1978). The number of structures generated was 3.9×10^4. Of these, 27 satisfy steric constraints while simultaneously maintaining the connectivity of the chain and permitting a disulfide bridge between residues Cys-58 and Cys-105. The five topological families of four helix bundles which remained are shown in Figure 1a. The right handed cylinder was reminiscent of other four helix bundle structures (Weber & Salemme, 1980) and was selected as the most likely structure (see Figure 1b). McKay and co-workers (Brandhuber et al., 1987) have solved the three-dimensional structure of Il-2 at 3.0Å resolution.

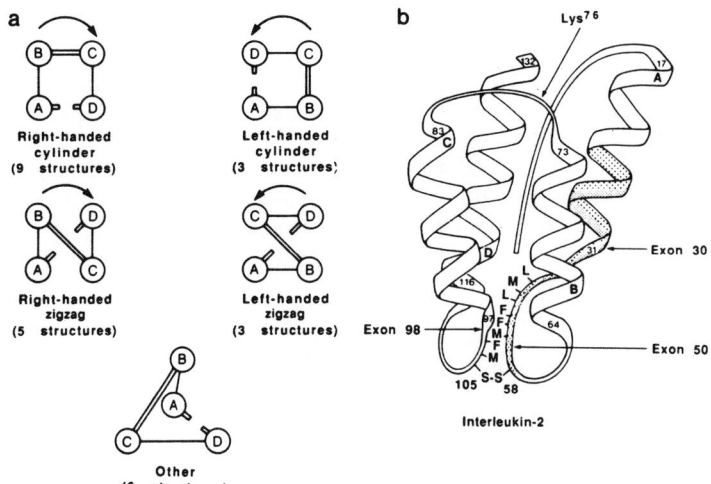

Figure 1a. Schematic representation of four helix bundle topologies generated by a combinatorial search of possible structures for Interleukin-2. Single lines show connections behind the page and double lines represent loops in front of the page. *b.* Ribbon diagram of the most likely model for Il-2. The disulfide bridge and exon boundaries are shown. A hydrophobic region in the AB and CD loops is a potential candidate for the ligand receptor binding site.

Six α-helices are noted including four core helices (B,C,D,F) which form a left-handed four fold α-helical bundle. The B helix is kinked at a proline. In comparing the predicted structures to the crystallographic model, the following conclusions can be reached (see Figure 2):

1) turn location using the pattern matching algorithm of Cohen *et al.* was able to accurately separate core α-helices within ± 2 residues. The kink in the B helix at Pro-47 was identified as a turn.

2) the α-helices C, D, and F were identified correctly C (66-78), D (83-101) and F (117-133) in the x-ray structure(64-73), (83-97) and (116-132) in the model structure. The B helix was completely missed. Instead of identifying residues 33-47 as helical, residues 17-31 were chosen. The prediction scheme relied upon the hydrophobic pattern of residues in positions $i, i+3, i+4$ and $i+7$ seen in the region 17-31 and absent in the region 33-47.

This pattern has been identified as important to helix-helix interactions. The energetic explanation for the predictive error remains obscure.

3) the combinatorial packing algorithm correctly identified the approximate packing angles (actual 26.6 °, predicted 20.0 °) and identified three possible left-handed structures from a list of 27 alternatives. We incorrectly identified the right-handed four helix bundle as the best model for Il-2 based on a survey of known structures all of which displayed the right handed topology. The correct structure is novel from the perspective of previously characterized four helix bundles. A better understanding of the structural determinants will hopefully follow from a detailed analysis of the refined atomic coordinates.

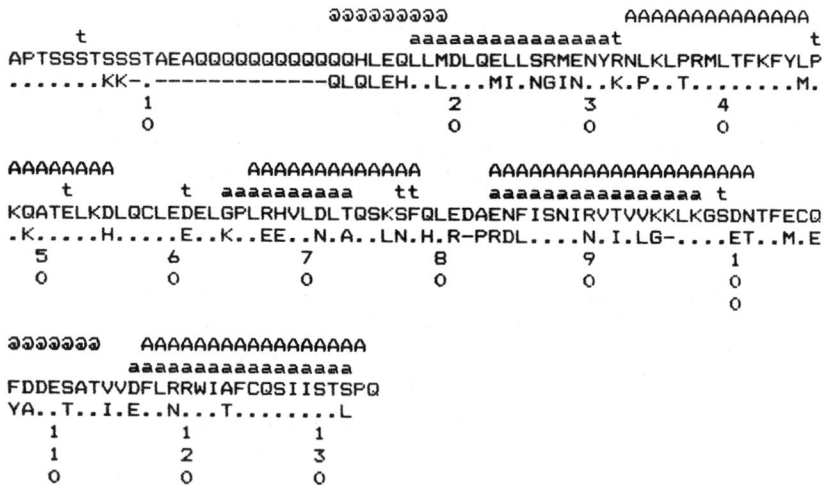

Figure 2. The sequences of mouse and human Interleukin-2 are presented in alignment. Dots (.) indicate identity and dashes (-) represent deletions. The predicted secondary structure, core α-helices (aaa) and turns (t), and the experimentally determined secondary structure, non-core (@@@) and core α-helices (AAA) are shown above the sequence.

HGH was studied in a similar fashion. Four helices were identified along the sequence and a most likely structure was selected from a list of 186 possibilities (see Figure 3a).

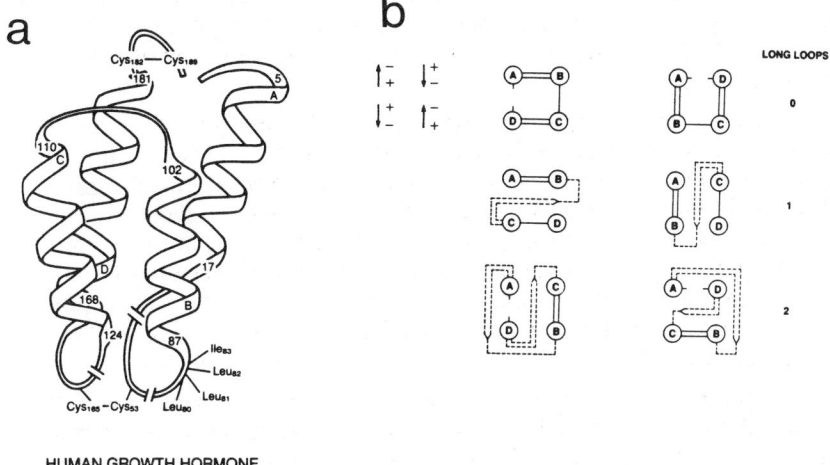

HUMAN GROWTH HORMONE

Figure 3a. A schematic diagram of the right handed four-fold α-helical structure predicted for human growth hormone. Helices are labelled sequentially A through D and the residue number of the N-terminii and C-terminii are marked. Loops are added for clarity and breaks are placed in the longer loops. Disulfide bridges are shown between Cys_{53} and Cys_{165} and between Cys_{182} and Cys_{189}. Four hydrophobic residues predicted to be important to receptor binding, Leu_{80}, Leu_{81}, Leu_{82}, and Ile_{83}, are marked. *3b.* A schematic representation of four α helix bundle topologies which are favored by helix dipole interactions. Helices are labelled sequentially A through D with double lines showing connections in front of the page, single lines showing connections behind the page and long loops which traverse the diameter of the molecule shown as transitions from single to double lines. For clarity, the relative directions of the chain for each helix is highlighted in the upper left hand corner of the figure.

Abdel-Meguid *et al.* (1987) have determined the crystal structure. Both are four helix bundles with a right handed sense. Helices are consistently shorter in the predicted structure however the major packing interfaces are located. Unfortunately, the topological arguments used to process the list of 186 structures again proved too limiting (Cohen & Kuntz, 1987). The crystal structure connectivity can be obtained from the model structure (Figure 3a) by interchanging the position and direction of the B and C helices. The crystal structure requires long loop connections between helices A and B and between helices C and D. This connectivity was generated by the combinatorial algorithm. However, in an attempt to find the most likely predicted structure, long loops were rejected on statistical grounds. As suggested by the breaks in the chain in Figure 3a, such long connections are sterically possible but had been thought to be kinetically unreasonable. Clearly a reevaluation of this presumption is indicated.

Baldwin and co-workers (Shoemaker et al., 1987) have shown that the helix dipole is important to the stability of isolated α-helices. If this effect is important for stabilizing helix pairs that adopt the relatively parallel orientation observed in four α-helical bundles, then six connectivities should be observed. These are shown schematically in Figure 3b. Obviously, those which require long loop connections would be possible only when sufficient chain is present between the helices. It is tempting to speculate that long loop connections have not be observed in the previously studied four α-helix bundles (Weber & Salemme, 1982). because they are consistently smaller than HGH.

More general schemes are needed to deal with the number of structures generated by combinatorial algorithms which satisfy connectivity and crude steric constraints. Energy calculations are insensitive to the gross structural errors observed in these structures (Novotny et al., 1984). A sorting procedure which could eliminate non-native-like structures from further consderation would be invaluable. One observed feature of native proteins is that they are well packed (Richards, 1977). Recently, we have developed a method for evaluating packing in proteins using a simplified representation. The protein chain is modeled as a series of spheres (residue side chains) attached to a polypeptide backbone (Levitt and Warshel, 1974). Each side chain sphere is located at the approximate center of mass of the side chain atoms. Every amino acid type is assigned an "ideal radius" related to its size, and, more subtly, its preference for the core or exterior of the protein. These ideal values are derived from a study of 72 proteins from the Brookhaven data bank. To test packing, the radii of all of the side chains are set equal to zero and increased at a rate proportional to their final "ideal" radius. Growth is terminated when the sum of the radii assigned to the two residues considered equals the separation between the residue centroids. The calculation is performed iteratively through the sequence with a step size of 1% of the ideal radius. until no residues can increase further in size. The size attained by each residue in a protein is judged relative to the "ideal radius".

We have compared some combinatorially-produced models of flavodoxin to the crystal structure. In general, the models are much less uniformly packed, with regions of loose packing (residues grew too big) and constrained packing (residues did not grow big enough.) The crystal structure of flavodoxin (Smith et al., 1977) is relatively homogeneous. We are developing methods to quantify this measure of packing quality.

Current heuristic methods are capable of generating plausible structural models which can be subjected to experimental testing. These methods fail to produce a unique answer. The HGH example clearly shows an inability of the method to distinguish the native structure from a reasonable, but incorrect, alternative. Significant progress in understanding protein stability is necessary before it will be possible to sort between alternative structures. Mutagenesis experiments may provide this insight. In the interim, a close collaboration between experimentalist and theoretician is vital if these modeling efforts are to have any utility.

Acknowledgements

This work was supported by the Defense Advanced Research Projects Agency under contract N00014-86-K-0757 administered by the Office of Naval Research, NIH RR01081 to the UCSF Computer Graphics Laboratory, and a grant from the Eli Lilly Company (FC). Ms. Gregoret is supported by an NSF fellowship. We thank Dr. Langridge for his ongoing help and encouragement.

REFERENCES

Abdel-Meguid SS, Shieh HS, Smith WW, Dayringer HE, Violand BN, Bentle LA (1987). Three-dimensional structure of a genetically engineered variant of porcine growth hormone. Proc Natl Acad Sci USA 84:6434-7.
Anfinsen CB, Haber E, Sela M, White FH (1961). The Kinetics of Formation of Native Ribonuclease During Oxidation of the Reduced Polypeptide Chain. Proc Natl Acad Sci USA 47:1309-1314.
Abarbanel RM (1984). Ph.D Thesis. Medical Information Sciences. University of California, San Francisco.
Baldwin RL (1980). The Mechanism of Folding of Ribonucleases A and S. In Jaenicke, R. (ed.) "Protein Folding," Amsterdam: Elsevier, p 369.
Beveridge DL, Jorgenson WL (1986). Computer Simulation of Chemical and Biological Systems. Ann NY Acad Sci 482.
Blundell TL, Sibanda BL, Sternberg MJE Thornton J.M. (1987). Prediction of protein structure from amino acid sequence. Nature 326:347-352.
Brandhuber BJ, Boone T, Kenney WC, McKay DB (1987). Three-dimensional structure of interleukin-2. Science 238:1707.
Chothia C, Levitt M, Richardson D (1977). Structure of proteins: packing of alpha helices and pleated sheets. Proc Natl Acad Sci USA 74:4130-4134.
Chou PY, Fasman GD (1974). Prediction of protein conformation. Biochemistry 13:222-245.
Chou PY, Fasman GD (1977). Beta-turns in proteins. J Mol Biol 115:135-175.
Cohen FE, Richmond TJ, Richards FM (1979) Protein Folding: Evaluation of Some Simple Rules for the Assembly of Helices into Tertiary Structure with Myoglobin as an Example. J Mol Biol 132:275-288.
Cohen FE, Sternberg MJE (1980a). On the Prediction of Protein Structure: The Significance of the Root-mean-square Deviation. J Mol Biol 138:321-333.

Cohen FE, Sternberg MJE (1980b). On the Use of Chemically Derived Distance Constraints in the Prediction of Protein Structure with Myoglobin as an Example. J Mol Biol 137:9-22.

Cohen FE, Sternberg MJE, Taylor WR (1980). Analysis and Prediction of Protein Beta-Sheet Structures by a Combinatorial Approach. Nature (London) 285: 378-382.

Cohen FE, Sternberg MJE, Taylor WR (1982). The Analysis and Prediction of the Packing of α-Helices against a β-Sheet in the Tertiary Structure of Globular Proteins. J Mol Biol 156:821-862.

Cohen FE, Abarbanel RA, Kuntz ID Fletterick RJ (1983). Secondary Structure Assignment for α/β Proteins by a Combinatorial Approach. Biochemistry 22:4894-4904.

Cohen FE, Abarbanel RA, Kuntz ID, Fletterick RJ (1986a). Turn Prediction in Proteins Using a Pattern Matching Approach. Biochemistry 25:266-275.

Cohen FE, Kosen PA, Kuntz ID, Epstein LB, Ciardelli TL, Smith KA (1986b). Structure Activity Studies of Interleukin-2. Science 234:349-352.

Cohen FE, Kuntz ID (1987). Prediction of the Three-Dimensional Structure of Human Growth Hormone. Proteins: Structure, Function and Genetics 2: 162-166.

Creighton TE (1978). Experimental Studies of Protein Folding and Unfolding. Prog Biophys Mol Biol 33:231-297.

Dayhoff M (1978). "Atlas of Protein Sequence and Structure." Silver Spring Md.: National Biomedical Research Foundation

Doolittle RF, Feng D, Johnson MS (1984). Computer-based characterization of epidermal growth factor precursor. Nature (London) 307:558-560.

Farrah T, Kuntz ID, Cohen FE, Abarbanel RA, Sloan KR (1988). Protein Secondary Structure Definition: An Approach Based on Local Inter-C^{α} Distances. In preparation.

Gillis S, Smith KA (1977). Long term culture of tumour-specific cytotoxic T cells. Nature 268:154-156.

Hagler AT, Honig B (1978). On the formation of protein tertiary structure on a computer. Proc Natl Acad Sci USA 75:554-558.

Molecular Dynamics and Protein Structure (1985) In Hermans J (ed): Proceedings of Workshop 13-18 May 1984, University of North Carolina. Western Springs, IL: Polycrystal Book Service.

Kabsch W, Sander C (1984). On the use of sequence homologies to predict protein Structure: Identical Pentapeptides can have completely different conformations. Proc Natl Acad Sci USA 81:1075-8.

Karplus M, Weaver DL (1979). Diffusion Controlled Model for Protein Folding. Biopolymers 18:1421-1437.

Karplus M, McCammon, JA (1983). Dynamics of Proteins: Elements and Function. Ann Rev Biochem 52:263-300.

Lee BK, Richards FM (1971). The interpretation of protein structures: estimation of static accessibility. J Mol Biol 55:379-400.

Levitt M, Warshel A (1975). Computer simulation of protein folding. Nature

253:694-698.
Levitt M, Chothia C (1976). Structural Patterns in Globular Proteins. Nature 216:552-557.
Lim VI (1974). Algorithms for prediction of alpha-helical and beta-structural regions in globular proteins. J Mol Biol 88:873-894.
McCammon JA, Gelin BR, Karplus M (1977). Dynamics of folded proteins. Nature (London) 267:585- 590.
Martinez HM, Katzung B, Farrah T (1983). "Sequence Analysis Program Manual." San Francisco Biomathematical Computation Laboratory, University of California.
Morgan DA, Ruscetti FW, Gallo R (1976). Selective in vitro growth of T lymphocytes from normal human bone marrows. Science 193:1007-1008.
Needleman S, Wunsch C (1970). A General Method Applicable to the Search for Similarities in the Amino Acid Sequence of Two Proteins. J Mol Biol 48:443-453.
Nemethy G, Scheraga HA (1977). Protein folding. Quart Rev Biophys 10:239.
Post CB, Brooks BR, Karplus M, Dobson CM, Artymiuk PJ, Cheetham JC, Phillips DC (1986). Molecular dynamics simulations of native and substrate-bound lysozyme. A study of the average structures and atomic fluctuations. J Mol Biol 190:455-479.
Richards FM, (1977). Areas, volumes, packing and protein structure. Ann Rev Biophys Bioeng 6:151-176.
Richardson JS (1981). The anatomy and taxonomy of protein structure. Adv Protein Chem 34:167-339.
Richmond TJ, Richards FM (1978). Packing of alpha-helices: geometrical constraints and contact areas. J Mol Biol 119:537-555.
Schulz GE, Schirmer, RH (1979). "Principles of Protein Structure". New York Springer-Verlag.
Sheriden RP, Dixon JS, Venkataraghavan R, Kuntz ID, Scott KP (1985). Amino acid composition and hydrophyobicity patterns of protein domains correlate with their structures. Biopolymers 24:1995-2023.
Shoemaker KR, Kim PS, York EJ, Stewart JM, Baldwin RL (1987). Tests of the helix dipole model for stabilization of alpha-helices. Nature 326:563-567.
Sternberg MJE, Thornton JM (1978). Prediction of protein structure from amino acid sequence. Nature 271:15-20.
Sternberg MJE (1983). The Analysis and Prediction of Protein Structure. In Geisow M, Barrett A (eds): "Computing in Biological Science" Amsterdam: Elsevier, p 143-177.
Taylor WR, Thornton JM (1984). Recognition of super-secondary structure in proteins. J Mol Biol 173:487-514.
Weber PC Salemme FR (1982). Structural and functional diversity in 4-alphahelical proteins. Nature (London) 287:82-84.
Weiner SJ, Kollman PA, Case DA, Singh UC, Ghio C, Alagona G, Profeta S, Weiner P (1984). A New Force Field for Molecular Mechanical Simulations of Nucleic Acids and Proteins. J Am Chem Soc 106:765-784.

NORMAL-MODE DYNAMICS AS A TOOL FOR PREDICTING ANTIGENIC SITES ON PROTEINS

Peter S. Stern

Chemical Physics Department, Weizmann Institute of Science, 76100 Rehovot, ISRAEL

INTRODUCTION

The importance of flexibility in the biological function of macromolecules is generally recognized. Examples include a variety of interactions between receptors and ligands, such as that between an enzyme and its substrate. More recently, antibody recognition of continuous antigenic determinants have been presumed to be correlated with the segmental mobility of these sites (Westhof et al., 1984; Tainer et al., 1984). As a result, there has been considerable interest in predicting the location of antigenic determinants in proteins by determining sites of segmental mobility in accessible surface regions. Despite the controversy as to whether atomic mobility is a significant factor in locating antigenic determinants (Jemmerson et al., 1985), groups synthesizing the peptides have successfully incorporated mobility into their prediction schemes. These groups have prepared antibodies from peptide sequences representing mobile regions of the protein, using x-ray crystallographic temperature factors to choose these regions. These antibodies react strongly with native protein and are useful for detecting proteins and studying their structure.

A method of modeling protein dynamics using normal-mode analysis in torsional space was previously developed and applied to four proteins (Levitt et al., 1983, 1985). The method of normal modes was developed as an alternative to the more costly method of molecular dynamics. In the further simplification of imposing rigid geometry and using only torsional degrees of freedom, this method seems well-suited to studying the collective, segmental motion of the lowest (and slowest) vibrational modes of a protein. Examples of these modes, when displayed with computer graphics, give a great deal of insight as to the nature of protein motion. The rms atomic fluctuations calculated from these modes agree well with experimental temperature factors (Levitt et al., 1985). In addition, the rms fluctuations for the α-carbon atoms of different conformations of the same protein are insensitive to rms differences of conformation on the order of 2-3 Å. There are a number of instances in which structural information on proteins is available, but temperature factors are not. These include lower resolution structures and cases where modeling of a similar protein is used to predict the structure.

The normal modes of proteins are analagous to the molecular vibrations of small organic molecules studied with IR and Raman spectroscopy. If we consider a ball of mass $\underline{m}$, connected to a stationary object by a spring of force constant $\underline{k}$, we can easily solve the equations of motion to give us the amplitudes and modes of vibration of this system. The frequency of vibration ω is equal to $\sqrt{k/m}$. For the many atoms in a molecule connected by spring-like bonds, the equations are more complicated, but when solved yield a spectrum of vibrational modes each with a different frequency and a characteristic coherent motion of all the atoms in the molecule moving together, in phase. Each atom, of course, may have a different amplitude and direction of motion, but all atoms reach their maximum amplitude together and return to their equilibrium starting position at the same time. This is somewhat of an approximation; because of the nature of the forces with which the atoms in a molecule interact, the motion will not be so harmonic. Here we have assumed harmonic motion about the equilibrium coordinates. Also, the true motion of the molecule will most certainly be some linear combination of different modes in random phases, so that the atoms will not be moving coherently. Nevertheless, by breaking down the true motion of the protein into its component parts one may get a much better feel for the different possible types of motion available to the molecule.

The vibrational frequencies of organic molecules range from 3000 cm^{-1} (a period of 10^{-14} s) for C-H stretching vibrations, down to a few hundred cm^{-1} (only an order of magnitude slower) for torsional modes. For the proteins we studied, the large reduced masses lower the range of the torsional frequencies from about 200 cm^{-1} down to as low as 2-3 cm^{-1}. The latter corresponds to periods of vibration in the 10 ps range. Unfortunately, this range is still to fast to be compared to most experiments on protein dynamics which are usually 10-100 times slower and only overall averaged results such as temperature factors from x-ray crystallography and low-frequency vibrations from Raman spectroscopy can be used to confirm that the results are reasonable. Hopefully, the introduction of picosecond lasers in future experiments will change this.

The method is applied to proteins with known antigenic sites and will be extended to predicting antigenic sites on proteins of interest.

METHOD

The method used to calculate the normal modes of proteins in torsional space has been described in detail previously (Levitt et al., 1983, 1985) and will only be outlined briefly. First, we use an empirical energy function to describe the interaction between all the atoms of the molecule at their x-ray crystallographic positions. This function is minimized in Cartesian space until the energy change is less than 10^{-4} kcal/mol. It is then minimized in torsional space and at the minimum, the equations of motion are solved and the normal modes determined.

The equations of motion to be solved reduce to an eigenvalue equation which may be written in matrix notation as:

$$FA = HA\Lambda$$

where F is the second derivative matrix of the potential energy, E_P, with respect to the $\underline{n}$ torsion angles, q ($F_{ij} = \partial E_P/\partial q_i \partial q_j$) and H is the second derivative matrix of the kinetic energy, E_K, with respect to the $\underline{n}$ torsional velocities, $\dot{q}$ ($H_{ij} = \partial E_K/\partial \dot{q}_i \partial \dot{q}_j$). Since

$$E_K = 1/2 \sum_{\ell}^{N} m_\ell \dot{r}_\ell^2$$

in N Cartesian coordinates r_ℓ with atomic masses m_ℓ,

$$H_{ij} = \sum_{\ell}^{N} m_\ell (\partial r_\ell / \partial q_i) \cdot (\partial r_\ell / \partial q_j).$$

The eigenvalue equation is readily solved to give a set of eigenvalues, $\Lambda_{ii} = \omega_{ii}^2$, the normal mode frequencies, and a set of eigenvectors, A_{ij}, the normal mode vectors, which give the relative motion of each torsion angle, q_j, at frequency ω_i.

This equation is usually a function of the 3N (3 times the number of atoms in the molecule) atomic postion coordinates, but since we vary only the n single-bond torsional angles, the complexity of the problem is considerable reduced. A protein with 2500 atoms, for example, would have $(7500)^2$ second derivative matrix elements in Cartesion space, but only $(600)^2$ matrix elements in torsional space.

The amplitude of a normal mode, $\alpha_i = (2k_B T / \omega_K^2)^{1/2}$, where k_B is Boltzmann's constant and T is the absolute temperature. The temperature factor, $B = 8\pi^2 \sigma^2 / 3$, where the atomic fluctuations, σ, are readily calculated from the amplitudes, α, the eigenvectors, A, and the transformation matrix, $\partial r / \partial q$ (see Levitt et al., 1985 for details).

RESULTS

In Table 1, the frequencies and atomic fluctuations for oxymyoglobin are compared with those of lysozyme, previously calculated by Levitt et al. (1985). The two proteins have similar values. The larger myoglobin has a low frequency of only 2.7 cm^{-1}, but has a larger mean frequency. The mean α-carbon fluctuations are similar in the two proteins, but the mean Φ and Ψ torsional fluctuations are significantly less in myoglobin. In Table 2, the experimental and calculated temperature factors (B values) are compared. These are presented graphically in Figures 1 and 2. In both proteins, the mean calculated values are lower than the observed values, indicating the contribution of factors other just thermal vibrations to the observed B values.

TABLE 1. Comparison of Frequencies and Fluctuations

	Frequency (cm^{-1})		Number of modes		rms fluctuations		
	Low	Mean	Total	<20 cm^{-1}	σ_α(Å)	σ_ϕ(°)	σ_ψ(°)
Lysozyme*	3.0	64.0	471	60	0.56	15.9	16.2
Myoglobin	2.7	70.0	592	63	0.54	11.7	12.0

* Taken from Levitt et al. (1985).

TABLE 2. Comparison of Experimental and Calculated B values

	Mean (C^α atoms)		Correlation coefficient*			
	Obs	Calc	All	C^α	Main	Side
Lysozyme†	18.0	9.2	–	–	0.56	0.60
Myoglobin	11.3	7.4	0.50	0.65	0.56	0.44

* The correlation coefficient between observed {o} and calculated {c} sets of data is given by $(\langle oc \rangle - \langle o \rangle \langle c \rangle)/\{(\langle o^2 \rangle - \langle o \rangle^2)(\langle c^2 \rangle - \langle c \rangle^2)\}^{1/2}$. The headings All, C^α, Main and Side denote all non-hydrogen atoms, the α-carbon atoms, the N, C^α, C^β, C and O atoms and the remaining atoms, respectively.
† The values for lysozyme were taken from Levitt et al. (1985); the experimental B values (M. Sternberg, personal communication) were not refined for individual atoms, but only for main-chain and side-chain atoms of each residue.

Lysozyme is known to have continuous epitopes with antigenic activity at the loops formed by residues 38-54 and 64-80 (Westhof et al. 1984). Both the curves in Figure 1 show peaks in these two regions with maxima at residues 47 and 70, correlating well with the known epitopes. A third peak (at 101 in the experimental curve and at 103 in the calculated curve) does not correspond to a region of known antigenic activity. It should be noted that the experimental curve is from the tetragonal crystal form of hen egg white lysozyme. A corresponding plot for the orthorhombic form does not does not have peaks at residues 47 and 70 because of intermolecular contacts of these residues in the crystal (Westhof et al., 1984).

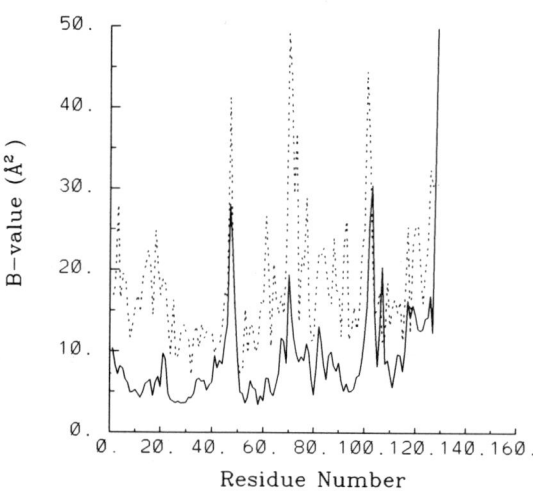

Figure 1. Comparing the experimental (dashed line) and calculated (solid line) variations with residue number of the main-chain B values at 300 K for lysozyme.

Myoglobin is known to have nine continuous epitopes at residues 1-6, 15-22, 22-55, 56-62, 72-89, 94-99, 113-119, 121-127 and 145-151 (Westhof et al. 1984). Except for the epitope at residues 56-62, for which both calculated and observed (Phillips 1980) temperature factors have a minimum, the other eight epitopes correspond to local maxima in both curves in Figure 2. Of course, the regions from residues 22-55 and 72-89 are too broad to identify precisely the location of the epitope, but the calculated curve gives maxima for these at 40 and 47 and at 80 and 84. The experimental curve has maxima in the same region.

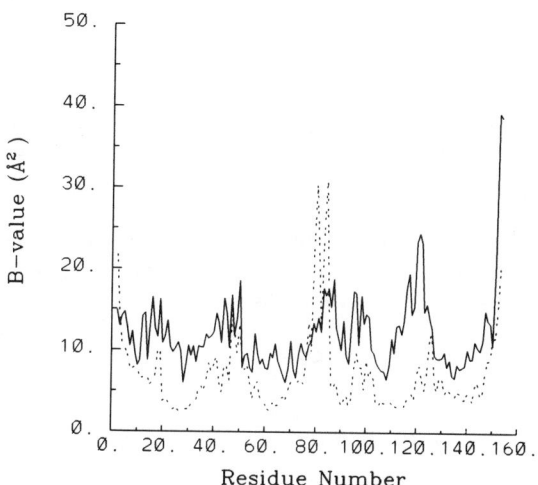

Figure 2. Comparing the experimental (solid line) and calculated (dashed line) variations with residue number of the main-chain B values at 300 K for oxymyoglobin.

CONCLUSIONS

Normal-mode dynamics appears to be a simpler and less costly method than molecular dynamics for narrowing down the number of peptide sequences which would have to be synthesized to try to find antigenic determinants. For a protein such as myoglobin with 153 residues, 1439 heavy and polar hydrogen atoms and 592 torsion angles, the entire calculation took less than 60 hours on a MicroVAX II. Although the question as to whether surface mobility is a factor in identifying antigenic sites is as yet unresolved, this method is a fast and relatively inexpensive way to identify such regions. It has the additional advantage of identifying regions of surface mobility which because of close crystal contacts have low temperature factors and are, therefore, not identified by x-ray crystallographic refinements.

ACKNOWLEDGEMENTS

The author is grateful to Professor Michael Levitt for helpful discussions and for the use of his program ENCAD.

REFERENCES

Jemmerson R, Paterson Y, Van Regenmortel MHV, Altschuh D, Tainer JA, Getzoff ED, Alexander H, Houghten RA, Olsen AJ, Lerner RA (1985). Mobility and evolutionary variability factors in protein antigenicity. Nature 317:89-90.

Levitt M, Sander C, Stern PS (1985). Protein Normal-mode Dynamics: Trypsin Inhibitor, Crambin, Ribonuclease and Lysozyme. J Mol Biol 181:423-447.

Levitt M, Sander C, Stern PS (1983). The Normal Modes of a Protein: Native Bovine Pancreatic Trypsin Inhibitor. Int J Quant Chem: Quant Biol Symp 10:181-199.

Phillips SEV (1980). Structure and Refinement of Oxymyoglobin at 1.6 Angstroms Resolution. J Mol Biol 141:531-544.

Tainer JA, Getzoff ED, Alexander H, Houghten RA, Olsen AJ, Lerner RA, Hendrickson WA (1984). The reactivity of anti-peptide antibodies is a function of the atomic mobility of sites in a protein. Nature 312:127-134.

Westhof E, Altschuh D, Moras D, Bloomer AC, Mondragon A, Klug A, Van Regenmortel MHV (1984). Correlation between segmental mobility and the location of antigenic determinants in proteins. Nature 311:123-126.

A THEORETICAL MODEL OF METAL BINDING SITES IN PROTEINS

Morris Krauss and Walter J. Stevens
Molecular Spectroscopy Division, National Bureau of Standards, Gaithersburg, Maryland 20899

INTRODUCTION

The binding of metal cations to a site consisting of neutral ligands has been studied exhaustively (1). If we restrict our attention to ions where there is no directed valence interaction, the binding is governed by the size of the cation and the electrostatic and inductive interactions between the cation and the ligands. Numerous models have been proposed for calculating the enthalpy of binding. Since the binding of the cation to first shell ligands substantially exceed water-water binding energies, a satisfactory model evaluates the first coordination shell energetics with explicit cation to ligand interactions (2) and uses the Born model (3) for the polarization interaction with the environment. Since we are concerned about the transfer of the cation from water to the protein binding site, the relative binding energy of the cations to water clusters is also required. This study will then emphasize the intrinsic energetics and size of the first coordination shell of idealized cluster models of the protein binding sites and the hydrated cations. The energetics of the clusters will be obtained with quantum calculations in order to account for the large polarization effects accurately.

In proteins the site invariably contains an anionic residue. The much stronger ionic interactions can alter the factors that have been proposed as the basis for the conformational behavior and selectivity among the cations when they are bound to neutral sites. The three most common binding ligands in a protein are the carboxylate moiety, the peptide-carbonyl moiety, and the water molecule (4). Even though coordination polyhedra with a coordination number of seven or greater are common for Ca, many sites are roughly octahedral with the larger polyhedra containing several water molecules. The carboxylate moiety can also bind in either a monodentate or bidentate mode. Although most of the Ca binding proteins have three, four, or even five Asp or Glu residues bound to the cation (4), the Subtilisin protease has only one Asp in each of two binding sites (5). In the site B at the surface of the protein, the binding is unidentate while the buried site A binds the carboxylate moiety to the cation in a bi-dentate fashion. The surface site has unusually large metal to oxygen bond distances. In the present study, we restrict the protein model calculations to approximate octahedral symmetry, one anionic ligand, and only oxygen binding sites. With the binding sites of Subtilisin BPN' in mind, three metal cations, Mg^{2+}, Ca^{2+}, and Na^+, were considered at this time. The ion transfer energies are calculated in the cluster model as a function of the cluster size and conformation. The accuracy with which we calculate the hydration enthalpies for these cations will support the cluster model as a means of estimating the transfer energetics.

AB INITIO CALCULATION

Ab initio molecular orbital studies of clusters of water (W), formate anion (F^-), and formamide (F) ligands were performed with the HONDO code (6). Compact effective potentials (CEP) are used in place of the chemically unimportant core electrons. CEP and their concomitant basis sets have been reported for the first two rows of the periodic table (7) and the CEP are available and basis sets have also been optimized for the elements from K to Kr. The interaction energy is dominated by charge-multipole and polarization contributions and the packing by the exchange repulsion interactions. The ligand geometry is chosen to be that of the isolated ligand. The accuracy of the self-consistent-field (SCF) calculations depends on the basis sets and the accuracy with which they represent the ligand moments and polarizabilities. Increasingly accurate basis sets were used to test the variation of the cluster binding energies and the accuracy of binding energy differences. The systematic examination of the clusters is described elsewhere (8). The basis set dependence of the SCF binding energy differences between the different metal complexes for a specific cluster for the double zeta (DZ) and various polarization basis sets (DZP, DZP+d) is found to be agree to about 5 kcal/mol in the worst case but of the order of 1 kcal/mol in most of the clusters tested. These relatively small differences will not affect the conclusions on the qualitative behavior of binding with respect to changing the charge and size of the cluster. The relevant cluster binding energies are given in Table 1.

TABLE 1.   **SCF Cluster Binding Energies**

Energies (kcal/mol)

Cluster	Na	Ca	Mg
W6	99.5	259.9	315.0
W9		292.0	320.2
F⁻ F4W (opt)	176.4	498.5	545.4
F⁻ F4W (2.7Å)	171.2	438.7	
F⁻ F4W (2.8Å)	166.7	422.7	
F⁻ F5 (bi)		490.7	530.9
F⁻ F5 (uni)		504.5	547.2

Since there are twelve nearest neighbor pairs for a cluster of six waters (W6) and twenty one for the nine coordinate cluster (W9), the total dispersion energy is of interest in estimating the first shell contribution to the enthalpy. The dispersion interaction between the water ligands was estimated by a second-order perturbation calculation (MP2) (9). The binding energy per pair in W6 was 0.58, 0.42, and 0.36 kcal/mole for the O-H distance of 2.1, 2.3, and 2.4Å, respectively. The correlation contribution to differences in binding for the different metals is small in the W6 cluster and of the order of at most 1 kcal/mol. The dispersion energy between nearest neighbor waters is much larger in the W9 cluster where the waters are more tightly packed. In fact, the dispersion energy stabilizes the W9 cluster relative to the W6 for the case of Ca. Correlation interactions will be larger for the F and F⁻ ligands, but the hydrogen bonding between these ligands is also far larger. Since the conformational surface is only roughly examined and the energy minima is not determined, there seems little point in calculating the correlation contribution to ligand binding except to note that it will not change substantially for the different metals as the protein model clusters are of similar size.

In order to determine the intrinsic energetically favorable orientations of the ligands with respect to the metal atom, calculations were done with just one ligand bound to the metal, (ML). For formamide the minimum energy structure is very close to a linear M·O·C bond reflecting the strong local interaction with and polarization of the CO bond rather than the overall interaction with the molecular dipole. This agrees with the theoretical literature on small clusters with alkali and alkaline earth cations (10,11). The conformations observed experimentally are considerably distributed over angle (4,12). The energy minimum in the case of the formate anion occurs when a bidentate bond is formed (13). However, the uni-dentate cluster of a metal bound to one formate and five formamide ligands (MF⁻F5) is calculated to be about 15 kcal/mol more stable than the cluster with bi-dentate binding of the carboxylate anion to the metal for both Mg and Ca. Experimentally, the uni-dentate bond to the anion seems to predominate (4). Evidently, the intrinsic binding to a single ligand of either F or F⁻ is not the determining factor in the ligand orientation. The variation of the cluster binding energy with the ligand orientation is found to be dependent on intra-ligand hydrogen-bonding or electrostatic interactions and steric crowding. The MF-F5 and MW6 clusters are illustrated in Figs. 1 and 2.

OUTLINE OF MODEL CALCULATION

The cluster reaction energetics for the following reaction will be determined:

A Model of Metal Binding Sites in Proteins / 99

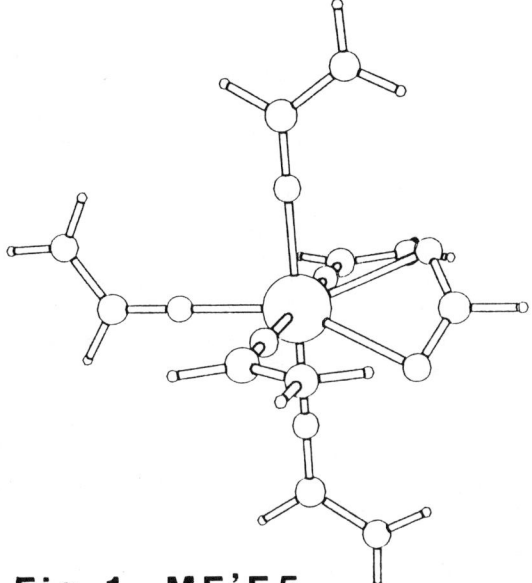

Fig. 1. MF'F5

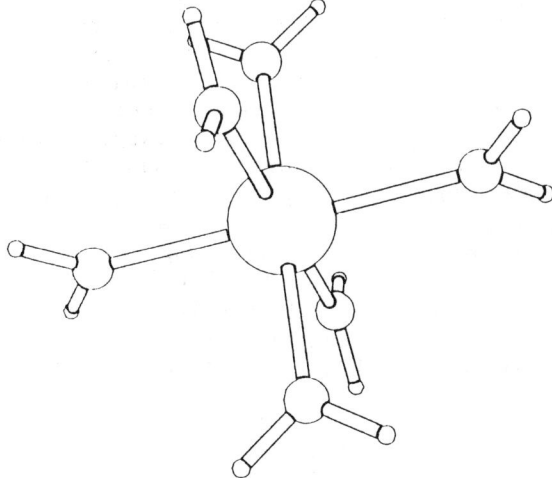

Fig. 2. MW 6

$$M_1W_1 + M_2P \longrightarrow M_1P + M_2W_2 + W_2 - W_1, \quad \Delta E1$$

where M_iW_i represents the metal and its first hydration shell, M_iP represents the metal bound to the protein cluster model, and W_i is the first shell water cluster. The difference in the number of first shell waters for the two metals, $W_2 - W_1$, is required to balance the total number of water molecules. Our model determines the binding of the cations primarily from the sum of the energies of the first coordination shell cluster and the Born polarization energy for this cluster. The procedure of Morf and Simon (1) is followed with the exception that the first shell cluster is evaluated by ab initio quantum chemical methods and the zero-point energy of forming the cluster is included.

Experimentally, only the heats of hydration of the cations are known (1,14,15). The calculated enthalpies of hydration for the cations are compared to the experimental values deduced by Rashin and Honig (15).

The first hydration shell of the Na and Mg cations is a six coordinate octahedral cluster (16) while the Ca shell may be nine coordinate (17). There is a dramatic difference between nine and six coordinate clusters regarding the dispersion energy. Because of the tight packing of the waters in W9, the dispersion energy is calculated to be 18 kcal/mol while for the six coordinate cluster it is about 7, 5, and 4 kcal/mol for Mg, Ca, and Na, respectively. It is evident that the dispersion contribution is significant for the larger cluster. With the inclusion of the dispersion, the W9 cluster is essentially as stable as the W6 for Ca. The zero-point energy is difficult to calculate because of the appreciable number of modes involving the metal that are not measured experimentally. Ab initio frequencies (18) suggest that the metal-oxygen stretches are not dominant and the sum of spectral frequencies is more than twice the stretches. Estimates of the stretch frequencies were calculated to suggest that the zero-point energies are 10, 7, and 6 kcal/mol for the W6 clusters for Mg, Ca, and Na, respectively, but is 12 kcal/mol for CaW9. A calculation of ab initio frequencies is in progress to obtain more accurate values.

In order to calculate the Born polarization energy, the radius of the first shell is required. This was obtained by calculating a two-shell cluster and optimizing the hydrogen-bonding distance between the first and second shell waters. The distance from the metal to the mid-point of the inter-shell hydrogen-bond was chosen as the radius of the first shell cluster. The radii were 3.87Å, 3.61Å, 3.71Å, and 4.36Å, respectively, for the Na, Mg, CaW6, and CaW9 clusters. The cation hydration enthalpies in Table 2 agree with the experimental values within the sizeable errors expected in calculating the various components of the model.

TABLE 2.   **Enthalpy of Model Clusters**
             $-\Delta H$ (kcal/mol)

Cluster	Mg	Ca	Na
W6	450 (462.4)[1]	392	100 (98.5)[1]
W9		387 (383.6)[1]	
F⁻ F5	570	527	
F⁻ F4W (opt)	568	521	176
F⁻ F4W (2.7Å)		460	171
F⁻ F4W (2.8Å)		443	167
F⁻ F4W2 (2.8Å)		453 - 463	

1. Experimental enthalpies from Ref. 15.

A number of small thermodynamic terms that are likely to cancel in the calculation of the transfer energetics are ignored because of the large uncertainties in the the terms given above.

The model protein binding energies are estimated from only the first coordination shell cluster binding energy and the Born polarization energy. The energy required to modify the protein in order to optimize metal binding is not considered at this time. The protein environment is considered to be immobile and the dielectric constant is assumed to be 4 and constant with temperature. The environmental interaction has a large hydrogen bonding contribution and the Born polarization energy as calculated must be considered a crude approximation. The ion cluster has a large dipole formed from the well-separated cationic and anionic charge distributions, which is not represented in the simple Born theory. The calculation of second shell binding energies in a cluster model can only be done for ideal hydrogen-bonded shells which have little relevance to the likely statistical behavior in the outer shells where the metal binding does not dominate. Although such calculations indicate a larger total binding than the Born model, the difference in binding for different metals between clusters containing second shell ligands is not calculated to be large.

The octahedral complex for MgP and CaP is defined differently for the A and B sites in Subtilisin. The Asp residues in the sites are bound to the cation by the carboxylate moiety and are modeled by the formate anion. For site A, the binding is bi-dentate to F⁻ and uni-dentate in site B. The binding to the amide side chain in Asn and Gln is represented by formamide, which also models the amide protein backbone. The size and stiffness of the cavity in surface site B is a question that will be explored in this study. It is evident that the experimental M-O distances in B shown in Table 3 exceed the optimum radial distances found in the present model.

TABLE 3. Metal Oxygen Distances in Subtilisin[1] (Å)

Site	Ca	Ca
A		
Q2E1	2.37	2.43
D41D1	2.42	2.40
D41D2	2.52	2.43
L75	2.29	2.36
N77D1	2.07	2.39
I79	2.29	2.27
V81	2.37	2.30
B	Ca	Na
G169	2.66	2.24
Y171	2.99	2.36
V174	2.66	2.41
E195	3.02	3.88
D197D1	2.84	3.63
W340	2.72	2.12
W372	2.97	2.77

1. Distances from Ref. 1 and unpublished data of Gilliland and Poulos.

The optimum radial distances are in approximate agreement with the distances in site A. Enthalpies are presented in Table 1 for radially optimum distances for both A and B sites, and the effect of a larger relatively stiff site is also examined for B. The site that is accessible to solvent, though, may have different numbers of waters coordinated to the cation and H-bonded to the other ligands. This is particularly important in stabilizing the Ca binding. The A site is then represented by the complex MF^-F_5 while the octahedral version of the B site is given by a cluster of one formate, four formamides, and one water molecule (MF^-F_4W). An upper bound of the binding energy of the seventh coordinated water in the first shell at the B site is obtained from the dissociation energy for water from the CaF^-W5 cluster at the larger distance of the B site. The seven-coordinate estimate for the enthalpy of binding to the B site is given in Table 1.

The water ligands in the octahedral cluster are replaced systematically by first one F^- ligand and then succesively F ligands. The bi-dentate complexes are still more stable for the MgF^-W5, CaF^-W5 and CaF^-FW4 systems but the uni-dentate systems are increasingly stable as the waters are replaced. The uni-dentate complex is increasingly more stable with the number of waters that are replaced reflecting the importance of ligand-ligand repulsion. A complete exploration of even the octahedral complex energy surface is not feasible with SCF calculations but should be possible with an accurately parametrized classical potential. The Mg complexes are more likely to be in the uni-dentate form reflecting the shorter radial metal to oxygen distances and the increased ligand crowding, but all the MF^-F5 clusters are most stable for uni-dentate binding of the metal to F^-.

The B or surface site in Subtilisin has radial metal-oxygen distances substantially in excess of any of the optimal complex distances (5) and one structure has been analyzed with the Ca cation replaced by another metal, Na. As a model of this interaction, the MF^-FW4 and MF^-F4W clusters were studied at larger distances to illustrate the dramatic narrowing of the binding energy difference between Mg and Ca. Increasing the radial distance about 0.5 Å uniformly leads to reduction in the difference in binding energies of 44 kcal/mole down to 14 kcal/mol from about 58 kcal/mole. For Na and Ca in the MF^-F4W cluster, the dependence on the size of the cluster is also dramatic. As the cluster size increases, the energy of the Ca cluster decreases much more rapidly than the Na cluster as expected for the interaction of a divalent cation with a cluster containing an anion.

DISCUSSION

The intrinsic energetics of binding to the first coordination shell is the basis of the specificity in favor of Ca over Mg in a single anion-containing cluster model of the A site. The cluster sizes for the optimum A site of F^-F5 for either Mg or Ca differ only slightly. The Born radii for the Mg and Ca clusters are 5.48Å and 5.58Å, respectively, which yield environmental polarization energies that differ by less than 1 kcal/mol. Direct calculations of the binding energies of second shell hydrogen-bonded ligands of either water or formamide yield energies within 0.1 kcal/mol for either Mg or Ca clusters. Thus the effect of ligands beyond the first shell on the cation transfer energetics is small. Howver, the
enthalpies of binding do not account for the modification of the protein due to the metal binding or the details of protein structure and hydrogen-binding so that the absolute differences must be considered cautiously. Desolvation energies determine the ion selectivity between Mg and Ca because the differences in binding energies to the protein, which are dominated by the first shell, are not as large as the differences in the enthalpies of binding to water. For the A site the energy of reaction is 20 kcal/mol favoring Ca binding to the protein.

A realistic and quantitative model for the protein requires explicit consideration of the second shell and the possible relaxation of the cluster conformation for each cation. Experimental data does not exist for the Mg protein. Site A is always occupied by a Ca cation, and while replacement of Ca has been observed in site B the structure has been analyzed only for a monovalent cation. If the conformations of the Mg and Ca proteins are comparable in the A site, then simple model calculations of hydrogen binding of a second shell formamide to the first shell formamide do not show sufficient difference to alter the selectivity of binding in favor of Ca. The binding energy for a second shell formamide is calculated to be within 0.1 kcal/mol for Mg and Ca in the cluster, MF^-F3W2.

For an optimum B site cluster, the enthalpy of reaction for replacing Mg by Ca would be -16 kcal/mol using the data for CaW9 in Table 1. The Ca protein model is larger than the Mg cluster and could easily be expanded to include an additional water in the first shell which would further enhance the selectivity of Ca over Mg. However, the B site is found to have unusually large metal to oxygen distances for the two metals that have been observed. However, if we assume that the binding distances are optimum, the enthalpy of reaction for displacing the Ca by Na is 58 kcal/mol. The variation of the binding energies with increasing size of the cluster shows that the Na rapidly becomes more favored, with the reaction enthalpy decreasing to about 2 kcal/mol for an octahedral cluster that is uniformly increased to 2.7Å. The Na transfer is favored by -11 kcal/mol if the cluster radius increases to 2.8Å. The rapid decrease in the binding of a divalent cation with the increasing size of the cluster determines this behavior. The Ca could remain bound by adding another water to the first coordination shell, which is not energetically favorable for a monovalent cation like Na. An upper bound to the binding of a seventh water ligand suggests that the Ca is still energetically preferred for the 2.8Å cluster, but the energy of reaction is close enough to zero to suggest that the preformed hole is designed not to bind Ca too strongly.

The rigidity of the site for divalent cations is maintained by a strong ion-pair interaction of Lys 170 and the side chain of Glu 195, which both bind to the side chain of Ser 163. Since the Lys and Glu residues are from opposite sides of the B site, this interaction helps to create a larger hole than the strong divalent cation interaction would normally permit. The rigidity of the anionic ligands also allows the Na to move closer to the amide ligands and permits waters to solvate the anionic site which is accessible to solvent. Apparently, this is energetically favorable for a univalent ion because of the relatively large hydration energy of the carboxylate anion (13).

Previous ab initio calculations of the cluster binding energies (19-21) that were done with smaller basis sets significantly over-estimate the binding energies. A polarization basis is required, as a minimum, that will accurately represent the molecular moments and dipole polarizability. The use of a DZP basis with MW_2 systems to estimate classical three-body terms (19) results in classical binding energies for the octahedral clusters that are good for Mg but high for Ca. The present calculations for the protein clusters are more inaccurate than the water clusters but the energy differences are not very sensitive to basis set variation. Nontheless the dipole moment of the formamide ligand is not as accurately determined as that for water for the best basis sets that were used. The polarizability of the formate anion is also unlikely to be well represented at the SCF level but damping effects at the short metal ligand distances may negate the lack of accuracy for asymptotic properties. The simplicity of the model, though, suggests that exploring more accurate estimates of the components would not be rewarding. The insight regarding the effect of charge and cavity size has on the transfer enegetics is supported by the reasonable hydration enthalpies that this simple model does obtain.

REFERENCES

1. Lehn, J-M. "Structure and Bonding" (J. D. Dunitz, P. Hemmerich, J. A. Ibers, C. K. Jorgensen, J. B. Neilands, D. Reinen, R. J. P. Williams, eds.) V. 16, (1973), pp. 1-69, Springer Verlag, Inc., New York.
2. Morf, W. E.; Simon, W. Helv. Chim. Acta (1971), 54,794; ibid (1971), 54,2683.
3. Born, M. Z. Phys. (1920), 1,45.
4. Einspahr, H.; Bugg, C. E. "Metal Ions in Biological Systems: V. 17, Calcium and Its Role in Biology" (H. Sigel, ed.), pp. 51-97, Marcel Dekker, Inc., New York.
5. Finzel, B. C.; Gilliland, G. L.; Howard, A. J.; Pantoliano, M. W.; Poulos, T. L. (to be published).
6. Dupuis, M.; King, H. F. Int. J. Quantum Chem. (1977), 11, 613, J. Chem. Phys. (1978), 68,3998.
7. Stevens, W. J.; Basch, H.; Krauss, M. J.Chem.Phys. (1984), 81,6026.
8. Krauss, M.; Stevens, W. J. manuscript in prep.
9. (a) Moller,C.; Plesset,M.S. Phys.Rev. (1934), 46,618. (b) Pople,J.A.; Binkley,J.S.; Seeger,R. Int.J.Quantum Chem., Quantum Chem.Symp. (1976), 10,1.
10. Fuchs, D. N.; Rode, B. M. Chem.Phys.Lett. (1981), 82,517.
11. Welti, M.; Pretsch, E.; Clementi, E.; Simon, W. Helv. Chim.Acta (1982), 65,1996.

12. Chakrabarti, P.; Venkatesan, K.; Rao, C. N. R. Proc. Roy.Soc.Lond. (1981), A375,127.
13. Berthod, H.; Pullman, A. J.Comp.Chem. (1981), 87.
14. Noyes, R. M. J. Am. Chem. Soc. (1962), 84,513.
15. Rashin, A. A.; Honig, B. J.Phys.Chem. (1985), 89,5588.
16. (a)Dietz, W.; Riede, W. O.; Heinzinger, K. Z. Naturforsch. (1982), 37a,1038,
    (b) Palinkas, G.; Radnai, T.; Dietz, W.; Szasz, Gy. I.; Heinzinger, K. ibid. 1982, 37a,1049.
17. (a)Probst, M. M.; Radnai, T.; Heinzinger, K.; Bopp, P.; Rode, B. M. J. Phys. Chem. (1985), 89, 753-759, Palinkas, G.; Heinzinger, K. Chem.Phys.Lett. 1986, 126,251.
18. Hashimoto,K.; Yoda,N.; Iwata,S. Chem.Phys. 1987, 116,193.
19. Ortega-Blake, I.; Novaro, O.; Les, A.; Rybak, S. J. Chem.Phys. 1982, 76,5405.
20. Sano, M.; Yamatera, H. "Ions and Molecules in Solution, Studies in Physical and Theoretical Chemistry, V.27" (H. Tanaka, H. Ohtaki, R. Tamamushi, eds.) 1982, pp.109-116, Elsevier Science Pub., Amsterdam.
21. (a)Gottschalk, K. E.; Hiskey, R. G.; Pedersen, L. G.; Koehler, K.A., J. Mol. Struct. 1982,90,265. (b) Maynard, A. T.; Hiskey, R. G.; Pedersen, L. G.; Koehler, K. A. ibid. 1985,124,213.

COMPUTER MODELING OF MEMBRANE-ANCHORED CELLULAR AND VIRAL PROTEINS: ORGANIZATION AND FUNCTION

Yechiel Becker

Department of Molecular Virology, Faculty of Medicine, Hebrew University, Jerusalem, Israel

INTRODUCTION

Recent developments in molecular biology and genetic engineering made possible the cloning and sequencing of a large number of cellular and viral genes. These sequences compiled in data banks can be translated into amino acid sequences in polypeptides coded for by the respective genes. The availability of the primary amino acid sequence of a polypeptide allows the prediction of its secondary and tertiary structures. Computer programs like those of Chou and Fasman (1978) and Garnier et al. (1978) were developed to predict the secondary structure from the primary amino acid sequence. The hydropathic properties of such polypeptides can also be determined (Kyte and Doolittle, 1982; Hopp and Woods, 1981).

The above computer algorithms are based on properties of soluble proteins which were analyzed by physical methods and their tertiary structure was elucidated by X-ray crystallography and NMR analyses. When compared to proteins with known properties the ability of the Chou and Fasman (1978) computer program to predict the secondary structure of a polypeptide from its primary amino acid sequence is estimated to be only 50% correct and that of Garnier et al. (1978) is estimated as 100% correct. Yet, ongoing studies on the chemical properties of soluble relatively short polypeptides, discussed in the present conference, could lead to improvement of the present programs for computer-assisted modeling of polypeptides (Greer, 1988). In this study the three-dimensional structure of the central portion of C5a, corresponding to residues 13 to 76, was determined from the published C3a crystal structure by comparative modeling techniques. Further knowledge on the conformational energy calculations of polypeptides (Scheraga, 1988) as well as additional information on their properties will eventually lead to improvements in the predictive capacity of computer programs used to study secondary and tertiary structure of proteins.

While basic studies on the properties of soluble proteins are advancing, much less is known about the conformational properties of membrane-bound proteins. However, it has been possible to dissolve at least some membrane-bound proteins with small chain detergents or amphiphiles to form a complex. By precipitation of the complex, three dimensional crystals were formed allowing X-ray crystallography analysis of the proteins (Eisenberg, 1984). Bacteriorhodopsin and the honey bee venom, militin and allamethicin, were analysed by the above method and their properties were investigated (reviewed by Eisenberg, 1984). Melithin (26 amino acids in length) is highly amphiphilic and is a bent α-helical rod which in aqua salt solution exists as a tetramer with a surface-seeking capacity related to its high amphiphilicity when the peptide is coiled as a helix. In attempts to determine the properties of amino acid sequences which span membranes, Sergest and Feldman (1974) and Rose (1978) reported that numerical hydrophobicities might be effective in detecting hydrophobic segments. Kyte and Doolittle (1982) and Argos et al. (1982) have developed algorithms for testing membrane-related proteins. Averaging hydrophobicity over segments of 19 residues was found by Kyte and Doolittle (1982) to be most effective in distinguishing membrane-spanning segments from globular conformation of soluble portions of the membrane-bound protein. An algorithm was also developed by Eisenberg (1984).

One of the major developments in the field of molecular virology has been the rapid cloning of genes from numerous viruses. The nucleotide sequences and the amino acid sequences of their putative protein products are now available in computerized banks. Many virus families (e.g. Poxviruses, Herpesviruses, Retroviruses) synthesize membrane-bound proteins as essential steps in their life cycle in infected cells. The use of computer programs to elucidate some properties of viral membrane-bound proteins might help in the development of new approaches to antiviral drugs. In the present study use was made of a computer program compiled by Dr. H. Wolf (The University of Munich, FRG; Wolf et al., 1988) based on the use of the Kyte and Doolittle (1982) algorithm together with the secondary structure analyses of Chou and Fasman (1978) and Garnier et al. (1978). The aims of the study were: a) to use the primary amino acid sequence of selected membrane-bound proteins of cellular or viral origin to determine their conformational and hydropathic properties, b) to detect of the amino acid sequences that interact with or anchor the protein in a membrane, and c) to study the predictive value of the computer analyses.

METHODOLOGY

The primary amino acid sequences of polypeptides were obtained from the NBRF data base. Secondary structure predictions of polypeptides were made according to Garnier et al. (1978), Chou and Fasman (1978), and Rawlings et al. (1983). Hydrophobicity and hydrophilicity determinations were according to Kyte and Doolittle (1982) or Hopp and Woods (1981). All programs are available in the University of Wisconsin Genetic Computer Group Software (Devereux et al., 1984), compiled by Wolf et al. (1988).

RESULTS

ANALYSIS OF PREPROMELITTIN AND MELITTIN.

Eisenberg (1984) summarized the studies on mellitin, a 26 amino acid (aa) peptide cleaved off a precursor polypeptide prepromelittin of 70 aa, coded by a gene which is active in the venom gland of the honey bee. Melittin contains aa 44-69 of prepromelittin cleaved from the precursor by a stepwise process (Vlasak et al., 1983). The hemolytic activity of melittin resides in both ends of the polypeptide while removal of the first and last 7 aa did not affect its action on the surface tension of aqueous solutions (Schröder et al., 1971). Melittin was found to exist in aqueous solutions as a tetramer and in the crystal the molecule is a bent helical rod, the bend being at Thr-11- Gly12. The NH_2 terminal 20 aa of melittin are arranged asymetrically about the bent rod according to their polarity (Terwilliger and Eisenberg, 1982; Terwilliger et al., 1982).

The prepromelittin polypeptide was analysed in this study by the Kyte and Doolittle (1982) program as well as by the Chou and Fasman (1978) and Garnier et al. (1978) programs. The first 22 aa of prepromelittin were found to be highly hydrophobic while in melittin the first 19 aa (aa 44-62) were hydrophobic while the 7 carboxy terminal aa (aa 63-69) were hydrophilic. The internal sequence of the precursor (aa 23-43) was hydrophilic. Thus the ability to interact with the membrane which is associated with the first and last 7 aa (Schroder et al., 1971) requires an amphiphilic property of the peptide while the property to affect the surface tension of an aqueous solution resides in a highly hydrophobic aa sequence.

The secondary structure prediction according to Chou and Fasman (1978) predicted two helical regions in melittin (aa 44-46 and aa 55-69) while aa 47-54 were calculated to be a β sheet. Since the bend in melittin was found by Terwilliger et al. (1982) to be in aa

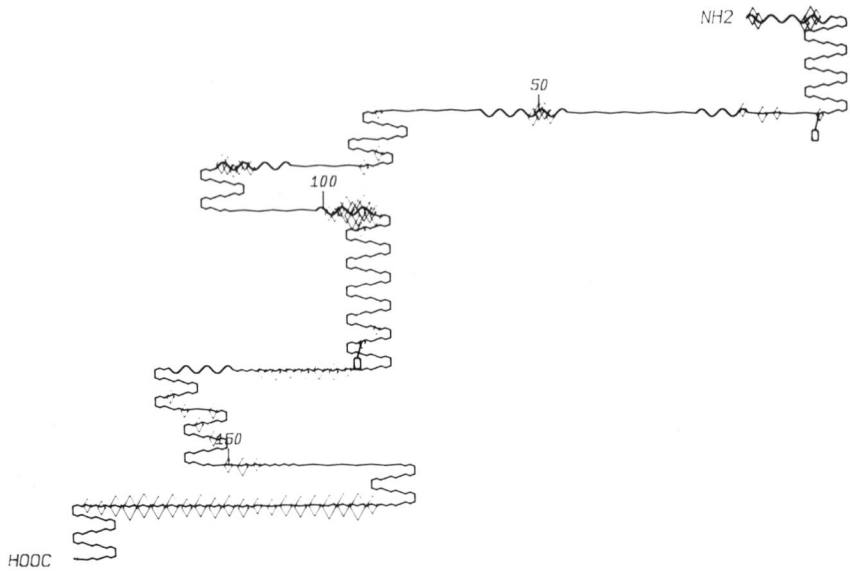

Figure 1. Conformation and hydropathy of EBV ORF BHRF1.

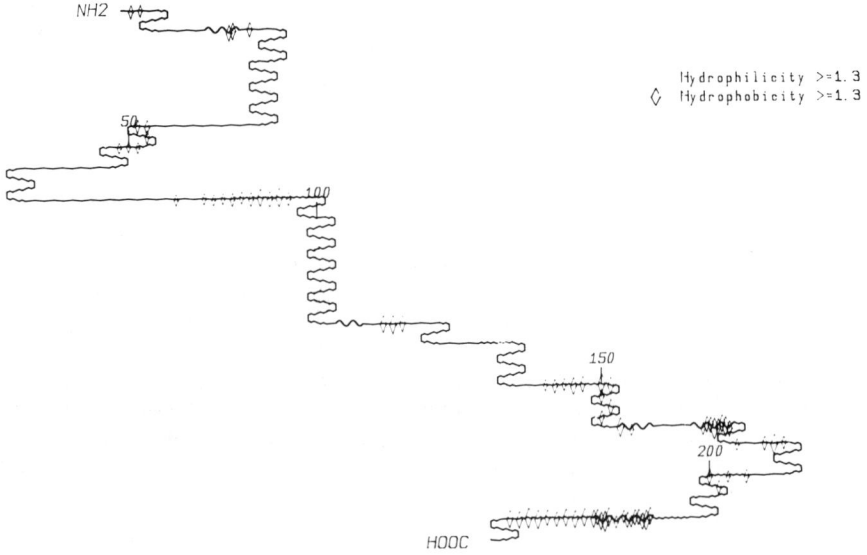

Figure 2. Conformation and hydropathy of bcl-2 protooncogene.

54-55 it seems that the Chou and Fasman (1978) algorithm extended the bend in the polypeptide over 8 aa rather than as in the X-ray crystallographic data over 2 aa. The Garnier et al. (1978) analysis predicted that the hydrophilic carboxy terminal aa had the conformation of a turn in the polypeptide. Thus, while the hydrophobicity of melittin calculated by the Kyte and Doolittle (1982) method agreed with the experimental data (Eisenberg, 1984), the secondary structure predictions by Chou and Fasman (1978) and Garnier et al. (1978) were unable to predict the correct tertiary structure of melittin.

PROTEINS THAT DO NOT HAVE A SIGNAL PEPTIDE BUT HAVE A MEMBRANE-ANCHORAGE DOMAIN

Viral and Cellular Proteins with no Signal Peptide but with a Membrane Domain.

The open reading frame (ORF) of Epstein-Barr virus (EBV) BHRF-1 can code for a putative polypeptide, the mRNA of which was found to be expressed in EBV-transformed cells (Becker et al., 1988). Computer analysis (Fig. 1) revealed one extended hydrophobic domain close to the carboxy terminus made up of 21 hydrophobic amino acids. The Garnier et al. (1978) conformational analysis predicted that the hydrophobic amino acids have a conformation of a β sheet. The N terminal aa sequence has an α-helical conformation with four hydrophobic aa. This analysis suggests that the polypeptide might have the ability to insert into a membrane but the lack of an N-terminal hydrophobic sequence (signal peptide) suggests that the polypeptide need not have the ability to penetrate through a cell membrane. Similar properties are displayed by the putative protein product of the bcl-2 protooncogene in malignant B cells. BHRF1 of EBV and the protooncogene bcl-2 have been found to be partly homologous (Cleary et al., 1986). However, the homology is in the aa sequence present in the hydrophobic domain of the two polypeptides. The conformation of this domain in bcl-2 protooncogene putative protein (Fig. 2) is partly α-helical and partly β-sheet conformation. Thus, although the two viral proteins have not yet been isolated the computer predicts that these proteins should be associated with intracellular membranes rather than with the cell surface membrane.

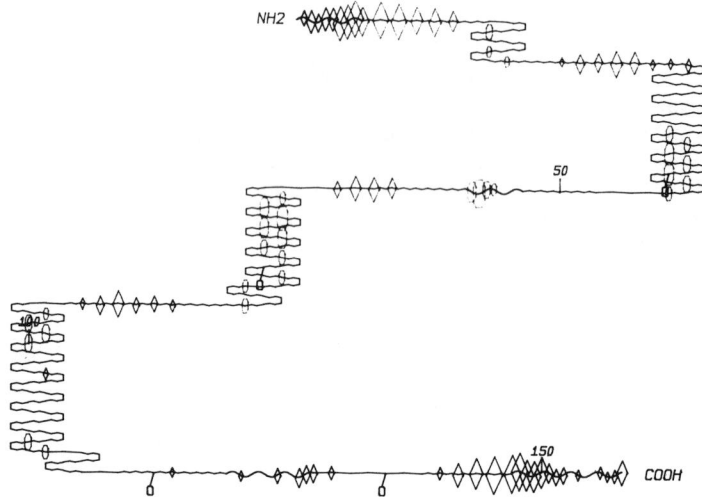

Figure 3. Conformation and hydropathy of Thy-1 membrane glycoprotein precursor.

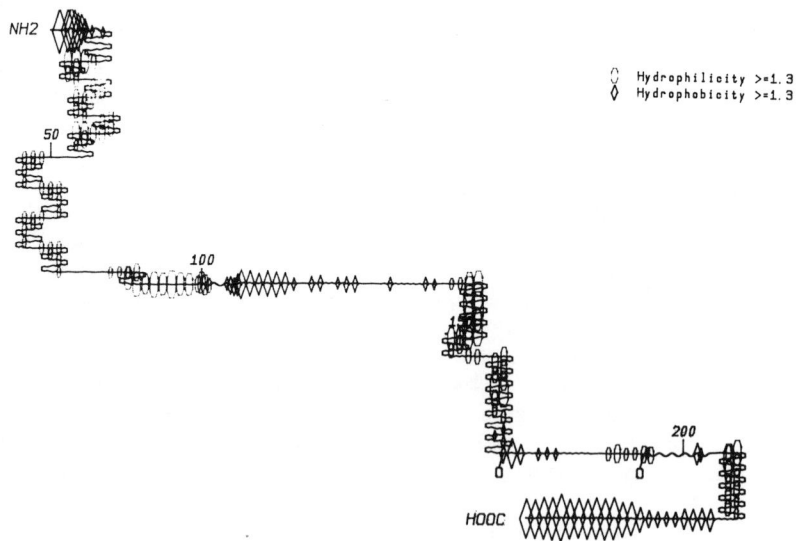

Figure 4. Conformation and hydropathy of human scrapie-like protein precursor.

Membrane Proteins Modified after Insertion into Outer Side of Cell Membrane.

The cellular membrane protein Thy-1. The computer analysis (Fig. 3) of the human Thy-1 glycoprotein precursor from human (Seki et al., 1985) or mouse (Williams and Gagnon, 1982) brain and T cells revealed that the protein has two hydrophobic domains: one at the N terminus (aa 2-16) and a second at the carboxy terminus (aa 130-138 and aa 142 to the end of the molecule, aa 161). It is assumed that the N-terminal hydrophobic sequence serves as a signal protein to lead the polypeptide through the cell membrane while the carboxy terminus serves as the anchorage domain inside the cell membrane. The Thy-1 protein is processed at the outer side of the membrane by cleavage of the signal peptide from the polypeptide as well as cleavage of the remaining external polypeptide leaving the anchorage domain in the membrane. The processed polypeptide is reattached to the outer side of the cell membrane via a phosphoinositol-lipid moiety attached to the carboxy terminus of the processed polypeptide while the lipid is inserted into the membrane.

Scrapie agent. Studies on the scrapie agent (Stahl et al., 1987) elucidated the structure and properties of the infectious protein involved in pathological processes in hamster brains. The scrapie agent was found to be a cleavage product of a precursor protein which at the final stage of its processing is anchored to the outer side of the cell membrane via a phosphoinositide lipid moiety. Computer analysis of the human scrapie protein precursor (Fig. 4) revealed a hydrophobic domain at the carboxy terminus (aa 223-245, end of molecule) as well as a signal peptide of 13 hydrophobic aa at the N terminus (aa 1-13).

The similarity in the organization of the Thy-1 and scrapie agent polypeptides led to the idea that the two might have common properties (Y. Becker, in preparation). Fig. 5 is a schematic representation of the states where these proteins are attached to the cell membrane.

Proteins Inserted into Inner Leaf of Cell Membrane.

Human immunodeficiency virus (HIV) gag protein. Rein et al. (1986) quoted the unpublished data of L. E. Henderson and S. Oroszlan that HIV gag protein is also myristylated. Computer analysis of the primary aa sequence of the group-specific antigen (gag) polypeptide (Ratner et al., 1985) presented in Fig. 6 revealed that it does not have any extended hydrophilic domain and thus cannot be classified as a membrane-bound protein. Yet, the addition of the myristic acid to the N-terminal glycine of the polypeptide (presumably after removal of

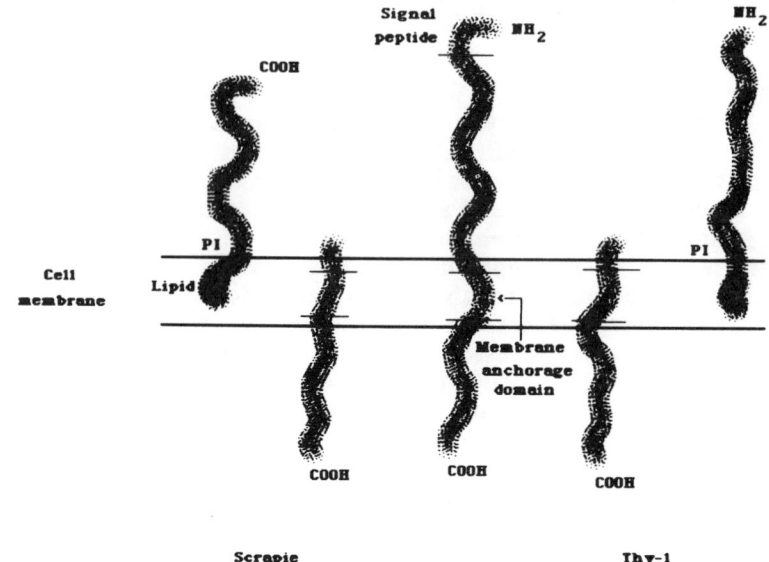

Figure 5. Schematic representation of the interaction of Thy-1 and the scrapie protein with the cell membrane.

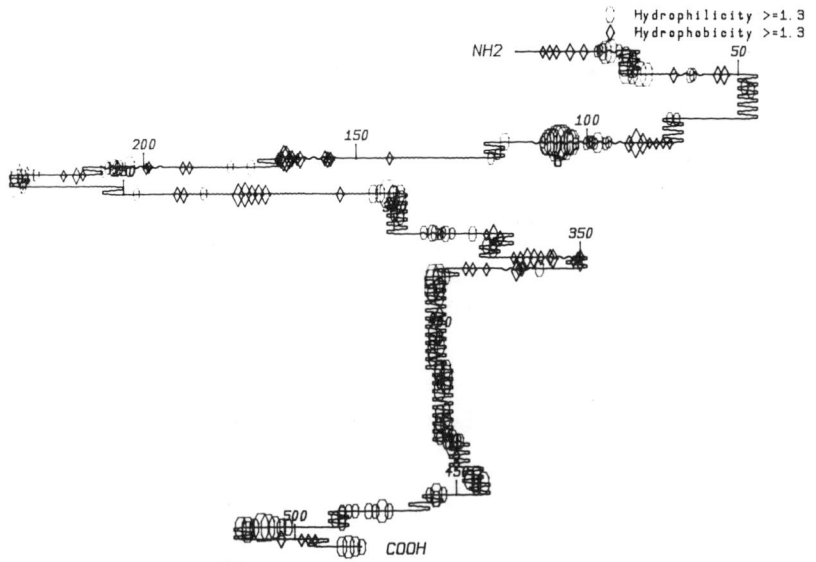

Figure 6. Conformation and hydropathy of the AIDS virus (HIV-1) gag protein.

the methionine) allows such a polypeptide to be positioned into the inner leaf of the cell membrane, a process followed by cleavage of the polypeptide into the core proteins p17 (aa 1-132), p24 (aa 133-391) and p15 (aa 392-512). The predicted comformation of p15 (extended turns) differs from that of the two other cleavage products (p17 and p24). p17 forms a shell underneath the virion envelope and p24 forms a capsid around the two viral RNA genomes and the reverse transcriptase. The p15 polypeptides seem to have the highly hydrophylic properties needed for attachment to nucleic acids and are thus involved in stabilization of the RNA genomes in the budding HIV virion. A schematic model is presented in Fig. 7.

Proteins with Multiple Hydrophobic Domains Inserted into Cellular Membranes.

EBV latent membrane protein.

Epstein-Barr virus (EBV), a member of the herpesvirus family, codes for a membrane-bound protein designated latent membrane protein (LMP) (Baer et al., 1984; Hudson et al., 1985). Computer analysis of this protein (Fig. 8) revealed that the N-terminal sequence (aa 1-21) is highly hydrophobic. This sequence is followed by 5 hydrophobic domains: 24 aa (aa 22-49), 25 aa (aa 50-75), 23 aa (aa 77-99), 57 aa (aa 103-159) and 19 aa (aa 166-184). The rest of the polypeptide is hydrophilic. The presence of the hydrophobic domains suggests that the LMP polypeptide has five membrane insertion domains with the hydrophilic N and carboxy terminus situated intra-cytoplasmically while only short peptide domains extrude from the cell membrane (Kieff et al., 1985). Although 4 out of the 5 hydrophobic domains conform in size to the membrane-anchorage domains described above, the presence of a 57 aa domain is a new feature that should be taken into account. Yet, it is possible that there are two hydrophobic domains. The conformation of the polypeptide in the hydrophobic domains is α-helical with a β-sheet extension. If the LMP contains 5 hydrophobic membrane-anchorage domains, it is possible to postulate that the hydrophobic N terminus of the polypeptide extends from the cell and the polypeptides crisscross into the membrane five times, allowing the hydrophilic carboxy terminal moiety to be situated in the cell cytoplasm. The studies by Kieff et al. (1985) suggest that the LMP might be a viral oncogene.

A cellular protein, masoncogene, product of the protooncogene.

A new oncogene designated masoncogene (Young et al., 1986) was found to have multiple hydrophobic domains as may be seen from the computer analysis (Fig. 9). It revealed multiple hydrophilic domains as well as the lack of a signal peptide at the N terminus. Six hydrophobic domains were detected by the Kyte and Doolittle

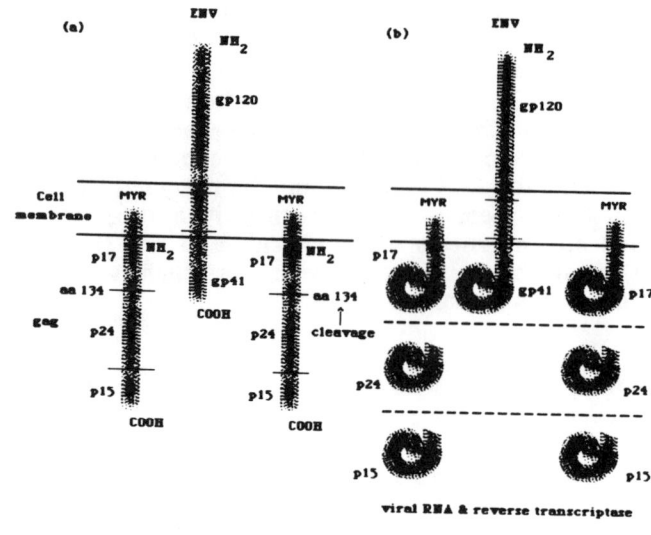

Figure 7. A schematic representation of the mode of membrane interaction and processing of the HIV-1 gag protein.

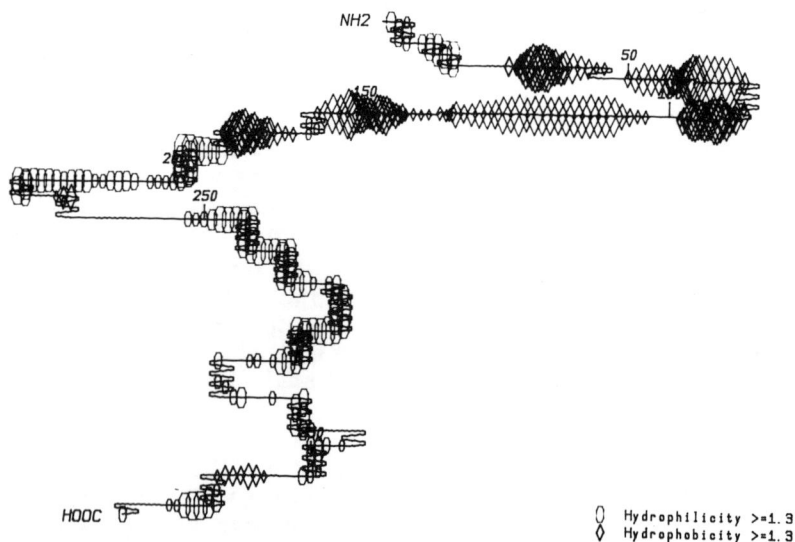

Figure 8. Conformation and hydropathy of EBV latent membrane protein (BNLF1 protein).

algorithm: 28 aa (aa 32-59), 26 aa (aa 67-92), 35 aa (aa 103-137), 24 aa (aa 149-172), 29 aa (aa 185-213) and 25 aa (aa 226-260). In addition, 3 short hydrophobic domains (11, 8 and 7 aa, respectively) were detected. The secondary structure predictions suggested α-helical conformation only in the hydrophobic domains 4 and 5.

A cellular protein functioning as a pump in the cell membrane capable of removing tetracycline. The gene coding for the protein responsible for resistance of E. coli to tetracycline is present in bacterial transposons and was used in the construction of the plasmid pBR322 (Peden, 1983). Computer analysis of the tetracycline resistance (tet^r) protein which is made of 386 aa (Fig. 10) revealed the presence of 10 main hydrophobic domains: 30 aa (aa 6-35), 23 aa (aa 44-66), 50 aa (aa 74-123), 47 aa (aa 135-181), 29 aa (aa 208-236), 24 aa (aa 245-268), 19 aa (aa 279-297), 23 aa (aa 303-325), 25 aa (aa 338-362) and 16 aa (aa 369-384). Two of the domains (Nos. 3 and 4) are made of 50 and 47 aa, respectively. The properties of such extended hydrophobic domains, which have more than twice the number of hydrophobic aa than in other domains, are not known. The Garnier et al. (1978) prediction suggested that the hydrophobic domains Nos 1, 3, 5, 9 and 10 are β-sheet while the other domains have the α-helical conformation. How the tet^r protein which is extensively hydrophobic is inserted and maintained in the cell membrane of E. coli is still to be determined, hopefully by X-ray crystallography, using small detergents to replace lipids to render the protein insoluble, thus obtaining crystals as suggested by Eisenberg (1984).

Proteins involved in cell fusion. Herpes simplex virus-1 (HSV-1) contains a gene coding for a protein (338 aa) involved in cell fusion (Debroy et al. 1985). This gene was found to be responsible for virus infectivity by the intracerebral route of inoculation into mouse brains (Ben-Hur et al. 1987). Computer analysis of the cell fusion protein (Fig. 11) revealed 5 major hydrophobic domains and several small domains: 23 aa (aa 10-32), 7 aa (aa 123-139), 23 aa (aa 154-176), 59 aa (aa 213-271) and 34 aa (aa 292-325). The viral cell-fusion proteins have a hydrophobic domain at the N terminus. The role of the hydrophobic domains in the process of fusion between membranes of two cells needs to be elucidated.

Human immunodeficiency virus (HIV) envelope protein: the ability of HIV infection to cause cell fusion is used as an assay for the detection of the virus in suspected infectious material. Thus, the computer analysis of this protein was used to predict some of its properties. The envelope (env) polypeptide (856 aa) is divided into the external gp120 region and the gp41 region which contains the membrane anchorage domain (Ratner et al., 1985). This protein has been the target of numerous studies on its antigenicity and was used

120 / Becker

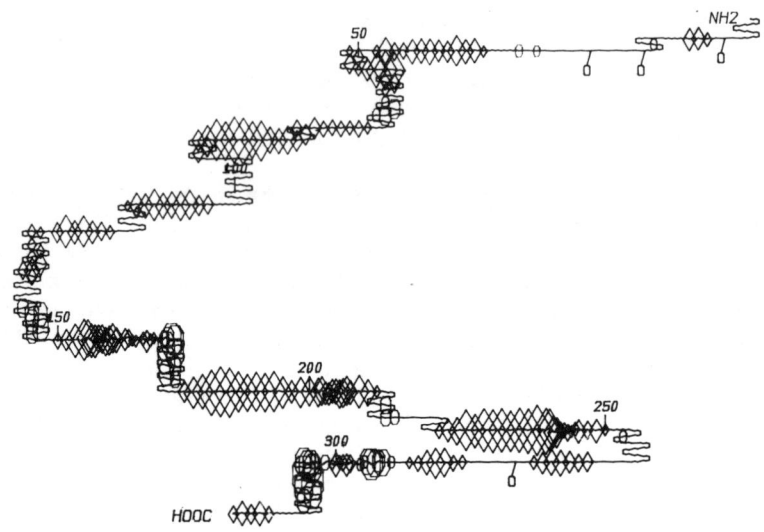

Figure 9. Conformation and hydropathy of the human masoncogene putative protein.

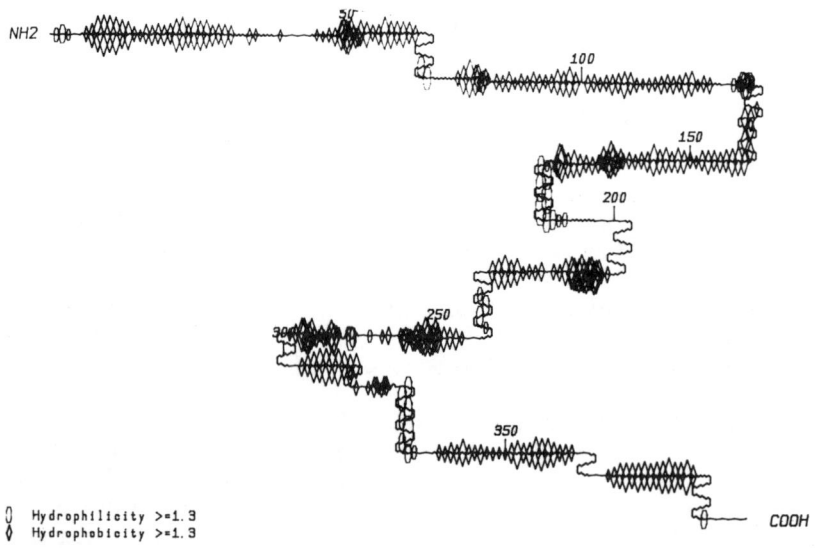

○ Hydrophilicity >=1.3
◊ Hydrophobicity >=1.3

Figure 10. Conformation and hydropathy of the tetracycline resistance (tet^r) protein.

to detect HIV antibodies in AIDS patients. Less is known about the reason why antibodies to epitopes on the env proteins do not protect HIV-infected AIDS patients in spite of the presence of large amounts of anti-env antibodies. Gelderblom et al. (1987) reported that the gp120 portion of the envelope protein detaches from the surface of infected cells and causes the formation of gp120-antibody complexes that are harmful to AIDS patients and require apheresis of the blood to remove such complexes.

Fig. 12 presents the Garnier et al. (1978) secondary structure analysis of the HIV envelope protein as well as the Kyte and Doolittle (1982) analysis. The gp120 sequence of the protein contains 14 short hydrophobic domains situated between hydrophilic domains. Since the gp120 is the external portion of the envelope it is possible to assume that the hydrophobic domains are situated internally and the hydrophilic domains externally when the polypeptide attains its globular conformation. A globular conformation is attained by the gp120 when released from the cells as reported by Gelderblom et al. (1987). The gp41 which is the membrane-bound portion of the env glycoprotein is attached to the cell membrane through an anchorage domain of 27 aa (aa 513-539). It is of interest that the computer analysis by Kyte and Doolittle (1982) revealed two additional hydrophobic domains further down the polypeptide chain: 19 aa (aa 592-604) and 25 aa (aa 681-705). Thus, the gp41 sequence of HIV envelope contains three hydrophobic domains (schematically shown in Fig. 13). The role of the three hydrophilic domains in cell fusion has not yet been elucidated. It is possible that the HIV envelope protein, inserted into the cell membrane, is used for the envelopment of the viral RNA during the formation of virions. Alternatively, when the HIV envelope is involved in cell fusion the three hydrophobic domains might be inserted into the membrane (as in Fig. 13) to cause cell fusion.

Analysis of the primary sequence of the HIV envelope (Ratner et al., 1988) revealed a relatively large number of cysteines in the gp120 moeity of the protein: between aa 53 and 250, 12 cysteines are located. An additional 7 cysteines are located between aa 250 and 511, the carboxy terminus of gp120 when detached from the membrane. When compared to the primary sequence of the insulin receptor (1370 aa) (Ullrich et al., 1985), it was noted that in this receptor polypeptide 20 cysteines are localized in the sequence between aa 153 and 360. Additional cysteines are localized in the carboxy terminus of the α chain of the insulin receptor which is cleaved after the precursor polypeptide is inserted into the cell membrane. The result of this process are α and β chains; the latter is inserted into the membrane with its carboxy terminus functioning as a tyrosine kinase. The α polypeptide chain is s-s bonded to the β polypeptide through the

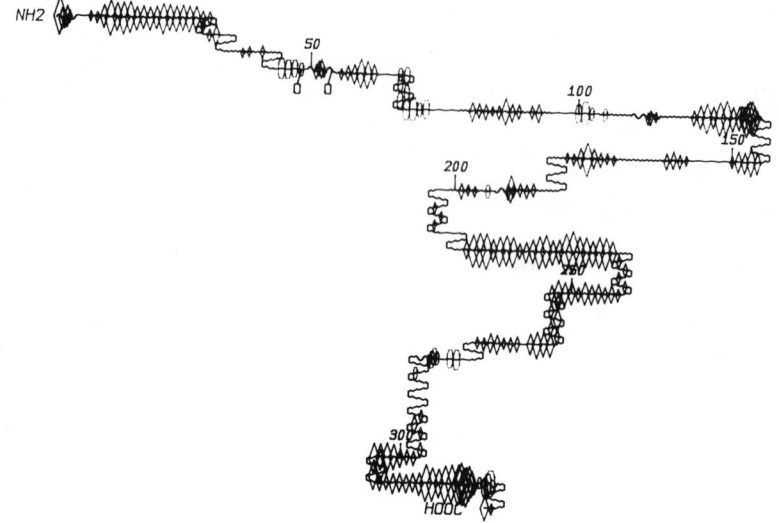

Figure 11. Conformation and hydropathy of HSV-1 cell fusion protein.

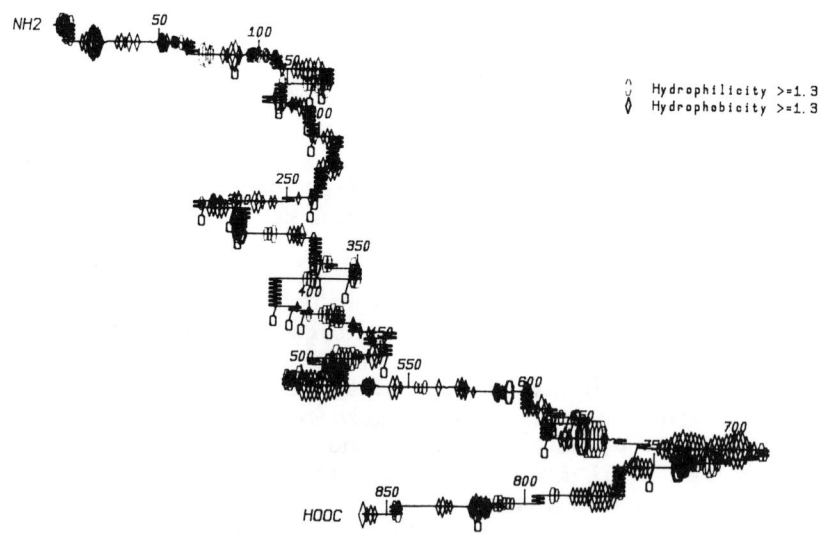

Figure 12. Conformation and hydropathy of AIDS virus (HIV-1) envelope glycoprotein.

cysteines in the two subunits of the receptor (Ullrich et al., 1985).

The ability of HIV gp120 to easily detach from the cell membranes (Gelderblom et al., 1987) might suggest that HIV envelope protein is processed similarly to the insulin receptor and the gp120 is released from the moiety gp41 which is inserted into the cell membrane. If this possibility does exist, gp120 in the HIV envelope could be viewed as a receptor for a factor that is capable of signalling the infected cells to function. Under such conditions the gp41 should be functioning as a tyrosine kinase, utilizing one of the three tyrosines as a target for phosphorylation.

It is of interest that the protein kinase (308 aa) of the fruitfly (Hoffman et al., 1983) when compared by the GAP program to the HIV envelope primary sequence resembles (12.5% similarity with 4 gaps), the HIV envelope gp41 sequence (aa 520-835), with three tyrosines of the fruitfly enzyme in the same position as in the p41 polypeptide (tyrosine in aa 628, 681 and 768). If the above idea describes a new function for the HIV envelope polypeptide it might be possible that antibodies to HIV gp120 are detrimental to the AIDS patients due to their ability to activate processes (e.g. cellular or viral DNA synthesis) in infected T cells and Langerhans cells.

DISCUSSION

The present study utilized the Kyte and Doolittle (1982) algorithm for determining the hydropathic properties of viral and cellular proteins known to function in association with cellular membranes. Proteins spanning the cell membrane were found to have a hydrophobic signal peptide and a hydrophobic anchorage domain. Proteins that are required for a specialized function in the membrane like pumping out of antibiotics from the cytoplasm to the cell environment (e.g. tetracycline-resistant proteins) as well as proteins which are involved in cell fusion (e.g. HSV-1 fusion protein or HIV envelope) have acquired more than one hydrophobic domain, some with α-helical conformations of the polypeptide in the hydrophobic domain to form, most probably, channels in the cell membrane. On the other hand, proteins which have a signal peptide and a membrane-anchorage domain (e.g. Thy-1 protein) can be modified through cleavage of the polypeptide and interaction with the outer cell membrane via phosphoinositol linkage. Other proteins (e.g. HIV gag) can be modified posttranslationally to interact with myristic acid followed by insertion into the inner leaf of the cell membrane. Such modifications can, a priori, be predicted from primary aa sequences when the structural rules for myristylation (e.g. need for glycine at the N terminus after removal of methionine) will be known.

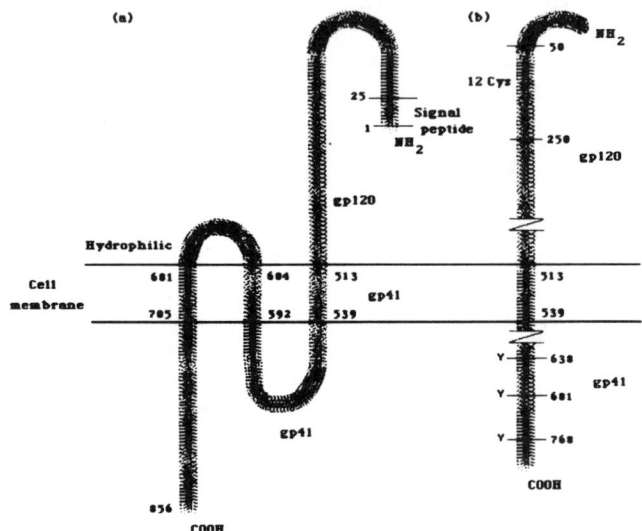

Figure 13. Schematic representation of HIV gp160 envelope protein inserted into the cell membrane: a) model for possible membrane insertion of three hydrophobic domains found in gp41, and b) model for insertion of gp160 into cell membrane using one hydrophobic domain of gp41 (aa 513-539). Amino acids 638, 681 and 768 are tyrosines (Y).

Table 1 summarizes the length of the hydrophobic domains of signal peptides and the membrane anchorage domains in the proteins analysed in this study. The number of hydrophobic amino acids in each domain is between 19 and 35 aa. Only in EBV LMP and HSV-1 cell fusion protein, one of the hydrophobic domains is 57 and 59 aa, respectively. The mode of insertion and function of such a long hydrophobic domain in the membrane is not yet known. With the help of site-specific deletions in the genes for the membrane-bound proteins it will be possible to determine the minimal hydrophobic domain to allow it to become a membrane-bound polypeptide. One example of such an approach is the study by Davis et al. (1985) on the fine structure of a membrane anchorage domain of 23 hydrophobic aa at the carboxy terminus of bacteriophage f1 gIII protein (designated PIII). They showed that deletion of 6 aa from the 23 aa hydrophobic domain resulted in no diminution of the capacity of the protein to anchor in the membrane. PIII derivatives with over half of the hydrophobic aa sequence deleted still retain substantial residual anchorage functions.

Further studies are needed to define the properties of the aa sequence in the domains of polypeptides involved in the insertion into and movement through membranes as detailed by Wickner and Lodish (1985). It is hoped that additional computer programs will be added to the available programs that will improve the analyses of membrane-bound proteins. Eisenberg (1984) had suggested that "computational analysis of amino acid sequences from membrane proteins can yield structural information of value . . . once several structures for membrane proteins are known. The code that links primary to three-dimensional structure may be more easily deciphered for membrane proteins." With such a code available, the understanding of membrane proteins will be enhanced.

SUMMARY AND CONCLUSIONS

1. Although secondary structure predictions are only 60% correct, the computer programs provide some idea as to peptide conformation.
2. The detection of hydrophobic domains in membrane-bound proteins allowed the identification of membrane insertional domains in the polypeptide.
3. A combination of computer analyses of protein conformation and the identification of functional domains in polypeptides must be substantiated by experimental data.
4. Computer analyses of membrane proteins may provide a theoretical approach to the selection and development of antibiotics capable of interfering with the insertion and function of viral proteins in cellular membranes.

Table 1. Hydrophobic domains in membrane-bound proteins

	Signal		Hydrophobic domain	
	No. of aa	(aa-aa)	No. of aa	(aa-aa)
Prepromelittin	22	(1-22)	19	(44-62)
BHRF1	None		21	
bcl-2	None		21	
Thy-1 precursor	15	(2-16)	19	(142-161)
Scrapie agent	13	(1-13)	22	(223-245)
HIV gag	None		None	
EBV LMP	21	(1-21)	24	(22-49)
			25	(50-75)
			23	(77-99)
			57	(103-159)
			19	(166-184)
Mason oncogene			28	(32-59)
			26	(67-92)
			35	(103-137)
			24	(149-172)
			29	(185-213)
			25	(226-260)
Tetracycline-resistance protein			30	(6-35)
			23	(44-66)
			50	(74-123)
			47	(135-181)
			29	(208-236)
			24	(245-268)
			19	(279-297)
			23	(303-325)
			25	(338-362)
			16	(369-384)
HSV-1 cell fusion protein	23	(1-32)	17	(123-139)
			23	(154-176)
			59	(213-271)
			34	(292-325)
HIV envelope	19	(10-28)	27	(513-539)
			19	(592-604)
			25	(681-705)

ACKNOWLEDGMENT

The study was supported by the Foundation for Molecular Virology and Cell Biology, Phoenix, Arizona, USA.

REFERENCES

Argos P, Rao JKM, Hargrave PA (1982). Structural prediction of membrane-bound proteins. Eur J Biochem 128: 565-575.

Baer R, Bankier AT, Biggin MD, Deininger PL Farrell PJ, Gibson TJ, Hatfull G, Hudson GS, Satchwell SC, Sequin C, Tuffnell PS, Battell BG (1984). DNA sequence and expression of the B95-8 Epstein-Barr virus genome. Nature 310: 207-211.

Becker Y, Asher Y, Tabor Y (1988). Differential expression of Epstein-Barr virus (EBV) genes BBRF3 BILF1 and BMRF2 in EBV-transformed lymphoblastoid lines from ataxia- telangiectasia patients. Leukemia (in press).

Ben-Hur T, Asher Y, Tabor E, Darai G, Becker Y (1987). HSV-1 virulence for mice by the intracerebral route is encoded by the BamHI-L DNA fragment containing the cell fusion gene. Arch Virol 96: 117-122.

Chou PY, Fasman GD (1978). Prediction of the secondary structure of proteins from their amino acid sequence. Adv Enzymol 47: 45-142.

Cleary ML, Smith SD, Sklar J (1986). Cloning and structural analysis of cDNAs for bcl-2 and a hybrid bcl-2 immunoglobulin transcript resulting from the t(14;18) translocation. Cell 47: 19-28.

Davis NG, Boeke JD, Model P (1985). Fine structure of a membrane anchor domain. J Mol Biol 181: 111-121.

Debroy C, Pederson N, Person S (1985). Nucleotide sequence of a herpes simplex virus type 1 gene that causes cell fusion. Virology 145: 36-48.

Devereux J, Haeberli P, Smithies O (1984). A comprehensive set of sequence analysis programs for the VAX. Nucl Acids Res 12: 387-395.

Eisenberg D (1984). Three dimensional structure of membrane and surface proteins. Ann Rev Biochem 53: 595-623.

Garnier J, Osgerthorpe DJ, Robson B (1978). Analysis of the accuracy and implications of simple methods for predicting the secondary structure of globulin proteins. J Mol Biol 120: 97-120.

Gelderblom HR, Hausman EHS, Ozel M, Pauli G, Kock MA (1987). Fine structure of human immunodeficiency virus (HIV) and immunolocalization of structural proteins. Virology 156: 171-176.

Greer J (1988). Comparative modeling of proteins in the complement pathway. (This volume).

Hoffman FM, Fresco LD, Hoffman-Falk H, Shilo BZ (1983). Nucleotide sequences of the Drosophila src and abl homologs; conservation and variability in the src family oncogenes. Cell 35: 393-401.

Hopp TP, Woods KR (1981). Prediction of protein antigen determinants from amino acid sequences. Proc Natl Acad Sci USA 78: 3824-3828.

Hudson GS, Farrell PJ, Barrell BG (1985). Two related but differentially expressed potential membrane proteins encoded by the EcoRI Dhet region of Epstein-Barr virus B95-8. J Virol 53: 528-535.

Kieff E, Hennessy K, van Gaugh T, Matsuo T, Fennewald S, Heller M, Petti L, Hummel M (1985). Persistence and expression of the Epstein-Barr virus genome in latent infection and growth transformation of lymphocytes. In Levine PH, Ablashi DV, Pearson GR, Kottaridis SD (eds): "Epstein-Barr virus and Associated Diseases," Developments in Medical Virology (series editor Y. Becker), Boston: Martinus Nijhoff Publishing, pp 221-247.

Kyte J, Doolittle RF (1982). A single method for displaying the hydropathic character of a protein. J Mol Biol 157: 105-132.

Peden KWC (1983). Revised sequence of tetracycline resistance gene of pBR322. Gene 22, 277-280.

Ratner L, Haseltine W, Patarca R, Livak KJ, Starcich B, Josephs SF, Doran ER, Rafalski A, Whitehorn EA, Baumeister K, Ivanoff L, Petteway SR, Jr., Pearson ML, Lautenberger JA, Papas TS, Ghrayeb J, Chang NT, Gallo RC, Wong-Staal F (1985). Complete nucleotide sequence of the AIDS virus, HTLV-III. Nature 313: 277-284.

Rawlings N, Ashman K, Wittman-Liebold B (1983). Computerised version of the Chou and Fasman protein secondary structure predictive method. J Peptide Protein Res 22: 515-524.

Rein A, McClure MR, Rice NR, Liftig RB, Schultz AM (1986). Myristylation site in Pr65 gag is essential for virus particle formation by Moloney murine leukemia virus. Proc Natl Acad Sci USA 83: 7246-7250.

Rose GD (1978). Prediction of chain turns in globular proteins on a hydrophobic basis. Nature 272: 586-590.

Scheraga HA (1988). Approaches to the multiple-Minima problem in conformational energy calculations on polypeptides and proteins (This volume).

Schröder E. Lubke K, Lehmann M, Bectz I (1971). Hemolytic activity and action on the surface tension of aqueous solutions of synthetic melittin. Experientia 27: 764-765.

Seki T, Chang HC, Moriuchi T, Denome R, Ploegh H, Silver J (1985). A hydrophobic transmembrane segment at the carboxyl terminus of Thy-1. Science 227: 649-651.

Segrest JP, Chung BH, Brouillette CG, Kanellis P, McGahan R (1983). Studies of synthetic peptide analogs of the amphipathic helix. J Biol Chem 258: 2290-2295.

Segrest JP, Feldman RJ (1974). Membrane proteins amino acid sequence and membrane penetration. J Mol Biol 87: 853-858.

Stahl N, Borchelt DR, Hsiao K, Prusiner SB (1987). Scrapie prion protein contains a phosphatidylinositol glycolipid. Cell 51: 229-240.

Terwilliger TC, Eisenberg D (1982). The structure of melittin II: interpretation of the structure. J Biol Chem 257: 6016-6022.

Terwilliger TC, Weismann L, Eisenberg D (1982). The structure of melittin in the form of crystals and its implication for melittin's lytic and surface activities. Biophys J 37: 353-361.
Ullrich A, Bell JR, Chen EY, Herrera R, Petruzzelli LM, Dull TJ, Gray A, Coussens L, Liao Y-C, Tsubokawa M, Mason A, Seeburg PH, Grunfeld C, Rosen OM, Ramachandran J (1985). Human insulin receptor and its relationship to the tyrosine kinase family of oncogenes. Nature 31: 756-761.
Vlasak R, Urger-Ullman C, Krest G, Frischanf AM (1983). Nucleotide sequence of cloned cDNA coding from honeybee prepromelittin. Eur J Biochem 135: 123-126.
Wickner WT, Lodish HF (1985). Multiple mechanisms of protein insertion into and across membranes. Science 230: 400-407.
Williams AF, Gagnon J (1982). Neuronal cell Thy-1 glycoprotein: homology with immunoglobin. Science 216: 696-703.
Wolf H, Modrow S, Motz M, Jameson B, Hermanu G, Fortsch B (1988). An integrated family of amino acid sequence analysis programs. CABIOS (in press).
Young, D. Waitches G, Birchmeier C, Fasano O, Wigler M (1986). Isolation and characterization of a new cellular oncogene encoding a protein with multiple potential transmembrane domains. Cell 45: 711-719.

Computer-Assisted Modeling of Receptor-Ligand Interactions:
Theoretical Aspects and Applications to Drug Design, pages 131–142

C-H...X HYDROGEN-BONDED PSEUDO-WATSON-CRICK BASE PAIRING WITH 7-DEAZANEBULARIN AND CANONICAL BASES IN DNA AND RNA

Rick L. Ornstein

Department of Biochemistry, Princeton University, Princeton, New Jersey 08544

INTRODUCTION

It is generally thought that stereochemical and energetic features of base pairing are essential to polymerase-mediated nucleic acid processes such as replication and transcription as well as ribosomal mediated codon-anticodon interactions. Such classical pairing interactions, i.e., Watson-Crick pairing, requires at least two classical hydrogen bonds; in the case of the canonical base moieties, the donor group is always N-H and the acceptor is nitrogen or oxygen. Each residue in a Watson-Crick pair contributes at least one donor and at least one acceptor. The purine analogue 7-deazanebularin (DN) is exceptional in that it partakes in these nucleic acid processes (Ward and Reich, 1972; Grunberger et al., 1972; Brdar and Reich, 1972), yet does not allow classical pairing interactions since it lacks any donor group. From inspection, DN should be capable of forming but a single hydrogen bond with thymidine or uracil in the course of Watson-Crick type interactions (see Fig. 1); no hydrogen bonds should be

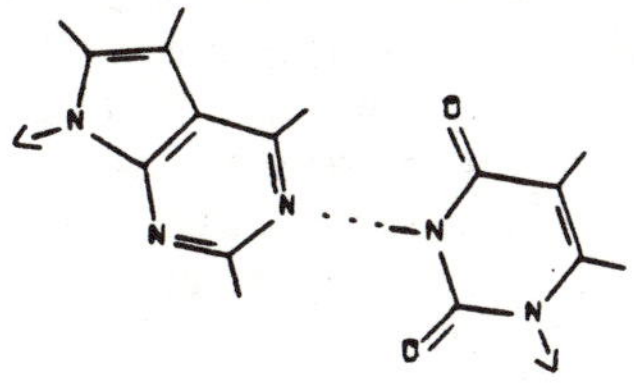

DN : U

Figure 1. 7-deazanebularin-uracil Watson-Crick type base pair with one hydrogen bond.

formed when DN attempts to pair with cytosine (Ward and Reich, 1972; Kornberg, 1980). Nevertheless, DN exhibits ambiguity in vitro when it functions as a substrate in transcription (Ward and Reich, 1972) and as a template in translation (Grunberger et al., 1972); in both cases, DN can replace adenine and guanine, but is more effective as an adenine analogue. The DN nucleoside is also rapidly phosphorylated and incorporated into cellular DNA and RNA in mammalian cell cultures and in viral nucleic acids (Brdar and Reich, 1972). In addition, poly[r(DN-U)] forms a stable double stranded structure with a thermostability (in the presence of Mg++) in the same range as that of poly[r(2-aminopurine-U)] (Ward and Reich, 1972).

In simple hydrocarbons, the constituent groups lack any significant capacity for electrostatic or hydrogen bonding interactions. Quantum chemical studies indicate, however, that a C-H moiety may become "activated" due to the presence of electronegative substituents (Bonchev and Cremaschi, 1974; Vishveshwara, 1978). Because of the heteroatoms and conjugated double bonds in nucleic acid base residues, their C-H moieties possess a significant electrostatic character intermediate between N-H and simple C-H groups (Ornstein and Fresco, 1983; 1984; 1988). Since hydrogen bonding is essentially an electrostatic phenomenon, it is not surprising to note that some observed distances and angles between C-H...X in x-ray and neutron crystal structures suggests that pyrimidine and purine C-H groups can serve as hydrogen-bond donors (Sundaralingam, 1966; Seeman et al., 1971; Rubin et al., 1972; Kvick et al., 1974; Parthasarathy and Soriano-Garcia, 1976; Srikrisknan and Parthasarathy, 1976; Takusagawa et al., 1979). This notion is also supported by NMR data indicating that nucleic acid base C-H group hydrogens are considerably acidic (Bullock and Jardetzky, 1964; Schweizer et al., 1964; Batterham et al., 1967; Lichtenberg and Bergmann, 1973; Bruskov et al., 1980). Generally, however, C-H donor hydrogen bond interactions involving nucleic acid bases are not likely to manifest in competition with N-H or O-H donors. The measured enthalpy of pairing in vacuum for G:C and A:U is -21.0 and -14.5 kcal/mol, respectively (Yanson et al., 1979), while in chloroform solution the enthalpies are reduced by half (Kyogoku et al., 1969) and in aqueous solution pairing of the corresponding nucleosides is almost below detection (Lord and Thomas Jr, 1967). Since nucleic acid base pairing in vivo requires a high level of fidelity, pairing interactions take place inside a sequestering polymerase or ribosomal complex, in order to reduce solvent competition and check for Watson-Crick type stereochemistry. Under such conditions and in the absence of classical donor groups, as on canonical bases, C-H donor interactions may considerably impact recognition specificity (Ornstein and Fresco, 1984; 1988; Ornstein, 1988).

The purpose of the work reported in this paper is to use a recently developed force-fit computational method (Ornstein and Fresco, 1988) to assess the base pairing potential of DN with the canonical bases. The results are found to be consistent with the trends observed for in vitro transcription, in vitro codon-anticodon interactions, and equilibrium helices. The present work indicates that "deleted" Watson-Crick donor sites at the purine ring 2 and 6 positions are capable of mediating recognition interactions when neither one has bonded to it an exo-ring amino group, as is the case for DN. Thus the experimental trends are rationalized with pseudo-Watson-Crick type pairing arrangements containing one classical N-H donor hydrogen bond and one (or more) C-H donor hydrogen bond(s). Some related predictions are offered.

METHODS

We previously developed an empirical potential-function (EPF) using CNDO/2 net atomic charges that reproduced experimental nucleic acid ("free") base hydrogen-bonded energies of interaction in chloroform solution (Ornstein and Fresco, 1981) and structures in the solid state (Ornstein and Fresco, 1983a); these interactions are dominated by classical hydrogen bonds involving only N-H donor groups. For reasons considered elsewhere (Ornstein and Fresco, 1988), STO-3G basis-set net atomic charges are more realistic for treating electrostatic interactions of "activated" C-H groups. Having chosen to switch the source of net charges, we were obligated to "refit" the EPF, in a manner similar to that used in our earlier work. We also took this opportunity to switch the Lennard-Jones type term from a 6-9 to a 6-12 expression because parameters are better defined for the latter case. No explicit hydrogen-bond component was considered. The electrostatic component is described below.

Using experimental chloroform solution base-pairing interaction energies for 10 base pairs, we force-fitted the EPF to the data (Ornstein and Fresco,1988) by varying the form of the dielectric and magnitude of charge chosen to represent "lone electron pairs", in a manner similar to that as in our previous nucleic acid studies (Ornstein and Fresco, 1981; 1983b,c). We believe that this most recent effort represents the most thorough and successful attempt to date to develop a consistent computational method for determining a wide range of possible base-pairing interaction energies. The energies and structures reported below were obtained by using this computational method.

RESULTS AND DISCUSSION

A. Energetic and Geometric Features of Pseudo-Watson-Crick or "Wobble" Type Base Pairs Containing a DN Residue:

Energy-minimized base pairs with pseudo-Watson-Crick or "wobble" (Crick, 1966) geometry containing DN and a canonical base are shown in Figure 2. The energies

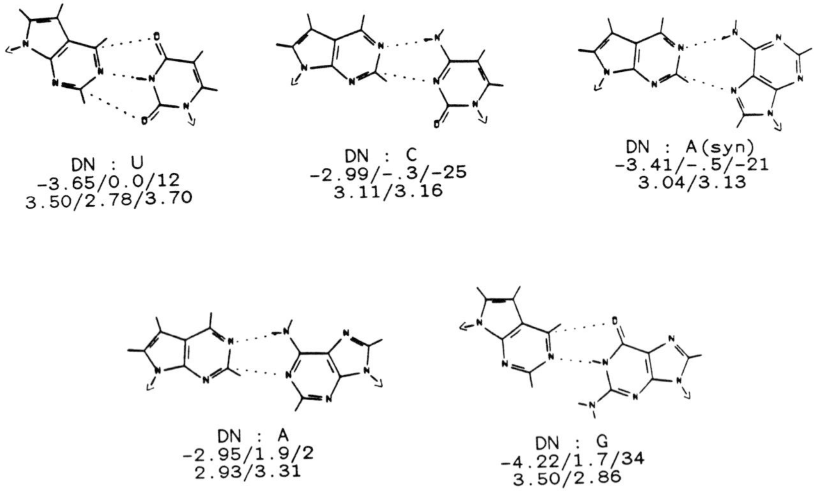

DN : U
-3.65/0.0/12
3.50/2.78/3.70

DN : C
-2.99/-.3/-25
3.11/3.16

DN : A(syn)
-3.41/-.5/-21
3.04/3.13

DN : A
-2.95/1.9/2
2.93/3.31

DN : G
-4.22/1.7/34
3.50/2.86

Figure 2. Computer-generated pseudo-Watson-Crick or

obtained for the pairs DN:U, DN:C, DN:A(syn), DN:A, and DN:G are -3.7, -3.0, -3.4, -3.0, and -4.2 kcal/mol, respectively. Each of these base pairs has one N-H donor hydrogen bond. Three of the five base pairs in Figure 2 have a second hydrogen bond of the C-H...N type, while the other two base pairs have at least one C-H...O type hydrogen bond. The heavy atom separation distance in the C-H...N bonds range from 3.11 to 3.31 angstroms with a deviation from linearity of less than 24 degrees; the sum of van der Waals radii for these heavy atoms is about 3.5 angstroms. Using classical geometric criterion these interactions would qualify as (weak) hydrogen bonds. The heavy atom separation distance in the C-H...O bonds are longer but with similar deviation from linearity; it is more subjective whether any of these interactions qualify as hydrogen bonds based on geometric arguments. Using 2.1 kcal/mol as the average stabilization afforded by a classical hydrogen in a base pair (Ornstein and Fresco, 1988), it is reasonable to conclude that the C-H...X interactions in each of the base pairs in Figure 2 give rise to extra stabilization. (It is noteworthy for comparative purposes that two of the 10 base pairs analyzed with the same computational method (Ornstein

---

"wobble" geometry base pairs between canonical bases and 7-deazanebularin (DN). On the first line below the identity of each pair are the calculated enthalpy in kcal/mol (left of first slash) and two geometric parameters of the energy minimized pair. The first geometric parameter (between the slashes) is the difference in the glycosyl-N to glycosyl-N separation distance in angstroms, d(NN), between the present pair and "idealized" Watson-Crick pairs. The second geometric parameter (right of second slash) is the difference in the angle made by the extension of the two glycosyl bonds and the angle in "idealized" Watson-Crick pairs, d(ANG). On the second line below the identity of each pair are indicated the hydrogen bond (heavy atom) separations (in angstroms). The separation for the upper hydrogen bond (indicated by dashes) is to the left of the slash; the separation for the lower hydrogen bond is to the right of the slash. The site of attachment of each base in a pair to the corresponding sugar-phosphate chain is indicated by an arrow pointing away from the glycosidic nitrogen.

and Fresco, 1988) but containing two classical hydrogen bonds have interaction energies of -3.74 and -3.92 kcal/mol, which is less than or similar to those energies reported in Figure 2. The Watson-Crick pairs A:U and G:C have pairing energies of -4.9 and -6.7 kcal/mol, respectively.)

The base pair DN:A(syn) in Figure 2 contains adenine in the syn-glycosyl configuration. The penalty for a purine base in a syn-configuration is 1.0 kcal/mol (Saenger, 1984). Although such isomerization is allowed for a substrate base residue, it is forbidden for a template base residue (Topal and Fresco, 1976).

The pairs shown in Figure 2 are stated as having pseudo-Watson-Crick or "wobble" geometry. Compared to the "idealized" Watson-Crick pairs, the glycosyl-N to glycosyl-N separation distance of the three pairs DN:U, DN:C and DN:A(syn) differ by no more than 0.5 angstroms (see Figure 2 for details); the remaining two pairs (DN:A and DN:G) differ by at least 1.7 angstroms and resemble wobble type pairs. The difference in the angle made by the extension of the two glycosyl bonds and the angle in "idealized" Watson-Crick pairs is no more than 25 degrees for the three non-wobble pairs. We therefore feel justified in referring to the three non-wobble pairs as pseudo Watson-Crick type pairs; all but the most stereochemically demanding polymerase and or proof-reading enzyme complex should tolerate these pairs on structural grounds.

B. RNA Polymerase Mediated Pairing with Substrate DN:

7-Deazanebularin nucleoside monophosphate (DNMP) is readily incorporated into RNA with natural and synthetic DNA templates using E. coli RNA polymerase (Ward and Reich, 1972). 7-Deazanebularin nucleoside triphosphate (DNTP) can substitute for either ATP or GTP (but not for CTP or UTP) with almost equal efficiency as an analogue of ATP or GTP. In competition with DNTP, however, ATP is incorporated 20-fold more readily (Ward and Reich, 1972).

When a nucleoside triphosphate is in solution the donor/acceptor sites will form hydrogen bonds with solvent.

Thus, upon incorporating a nucleotide during template-directed synthesis, some of the base donor/acceptor sites must be stripped of solvent. This thermodynamically costly step will depend on the base residue and can be compensated for in large measure if the incorporating base forms hydrogen bonds with the template base. It is therefore obvious that template and incorporating bases which do not form hydrogen bonds will be at considerable thermodynamic disadvantage compared to those that can; at the same time, however, a near Watson-Crick geometry is required.

At least part of the cost of solvent stripping can be recovered by the formation of a base pair containing one classical N-H donor hydrogen bond and a second C-H donor hydrogen bond. In fact, in some cases (see above), the C-H...X interactions are comparable in stabilization afforded to that of a classical interaction. As discussed above, DN can form pseudo-Watson-Crick geometry pairs with uracil, cytosine and adenine bases. The pair with adenine, however, requires adenine to be in the syn-glycosyl orientation which for a template strand residue is not a permitted isomerization. Therefore, the presently suggested pairing schemes involving C-H...X type interactions are thermodynamically and stereochemically plausible and in agreement with the observed incorporation specificity of DNTP.

C. Binding of Aminoacyl-tRNA to Ribosomes Directed by Polymers or Trinucleoside Diphosphates Containing Template DN Residues:

The coding properties of DN-containing polymers during translation *in vitro* have been studied by template-mediated binding of aminoacyl-tRNA to ribosomes (Grunberger et al., 1972). Homopolymer poly[r(DN)], although less effective than poly[r(A)] is a template for binding of lysyl-tRNA and synthesis of polylysine (anticodon UUU). Copolymers poly[r(DN,A)] and poly[r(DN,A3)] produce complexing to ribosomes of arginyl-tRNA; since all known arginine codons contain a guanine residue in the second position and since poly[r(A)] does not bind arginyl-tRNA, the guanine analogue in the copolymer must be DN (Grunberger et al.,1972). DN in various positions of different trinucleoside diphosphates also has been found to stimulate aminoacyl-tRNAs to ribosomes acting as an analogue of adenine and guanine (Grunberger et al., 1972). DN is less effective as an analogue of guanine than of adenine.

In simulating the relative extent of the relevant codon-anticodon interactions computationally, it is not necessary to take account of dehydration effects nor to limit anti/syn isomerization to one strand or the other since RNA has significant random coil character. Therefore the appropriate computed energies for DN:U and DN:C, both of which are pseudo-Watson-Crick type pairs, are -3.7 and -3.0 kcal/mol, respectively. This result is consistent with the observation that DN is a better analog of adenine than of guanine. The computed interaction energy for A:U is 1.2 kcal/mol more stable than for DN:U and this also agrees with the somewhat better binding of templates containing adenine compared to DN.

The present calculations lead to the prediction that a DN template third position codon base should be capable of pairing with a third position anti-codon adenine or guanine base residue; the pairs DN:A and DN:G are wobble type pairs and should be limited to the third position. In the case of DN pairing with adenine, it is also possible to have a pseudo-Watson-Crick type pair but adenine must now adopt the syn-glycosyl configuration and this pairing arrangement is less stable. Thus DN can serve as an analogue of cytosine and uracil in the third position, while DN may somewhat less readily serve as an analogue of uracil in the first and second positions as well.

D. Equilibrium Properties of Polymers Containing DN:

Synthetic poly[r(DN-U)] has a thermal denaturation profile that is sharp, cooperative, and reversible, just as in the case of double stranded poly[r(A-U)]; though the melting temperature of poly[r(A-U)] is about 50 C higher (Ward and Reich,1972). A double stranded structure for poly[r(A-U)] is also supported by nuclease digestion and other physical evidence (Ward and Reich, 1972).
A large part of the stability of nucleic acid double stranded helics is due to base stacking (Ornstein and Fresco, 1983b,c). It is thus interesting to note that DN does not stack well in comparison to adenine. Since

poly[r(2,6-diaminopurine-U)] has a melting temperature (Howard et el., 1966) about 45 C higher than poly[r(A-U)] but has three hydrogen bonds per base pair, it appears that one can neatly explain the 50 C melting temperature difference between poly[r(DN-U)] and poly[r(A-U)] as being due to a difference of one hydrogen bond per base pair. But since 2,6-diaminopurine has one more exo-ring amino group than does adenine, poly[r(2,6-diaminopurine-U)] is expected to be stabilized by base stacking to a greater degree than poly[r(A-U)]. So the question becomes where does the stability of DN:U in helices come from? On the one hand, extra stabilization of DN:U comes from the nonclassical C-H donor hydrogen bonds. On the other hand, extra stabilization of the double-stranded helix is afforded by the reduced thermodynamic drive of solvent water molecules to hydrogen bond with the donor moieties of DN versus adenine in the random coil state. Thus it appears that C-H donor hydrogen bonds form between DN and uracil in equilibrium helices.

CONCLUSION

It has been shown that 7-deazanebularin, a nucleic acid base analogue with a substituent "deleting" a normal hydrogen-bonding site required for Watson-Crick base pairing, is capable of forming pseudo-Watson-Crick and "wobble" geometry base pairs by virtue of a C-H...X hydrogen bond. Energy minimized pairing schemes are presented as obtained with a recently force-fit computational method. The energies and stereochemical features of base pairs involving 7-deazanebularin with the canonical bases are consistent with specificities observed in transcription and translation experiments and in equilibrium helices.

ACKNOWLEDGEMENT

Supported by NSF grant PCM-8023706 to J.R. Fresco (Princeton University); I thank Dr. Fresco for advice, encouragement and support. Part of this paper was written while I was employed at Eastman Kodak, Research Laboratories, Kingsport, Tennessee as were some of the calculations. My present address is Battelle, Pacific Northwest Laboratories, Molecular Science Research Center, Richland, WA 99352.

REFERENCES

Batterham TJ, Brown DJ, Paddon-Row MN (1967). Simple pyrimidines. IX. Deuterium exchange of C-methyl protons. J Chem Soc (B) 171-3.

Bonchev D, Cremaschi P (1974). C-H group as proton donor by formation of a weak hydrogen bond. Theor Chim Acta 35: 69-80.

Brdar B, Reich E (1972). 7-Deazanebularin: Metabolism in cultures of mouse fibroblasts and incorporation into cellular and viral nucleic acids. J Biol Chem 247: 725-30.

Bruskov VI, Bushuev VN, Poltev VI (1980). Nuclear magnetic resonance study of C-H...O hydrogen bonds in nucleic acid base analogs. Mol Biol 14: 245-51 (Eng.)

Bullock FJ, Jardetzky O (1964). Proton magnetic resonance studies of purines and pyrimidines. XII. An experimental assignment of peaks in purine derivatives. J Org Chem. 29: 1988-90.

Crick FHC (1966). Codon-anticodon pairing: the wobble hypothesis. J Mol Biol 19: 548-55.

Cullis PM, Wolfenden R (1981). Affinities of nucleic acid bases for solvent water. Biochemistry 20: 3024-8.

Grunberger D, Ward DC, Reich E (1972). 7-Deazanebularin: Coding properties of triplets and polynucleotides. J Biol Chem 247: 720-4.

Howard FB, Frazier J, Singer MF, Miles HT (1966). Helix formation between polyribonucleotides and purines, purine nucleosides, and nucleotides. J Mol Biol 16: 440-53.

Kornberg A (1980). "DNA Replication." San Francisco: Freeman, p426.

Kvick A, Koetzle TF, Thomas R (1974). Hydrogen bond studies. 89. Neutron diffraction study of hydrogen bonding in 1-methylthymine. J Chem Phys 61: 2711-9.

Kyogoku Y, Lord RC, Rich A (1969). Infrared study of the hydrogen-bonding specificity of hypoxanthine and other nucleic acid derivatives. Biochim Biophys Acta 142: 10-17.

Lichtenberg D, Bergmann F (1973). Mechanism of hydrogen-deuterium exchange in hypoxanthines. J Chem Soc Perk Trans 1: 789-93.

Lord RC, Thomas, Jr GJ (1967). Raman studies of nucleic

acids. II. Aqueous purine and pyrimidine mixtures. Biochim Biophys Acta 142: 1-11.

Ornstein RL, Fresco JR (1981). Successful force-fitting of nucleic acid base dimerization energies and distances with an empirical-potential function that uses an adjustable dielectric and explicitly accounts for lone-pair electrons with an adjustable set of charges. In Sarma RH (ed): "Biomolecular Stereodynamics", Vol 1, Albany, NY: Adenine, pp 151-62.

Ornstein RL, Fresco JR (1983a). Correlation of crystallographically determined and computationally predicted hydrogen-bonded pairing configurations of nucleic acid bases. Proc Natl Acad Sci USA 80: 5171-5.

Ornstein RL, Fresco JR (1983b). Correlation of Tm and sequence of DNA duplexes with enthalpies computed by an improved empirical method. Biopolymers 22: 1979-2000.

Ornstein RL, Fresco JR (1983c). Correlation of Tm, sequence and enthalpies of complementary RNA helices and comparison with DNA helices. Biopolymers 22: 2001-16.

Ornstein RL, Fresco JR (1984). Alkylated template bases in polymerase-catalyzed RNA synthesis pair to normal NTP bases with a C-H donor containing hydrogen bond. 18th Middle Atlantic Regional Meeting ACS, May 21-23, Abstract #47, p49.

Ornstein RL, Fresco JR (1988). Mediation of base pairing in nucleic acids by CH-donor hydrogen bonds. Science (in press).

Ornstein RL (1988). Novel base pairing in nucleic acids: C-H donor group mediated base pairing involving 3-methyluracil. J Mol Structure: Theochem

Parthasarathy R, Soriano-Garcia M (1976). Bifurcated hydrogen bonds and flip-flop conformation in a modified nucleic acid base, gc6Ade. Nature 260: 807-8.

Rubin J, Brennan T, Sundaralingam M (1972). Crystal and molecular structure of a naturally occurring dinucleoside monophosphate. Uridylyl-(3'-5')-adenosine hemihydrate. Conformational rigidity of the nucleotide unit and models for polynucleotide chain folding. Biochemistry 11: 3112-28.

Saenger W (1984). "Principles of Nucleic Acid Structure." NY: Springer-Verlag, pp 69-78.

Schweizer MP, Chan SI, Helmkamp GK, Ts'o POP (1964). An experimental assignment of the proton magnetic resonance spectrum of purine. J Am Chem Soc 86: 696-700.

Seeman ND, Sussman JL, Berman HM, Kim SH (1971). Nucleic acid conformation. Crystal structure of a naturally-occurring dinucleoside phosphate (UpA). Nature New Biol 233: 90-2.

Srikrisknan T, Parthasarathy R (1976). 'Sandwiched' water molecule between pyrimidine bases and intramolecular C-H...O hydrogen bonding in 5-nitro-1-(.beta.-D-ribosyluranic acid)-uracil monohydrate. Nature 264: 379-80.

Sundaralingam M (1966). Stereochemistry of nucleic acid constituents. III. Crystal and molecular structure of adenosine 3'-phosphate dehydrate (adenylic acid b). Acta Cryst 21: 495-506.

Topal MD, Fresco JR (1976). Complementary base pairing and the origin of substitution mutations. Nature 263: 285-9.

Takusagawa F, Koetzle TF, Srikrishnan T, Parthasarathy, R (1979). C-H...O interactions and stacking of water molecules between pyrimidine bases in 5-nitro-1-(.beta.-D-ribosyluronic acid)-uracil monohydrate [1-(5-nitro-2,4,di-oxopyrimidinyl)-.beta.-D-ribofuranoic acid monohydrate]: a neutron diffraction study at 80 K. Acta Cryst B35: 1388-94.

Vishveshwara S (1978). Ab-initio molecular orbital studies on carbon-hydrogen-x hydrogen bonded systems. Chem Phys Lett 59: 26-29.

Ward DC, Reich E (1972). Fluorescence Studies of Nucleotides and Polynucleotides II. 7-Deazanebularin: coding ambiguity in transcription with base pairs containing fewer than two hydrogen bonds. J Biol Chem 247: 705-19.

Yanson IK, Teplitsky AB, Sukhodub LF (1979). Experimental studies of molecular interactions between nitrogen bases of nucleic acids. Biopolymers 18: 1149-70.

SECTION II. PHYSICAL METHODS IN DRUG DESIGN

THE STRUCTURE OF PROTEINS AND THEIR BINDING SITES: NMR AND ARTIFICIAL INTELLIGENCE

Oleg Jardetzky[1], Olivier Lichtarge[1], James Brinkley[2], and Marcela Madrid[1]

[1]Stanford Magnetic Resonance Laboratory and
[2]Knowledge Systems Laboratory, Stanford University, Stanford, CA 94305-5055

INTRODUCTION

It has been known for some time that the NMR spectra of proteins and protein-ligand complexes contain a wealth of structural information (Roberts and Jardetzky, 1970). Strategies for the determination of active site structures have been worked out and tested on proteins of a wide range of molecular weights (Jardetzky and Roberts, 1981, Ch. IX and X). In recent years the increasing availability of high resolution instruments and efficient techniques of data collection has made it possible to obtain sufficient NMR data to make complete sequential assignments and to propose the three dimensional structure of small proteins (MW up to 10,000) from them (Wüthrich, 1986). At the same time this development has brought into focus the fundamental problems of interpreting spectroscopic data in structural terms (Jardetzky and Lane, 1987).

Over the past five years we have analyzed the issues involved in deriving structures from NMR data and the different options available, in some detail, and have developed a methodology that avoids most of the pitfalls. This has been extensively discussed elsewhere (ref. cited) and only a brief synopsis is given here.

THEORETICAL CONSIDERATIONS

The major source of structural information in the NMR spectra of proteins and nucleic acids is to be found in the

Nuclear Overhauser Enhancement, which reflects the transfer of magnetization between nuclei in the structure, which, in turn, is strongly dependent on internuclear distances (Ernst et al., 1987). Magnetization transfer is described by the Bloch equations, or, in its most general form by the Redfield density matrix (Abragam, 1961). An inspection of these equations (Eq. 1a-c) shows that internuclear distances can be obtained from the individual cross-relaxation terms, provided the spectral density functions are known.

$$dm_i/dt = -\rho_i m_i - \sum_{j=1}^{N} \sigma_{ij} m_j \tag{1a}$$

$$\rho_i = \gamma_I^2 \gamma_S^2 \hbar^2 S(S+1) \sum_k \{1/12\, J_{ik}^0(\omega_I-\omega_S)+3/2\, J_{ik}^1(\omega_I)+3/4\, J_{ik}^2(\omega_I+\omega_S)\} \tag{1b}$$

$$\sigma_{IS} = \gamma_I^2 \gamma_S^2 \hbar^2 I(I+1)\{-1/12\, J_{IS}^0(\omega_I-\omega_S)+3/4\, J_{IS}^2(\omega_I+\omega_S)\} \tag{1c}$$

Although reasonable assumptions about the magnitude of the spectral density functions can be made in many cases, lack of precise knowledge introduces an inherent inaccuracy in the distances calculated from cross-relaxation rates. A second important source of inaccuracy arises from the difficulty of measuring the individual cross relaxation rates (Jardetzky and Lane, 1987). This difficulty stems from the existence of multiple pathways for rapid magnetization transfer in macromolecules ("spin diffusion"). The magnitude of the resulting inaccuracy depends on several variables, such as the molecular weight of the macromolecule, temperature, length of the mixing time used in the 2D NOE experiment, etc. This is shown in Fig 1.

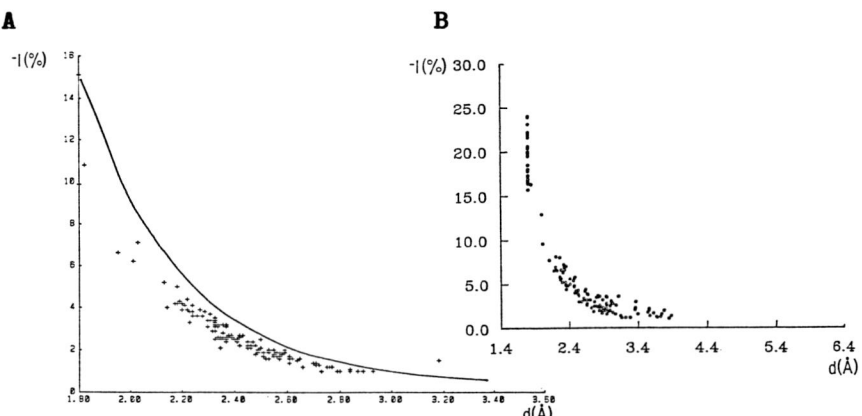

Figure 1

Figure 1A shows the distribution of NOEs as a function of internuclear distance obtained by solving the Bloch equations for the Bovine Pancreatic Trypsin Inhibitor (BPTI, MW 6000) at 100msec mixing time and 68°C (Madrid and Jardetzky, 1988). In general the $1/r^6$ relationship between the cross relaxation rate predicted by Equation 1c is apparent. However, comparison of the NOE crosspeak intensities obtained from the solution of the (complete) equation 1a (crosses) to those obtained assuming a two-spin approximation (solid line) makes it apparent that neglecting spin diffusion implicit in equating the observed NOE to an individual cross relaxation rate (the two-spin approximation) introduces a systematic error into the distances estimated from the NOEs. The magnitude of the error depends on the size of the protein and the experimental conditions.

Figure 1B shows a similar calculation for T4 lysozyme (MW 30084), with a mixing time of 50 msec and a temperature of 50°C. For the larger protein under reasonable experimental conditions the scatter is larger (the systematic error is not shown). Such calculations show (cf also Jardetzky et al., 1986) that the experimental data can be translated into structural parameters only with considerable caution and that the estimated distances must be treated as distance ranges of somewhat uncertain accuracy.

Several methods of data analysis to yield structures from the inexact NMR data have been proposed. All require prior knowledge of the amino acid (or nucleotide) sequence and all search the conformational space checking for the satisfaction of constraints. Differences between methods stem from the different optimizing functions used to guide the search and the different degrees of completeness with which the entire conformational space is explored. In distance geometry (Crippen, 1977) the exploration is guided by a global error function to find the best structure. In restrained molecular dynamics (Kaptein et al., 1985) the target function is given by the equations of motion for the macromolecular structure and assumed potentials are introduced as additional nonexperimental constraints. In the heuristic refinement method (Jardetzky, 1984, Lichtarge, 1986, Altman and Jardetzky, 1986, Lichtarge et al., 1987) no target function is used, rather the entire conformational space is sampled and structures are systematically checked for their ability to satisfy a given set of constraints. All structures that satisfy the constraints are retained as valid within the limits of the data set, all others are rejected. The different methods therefore also reflect different methodological paradigms. Distance geometry and molecular dynamics fall within the adjustment or optimization paradigm in which the generated starting structures are adjusted until a minimum error in the target function is obtained. Heuristic refinement in contrast falls within the exclusion paradigm, in which the generated starting structures are simply included if they are found to satisfy the constraints and excluded otherwise.

The importance of the exclusion paradigm to the interpretation of inherently inexact data is apparent from the following considerations:

The calculations cited above show that the internuclear distance which can be estimated from NOE data (without inverting the Redfield density matrix, which would require independent and precise knowledge of the spectral density functions) contain a systematic error. Thus the accuracy of the estimates is much lower than the apparent precision of the distances which can be calculated using equation 1c. Minimizing a distance geometry or molecular dynamics (or any other) global target function using precise, but in reality inaccurate constraints will drive the calculation into a local minimum, which will not necessarily correspond to the correct structure. Relaxing the precision of the constraints makes convergence less probable, and the sample of converging structures will not necessarily be more representative. It is therefore important to define the limits within which the structure MUST lie to satisfy the experimental constraints. It is equally important that the sampling of the conformational space be unbiased. Although some bias is inevitable in all methods of sampling, the exclusion paradigm offers two important advantages: (1) it is free of the relatively strong favorable bias implicit in any target function, which drives the calculation toward the (apparently) best case, and (2) it only rejects clearly unacceptable conformations so that by nature it is a cautious method which includes both best and worst case interpretations of the data. As a result the family of structures generated using the exclusion paradigm will more accurately reflect the range of possibilities defined by the initial data set. The accuracy of any structure calculated by global optimization methods remains indeterminate, even

if the precision of the structure (small RMSD from the mean)
is high, because overinterpreted constraints processed by a
global optimization program can produce precise, but
inaccurate, structures. With the use of the exclusion
paradigm it is at least possible to assert with confidence
that the real structure lies within the defined bounds of
conformational space.

THE PROTEAN EXPERT SYSTEM AND ITS STRUCTURE

An unbiased exploration of the entire conformational
space available to a macromolecule is, a priori, a limitless
task, beyond the capacity of current computers. The
approach of distance geometry or molecular dynamics is to
use a guiding (or target) function to explore only a limited
portion of space by threading a path towards a minimum. Our
alternative is to use hierarchical problem solving
(Jardetzky, 1984, Lichtarge, 1986). In this approach:
(a) the macromolecule is assembled piece by piece from
building blocks; (b) the building blocks can be represented
in varying degrees of detail, from a simple geometric solids
to full blown atomic representations; (c) the assembly
process is controlled by a strategy which usually consists
of defining first those regions of the macromolecule which
are subject to the largest number of constraints. This
approach is described in more detail below. For proteins
the natural hierarchy of primary-secondary-tertiary
structure parallels the severity of the constraints and
permits the abstraction of units of secondary structure into
geometric solids (e.g. a cylinder for an alpha helix, or for
a short stretch of beta sheet). While cylinders are
admittedly crude representations of helices or beta-sheets,
they are easily manipulated and permit an efficient search
through and a rejection of a large portion of the protein
conformation space. This results in a significant reduction
of computational time, and makes it possible to construct
topologically accurate, though "fuzzy" structures, the
"fuzziness" reflecting the uncertainties in the original
data base and the approximations inherent in the
abstractions as well as the intrinsic mobility within the
structure. Representation of the molecular structure at the
atomic level is a subsequent step, and can be derived from
the positions of the abstracted solids.

Placement of ligands by this technique proceeds in the same order as that used in the placement of any other coherent units - with the coarse placement of the abstracted ligand preceding any atomic refinement. This is illustrated in Fig. 2 for the placement of the heme in myoglobin (Lichtarge et al., 1987). Fig. 2A is the abstract representation of the myoglobin crystal structure with eight helices and the heme. Fig. 2B is the halo representing the uncertainty in the position of the heme, placed on the basis of those distance constraints which could be expected in an NOE spectrum, represented as distance ranges appropriate to an NOE experiment run under the conditions defined in Fig. 1B.

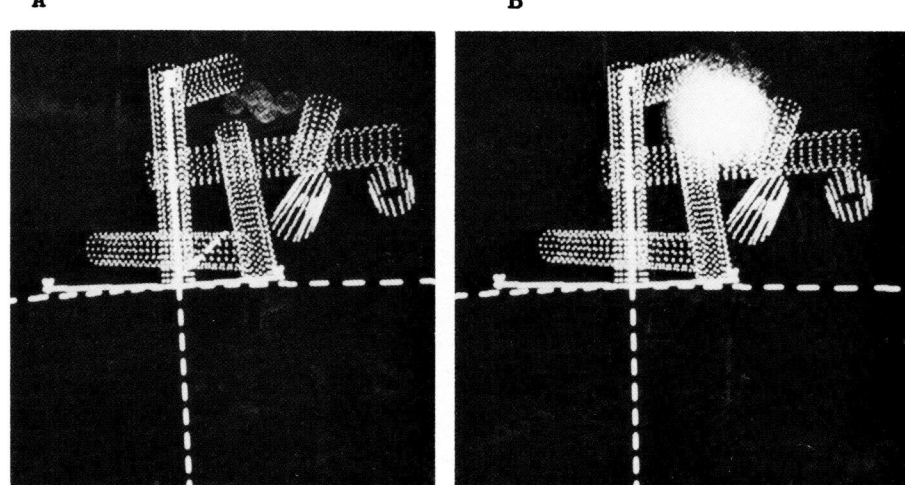

Figure 2

The computational cost of sampling all possible conformations grows with the number of atoms and with the number of different locations each atom can be in, where the set of these locations defines the <u>accessible volume</u> of the atom. PROTEAN uses four basic techniques to reduce computational demands:

Problem decomposition. Reduce the number of atoms that have to be simultaneously handled by breaking the overall protein into smaller units called <u>partial arrangements</u>, "solve" each partial arrangement separately, then combine them into the overall solution.

Problem abstraction. Reduce the number of atoms by grouping locally constrained sets of atoms, such as those forming sidechains or secondary structures, and consider the entire group as an abstract geometric object before considering each component atom.

Local satisfaction of constraints. Reduce the size of the accessible volume for each atom by sequentially applying constraints between pairs of geometric objects rather than all objects at once.

Heuristic control. At each point in the problem solving, choose that action which is likely to exclude the largest number of potential structures.

The heuristic refinement method therefore refines the protein along two main dimensions, that of structural representation and that of accessible volume and it uses heuristics to control the order of refinement operations to obtain the greatest efficiency.

INPUT/OUTPUT CHARACTERISTICS

PROTEAN currently accepts as input experimental data, standard parameters of chemical structure and method-specific parameters.

The experimental data consists of the primary structure, short and long range NOEs, hydrogen exchange rates and spin-spin coupling constants and volume and surface information. Any other data that can be expressed as distance constraints, or as constraints on the relative orientation of two geometric objects, is also acceptable to the program. The standard parameters of chemical structure arise from the assumption of standard backbone and sidechain geometry for all fixed bond lengths and angles, and hard-sphere radii for van der Waals forces.

The major method specific parameter is the desired resolution of the result, both in terms of the spatial resolution and the level of abstraction of the protein. For example, for highly underconstrained systems, coarse placement of secondary structures at 4 Å spatial resolution may be all that is justified, while for highly constrained systems, placement of the atoms at 1 Å resolution is desirable.

The output of PROTEAN is a set of covalently legal backbones which nearly satisfy the constraints and which are a representative sample of the entire family of structures satisfying the constraints. These structures may then be refined by any of the current adjustment methods.

GEOMETRIC REPRESENTATIONS

As discussed above, in order to reduce the number of atoms, PROTEAN groups related atoms together as more abstract geometric objects such as secondary structures. A geometric object may be composed of atoms or it may be composed of other geometric objects which are themselves composed of atoms. In this way a hierarchy of structure is maintained, from the most abstract protein level (a single geometric object such as an enclosing ellipsoid used for volume checks) to the most detailed atomic level.

Objects are related to each other by means of standard homogeneous coordinate transforms (Brinkley et al., 1988) used in computer graphics and robotics. These transforms are 4x4 matrices, equivalent to a set of six parameters describing the location and orientation of an object. The accessible volume of one object with respect to another is represented as a discrete list of locations sampled at some spatial resolution within a (typically) 64 Å cube (for example, 1 Å for the three position components and $10°$ for the three orientation components).

GEOMETRIC OPERATIONS

PROTEAN uses five basic operations to manipulate and assemble these geometric objects in order to generate the family of structures compatible with the constraints. These are ANCHOR, YOKE, APPEND, PRUNE and Coherent Instance Generation.

ANCHOR generates the initial explicit list of locations for an object. One object, called the <u>anchor</u>, is held fixed, and another object, called the <u>anchoree</u>, is systematically moved through different locations and orientations at some sampling resolution. The locations that satisfy the constraints between the anchor and the anchoree are retained and define the initial accessible volume of the anchoree in the partial arrangement.

The YOKE operation is used to apply constraints between two anchorees having constraints between them. At the end of this operation, a location is retained in each of the anchoree accessible volumes only if it is compatible (satisfies the constraints) with at least one location in the other accessible volume. The YOKE operation is the primary mechanism by which constraints between pairs of objects are able to propagate their influence to the overall structure.

The APPEND operation is used to combine partial arrangements. For example, it may be computationally advantageous to separately reduce the accessible volumes of the objects in two partial arrangements (such as two helices, each composed of peptide units and sidechains) before combining them into a single partial arrangment. The APPEND operation determines the accessible volume of an object C with respect to an object A, given that C has an accessible volume with respect to a third object B, and B has an accessible volume with respect to A. The derived accessible volume is simply the matrix composition of the cross-product of all the locations defining the relevant accessible volumes.

The PRUNE operation is a modified YOKE where constraints can be applied between two objects regardless of whether they are both anchorees of the same anchor. This situation arises when two partial arrangements are fused by APPEND.

All these operations define lists of locations for each object in a partial arrangement, with the guarantee that given constraints between two objects, at least one location in the list of the first object will satisfy these constraints to some location in the list of the second object. This defines _pairwise_ or _local simultaneity_ (Lichtarge, 1986).

The Coherent Instance Generator extends constraint satisfaction to the entire structure by exhaustively generating all possible combinations of locations of each object and retaining only those which simultaneously satisfy all the constraints on the structure. This process insures <u>global simultaneity</u> of constraint satisfaction. A representative sample of abstract secondary structure coherent instance defines the general topology of the molecule (Lichtarge et al., 1987). If there is no more information in the constraints, then this may be all that can be learned about the structure of the molecule. If there is more information, then each of these coherent instances can be expanded into its constituent parts, each of which then becomes a separate partial arrangement. This use of hierarchical structure representations allows an efficient use of the coherent instance generator to obtain a representative set of the overall family of structures.

CONCLUSION

The heuristic refinement method is well suited to generate structures of any macromolecule or set of interacting molecules for which there exist experimentally determined structural constraints. Furthermore, it has been shown that the computation is capable of reproducing known structures accurately, and that the precision of the structure depends on the size of the data set used as a basis of computation (Lichtarge, 1986, Lichtarge et al., 1987, Altman et al., 1988). The remaining sources of uncertainty can be measured by the extent to which alternate coherent instances differ from one another. They are: (a) the general paucity of constraints, (b) the low degree of information which the available constraints do provide and (c) the inherent conformational flexibility of chemical structures. What their relative contributions are is an important question which can now be tackled through the use of PROTEAN.

ACKNOWLEDGEMENTS

We acknowledge support from NSF grant DMB 8402348 and NIH grant RR02300.

REFERENCES

Abragam A (1961). "The Principles of Nuclear Magnetism." Oxford: Oxford University Press.

Altman R, Duncan B, Brinkley J, Buchanan B, Jardetzky O (1988). Determination of the Spatial Distribution of Protein Structure Using Solution Data. In Jaroszewski JW, Schaumburg K, Kofod H, (eds): "NMR Spectroscopy in Drug Research," Copenhagen: Monksgaard, in press.

Altman RA, Jardetzky O (1986). New Strategies for the Determination of Macromolecular Structure in Solution. J Biochem 100:1403-1423.

Brinkley JF, Altman RB, Duncan BS, Buchanan BG, Jardetzky O (1988). The Heuristic Refinement Method for the Derivation of Protein Solution Structures: Validation on Cytochrome-b562. Submitted to J Chemical Info & Comput Sci.

Crippen GM (1977). A Novel Approach to the Calculation of Conformation: Distance Geometry. J Comp Phys 24:96-107.

Ernst RR, Bodenhausen G, Wokaun A (1987). "Principles of Nuclear Magnetic Resonance in One and Two Dimensions." Oxford: Clarendon Press.

Jardetzky O (1984). A Method for the Definition of the Solution Structure of Proteins from NMR and Other Physical Measurements: The Lac-Repressor Headpiece. In Ovchinnikov YA (ed): "Progress in Bioorganic Chemistry and Molecular Biology," Amsterdam: Elsevier Science Publishers BV, pp. 55-63.

Jardetzky O, Lane A, Lefèvre J-F, Lichtarge O, Hayes-Roth B, Altman R, Buchanan B (1986). A New Method for the Determination of Protein Structures in Solution from NMR. In Maraviglia B, De Luca F, Campanella R, (eds): "Proc. XXIII Congress Ampere on Magnetic Resonance," Rome, Italy, pp. 64-69.

Jardetzky O, Lane AN (1987). Determination of the Solution Structure of Proteins from NMR. In "Proc. Int'l. School of Physics, Enrico Fermi," Bologna, Italy: Il Nuovo Cimento, in press.

Jardetzky O, Roberts, GCK (1981). "NMR in Molecular Biology." New York: Academic Press, Ch. IX & X.

Kaptein R, Zuiderweg ERP, Scheek RM, Boelens R, Van Gunstern WF (1985). A Protein Structure from Nuclear Magnetic Resonance Data Lac Repressor Headpiece. J Mol Biol 182:170-182.

Lichtarge, O (1986). "Structure Determination of Proteins in Solution by NMR." Stanford, CA: Stanford University, Ph.D. Thesis.

Lichtarge O, Cornelius CW, Buchanan BG, Jardetzky O (1987). Validation of the First Step of the Heuristic Refinement Method for the Derivation of Solution Structures of Proteins from NMR Data. PROTEINS, Structure Function and Genetics 2:340-358.

Madrid M, Jardetzky O (1988). Comparison of Experimentally Determined Protein Structures by Solution of Bloch Equations. Biochim Biophys Acta 953:61-69.

Roberts GCK, Jardetzky O (1970). Nuclear Magnetic Resonance Spectroscopy of Amino Acids, Peptides and Proteins. In "Advances in Protein Chemistry 24," New York: Academic Press, pp. 447-545.

Wüthrich K (1986). "NMR of Proteins and Nucleic Acid." New York: John Wiley and Sons.

New address for Dr. Marcela Madrid: Department of Biological Sciences, Carnegie Mellon University, Pittsburgh, PA

FOLDING AND DYNAMICS OF GLOBULAR PROTEINS STUDIES BY TIME RESOLVED FLUORESCENCE SPECTROSCOPY.

Elisha Haas
Department of Life Sciences, Bar-Ilan University Ramat Gan 52100; and Departments of Chemical Physics and Biophysics, Weizmann Institute of Science, Rehovot 76100, ISRAEL.

The design of modified proteins with altered specificities and affinities in ligand binding, as well as the design of novel proteins tailored for specific functions, is no more science fiction, it is a technological practice. The main reason for the present limited scope of successful applications of the available technology of production of engineered proteins is the yet unbridged gap in our understanding of the rules of the translation from the primary structure to the three dimensional structure. This is a translation of a one dimensional array of information to three (more correctly four) dimensional structure. The code for this translation is complex. Deciphering this "second genetic code" (Glodberg 1985) is a prerequisite for rational applications of the enormous potential of available biotechnologies of protein engineering and production.

The Anfinsen experiment (Anfinsen et al 1961) tells us two principles: (1) The free energy of folding provides the driving force for correct folding (and pairing of the disulfide bonds), (2) No auxiliary factors are necessary, the entire information needed for correct folding is contained in the amino acid sequence. In other words, the cell does not have a "translation machine". The high fidelity of the folded conformations and the fast transition rates lead to the concept of folding pathways (Matheson and Scherage 1978, Karplus and Weaver 1976 and Ptitsyn 1973, Kim and Baldwin 1982). The pathway can be viewed as a succession of intermediate structures which lead from the unfolded state to the unique native structure. Three features make the folding by a pathway an attractive model: (a) Local folding of segments of the protein, which reduces the complexity of the folding process and facilitates 'parallel processing'. (b) Intermediate local structures (including non-native structures) which are then rearranged. (c) Dynamic flexibility of the intermediates. High amplitude Brownian fluctuations facilitate constant editing of abortive conformations and rescue of molecules trapped in local minima.

The above concept of a pathway of folding leads to a working hypothesis that the coded sequence information includes instructions for the pathway of folding. Hence understanding the mechanism of folding of a protein depends on determining the intermediate structures. A test of this hypothesis is a major challenge for experimentalists in two respects: (a) The intermediates are only poorly populated (due to the cooperative nature of the folding transition), (b) Most intermediate structures are very unordered and dynamic and should be characterized in terms of distributions, averages, and rates of fast fluctuations.

The first challenge is commonly met by either trapping or perturbation approaches. (Creighton 1978, Kim and Baldwin 1982). In order to meet the second challenge methods based on combination of site-directed fluorescence labeling and time resolved fluorescence measurements of nonradiative excitation energy transfer (Forster 1948, Stryer and Haugland 1967 Stryer 1978, Steinberg 1971) are being developed. In the following I briefly describe the principles of the methods and results obtained for early steps in the folding pathway of bovine pancreatic trypsin inhibitor (BPTI).

Nonradiative Excitation Energy Transfer

It has been shown that the time dependence of the rates of non-radiative excitation energy transfer between probes attached to well defined sites on peptides and proteins is a function of both the distribution of distances between the labeled sites and the rates of conformational interconversions (Haas et al. 1975, 1978a, Haas and Steinberg 1984). The probability, $n_{D \to A}$, of energy transfer from a donor "D" to an acceptor "A" is given by (Förster 1948):

$$n_{D \to A} = \frac{9000(\ln 10)\kappa^2 \eta_0}{128\pi^5 n^4 N r^6 \tau} \int_0^\infty \frac{f(\bar{v})\varepsilon(\bar{v})}{\bar{v}^4} d\bar{v} = 1/\tau (R_0/r)^6 \quad (1)$$

where η_0 is the quantum yield of the donor in the absence of an acceptor, n is the refractive index of the medium, N is Avogadro's number, τ is the lifetime of the donor molecule, r is the distance between the donor and the acceptor, $f(\bar{v})d\bar{v}$ is the normalized fluorescence intensity of the donor in the wavenumber $\bar{v}$, R_0 (as defined by Eq. 1) is the distance between D and A when transfer efficiency is 50%, and κ^2 is a factor that expresses the orientational dependence of the probability of energy transfer. It has been shown that the orientational dependence of the transfer probability κ^2 can be made weak or insignificant, by selecting for probes which exhibit low limiting polarization properties for the electronic transitions involved in the transfer process (Haas et al. 1978b).

The efficiency of energy transfer measured by <u>steady-state</u> methods represents an average quantity which <u>yields</u> the equilibrium interprobe distance, when all molecules share the same distance. Analysis of the <u>kinetics</u> of the fluorescence decay of the donor or the acceptor, $F_d(t)$ or $F_a(t)$ respectively, contain enough information for the determination of the equilibrium interprobe distance distribution, $N_0(r)$. The impulse response of the donor's fluorescence can be written as:

$$F_d(t) = \int_0^\infty \sum_{i=0}^n N_0(r) \exp[\frac{-t}{\tau_i}(1+(R_0/r)^6)] dr, \quad (2)$$

where there are n donor lifetimes in the absence of an acceptor. Analysis of the donor fluorescence decay and calculation of the equilibrium distance distribution, $N_0(r)$, is achieved by methods based on least squares analysis (Haas et al. 1975). Equation (2) does <u>not</u> apply to cases in which the interprobe distances change during the lifetime of the excited state.

Any Brownian motion of the chain ends would enhance the rate of decay of the donor emission relative to that indicated in Eq. 2. This happens because the distribution of interprobe distances of the molecules that have an excited donor, $N^*(r,t)$, starts to deviate from the equilibrium distribution, $N_o(r)$, as time increases after excitation, because molecules with small r values have a faster rate of decay of excited state than molecules with a larger r values. The subsequent rearrangement towards the equilibrium distribution by diffusion of the labeled sites, can enhance the efficiency of energy transfer. Obviously, this enhancement in transfer efficiency contains information about the Brownian motion of the labeled sites relative to one another. The combination of fast optical excitation and distance dependent transfer rates, generates a perturbation of the interprobe distance distribution in the population of molecules with an excited donor. No conformational perturbation is involved in this process, the conformation of the population of the molecules is maintained at equilibrium.

The simultaneous analysis of multiple fluorescence decay measurements of the donor and the acceptor by the <u>global</u> approach (Beechem and Haas 1988), allows for simultaneous determination of the segmental diffusion parameter and the distance distribution parameters, without changing the physical conditions of the experiment. This method of analysis improves the accuracy and statistical significance of the determined parameters.

Fluctuations occurring on a time scale slower than the lifetime of the excited state of the donor, do not affect the fluorescence decay kinetics of the probes. Determination of slower conformational fluctuations is possible by autocorrelation analysis of donor emission in a very dilute solution under constant illumination (Haas and Steinberg, 1984). Thus the dynamic range of the conformational transitions which can be determined by excitation energy transfer measurements extends from the subnanoseconds to the subseconds time scales.

Intramolecular distance distributions in BPTI.

The rational for application of the above methods in the investigation of the pathway of protein folding is quite straightforward. A series of intramolecular distance distribution gives the overall conformation of the molecule, local structural states in the labeled segments and their dynamics. This is the type of data needed for determination of the conformational state of folding intermediates. The measurements can be repeated throughout the folding pathway by combination with either trapping techniques or with fast mixing experiments. The larger the number of derivatives, the higher the spatial resolution and localization of subdomain structures which may function in the folding pathway.

A necessary condition for application of the above methods is the availability of protein derivatives labeled each by a single donor and a single acceptor at well defined sites. Our approach to the labeling problem is based on attachment of the probes, by mild chemical reactions, to side chains of pairs of residues selected to probe specific local structures which may be functional in the folding transition. Only surface exposed side chains are modified, in order to minimize possible interference with the folding pathway. Labeling is done by modification of reactive side chains of either natural or engineered residues (Amir and Haas 1987).

The bovine pancreatic trypsin inhibitor, BPTI, is the first protein studied by the present methods. It is a single domain globular protein with elements of β sheet and α helix and three disulfide bonds (Deisenhofer and Steigmann 1975, Wlodawer et al. 1984, Figure 1). Procedures for the attachment of donor and acceptor probes to any one of the five amino groups of BPTI, (four ϵ-amino groups of lysine side chains and the α-amino group of arginine-1 at the N-terminus) were developed (Amir and Haas 1987). In the following I describe results obtained using the derivatives: N-α-MNA-arg^1-N-ϵ-DA-coum-lys^n-BPTI (n=15,26,41,46) (abbreviated (1-n)BPTI). Each derivative is labeled by a donor at the N-terminus and an acceptor at one of its lysine residues. All four BPTI derivatives retain full inhibitory capacity towards trypsin (except for (1-15)BPTI which is labeled at the active site), and can reversibly refold after unfolding, with 90% recovery of activities. This indicates that the probes attached to the amino groups do not perturb the capacity of the labeled derivatives to refold to the active structure and, hence, they are suitable models for investigation of protein folding.

162 / Haas

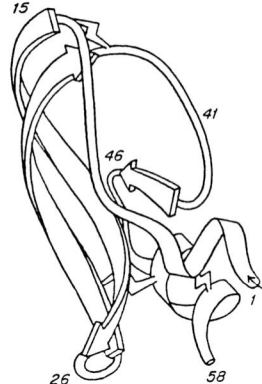

Figure 1. Native structure of BPTI schematically drown according to J. Richardson (Richardson 1985). Labeling sites are marked by residue number.

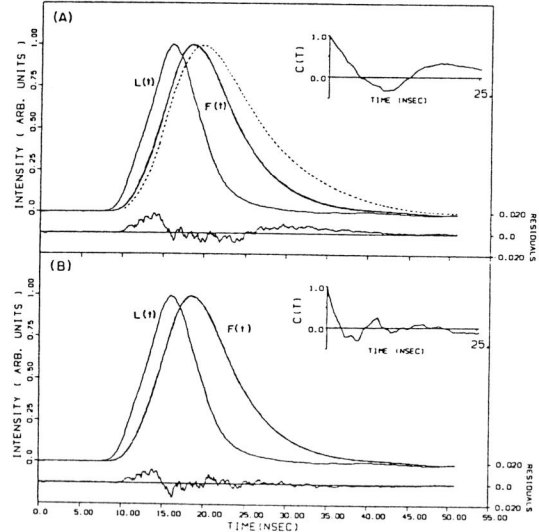

Figure 2. Fluorescence decay of the donor in (1-15)BPTI, in 0.05 M bicine buffer, pH 7.5, at room temperature. (A) Analysis assuming monoexponential decay. The distribution of the residuals and their autocorrelation function show a small deviation from monoexponential decay (RMS=0.0050). (B) The same decay curve shown in (A) was reanalyzed by using eqs. 2 and 3. The best fit, obtained with the parameters a=0.043 and b=30.5, is improved in comparison to the monoexponential fit shown in (A). RMS of fit equals

0.0035. The corresponding distribution function is shown in Figure 3. (---) Trace of donor fluorescence decay in the absence of an acceptor (τ=6.8 ns).

The donor fluorescence decay curves for each derivative in 0.05M Bicine pH 8.0 are analyzed with a least squares procedure (Fig. 2b) using Eq. (2) and a skewed Gaussians for $N_o(r)$:

$$N_o(r) = \sum_{i=1}^{2} 4\pi r^2 c_i \exp[-a_i(r-b_i)^2]. \qquad (3)$$

This function is flexible enough to adopt most experimental distributions, within the experimental noise, by variation of the two free parameters a_i and b_i. In Fig. 2b, note the improved fit, which is indicated by the randomness of the deviations, the low autocorrelation of the residuals and the lower RMS, which is 0.0035.

The average distances determined by the analysis of donor fluorescence decay curves are very similar to those determined from the crystal structure by X-ray crystallography (Wlodawer et al. 1984). The widths of the calculated interchromophoric distance distributions are obtained from the extent of deviation from monoexponential decay of the donor. This, in turn, includes contributions from experimental noise; segmental flexibility of the protein and mostly from the rotational freedom of probes, the side chain and the "arm" connecting it to the main chain. Based on these results, one can proceed using these derivatives for studying the unfolding of the protein.

Reduced Unfolded BPTI

BPTI is unfolded by reduction with 10-20 mM DTT in 6M GuHCl, 26% glycerol, 0.05M Bicin Buffer pH 8.0 (Amir and Haas 1988). Analysis of the donor fluorescence decay curves for the reduced unfolded derivatives under these conditions reveal the local changes. The average interprobe distance between residues 1 and 15 in the reduced unfolded state (in 6M GuHCl) is 31±4Å, shorter than that of the native state, with almost twice the width (Fig. 3). This result shows that the N-terminal segment of BPTI, residues 1 to 15, which is fixed at an extended conformation in the native state, relaxes upon unfolding, to a conformation similar to that of a random coil.

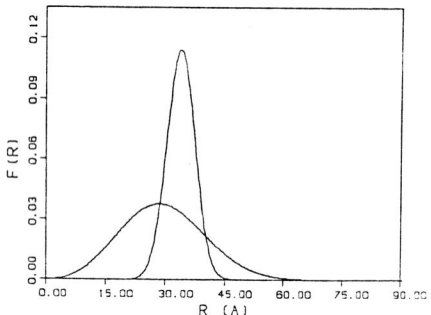

Figure 3. Distributions of distances between residues 1 and 15 in BPTI in the native state (narrow distribution) and in the reduced unfolded state (broad distribution).

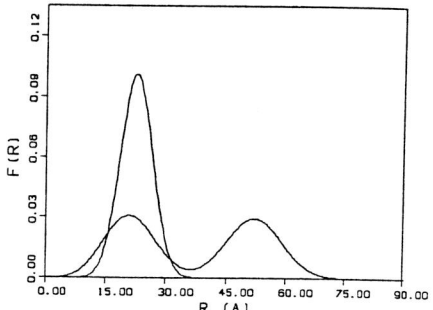

Figure 4. Distributions of distances between residues 1 and 26 in BPTI in the native (narrow distribution) and reduced unfolded state (bimodal distribution).

The distance between residues 1 and 26 in the reduced unfolded state is distributed in a bimodal form consisting of two subpopulations (Fig. 4). The best fit of the experimental decay curve is obtained when the first subpopulation, accounting for 50±20% of the overall population of molecules is averaged at 20±1.5Å, similar to that of the native state. The second subpopulation distance (50±20% of the molecules) assumed an average distance of 50±15Å and a large width of distribution, 20±10Å, which seems to be corresponding to fully unfolded random coil conformation.

Lower transfer efficiencies are found for all the derivatives of BPTI when they are reduced in 5.4M guanidium isothiocyanate (GuSCN). This shows that a more extended conformation is obtained i.e. the residual structures still present in 6M GuHCl are further unfolded by 5.4M GuSCN.

Reduced BPTI under folding conditions.

The intramolecular distances in the reduced unfolded BPTI show that in conditions considered as denaturing, The protein is not fully unfolded. GuHCl does not fully unfold reduced BPTI to the initial state of the folding pathway, the random coil conformation. A refolding experiment initiated with reduced BPTI unfolded in 6M GuHCl, actually starts at an intermediate state.

The next step in the investigation of the folding pathway is a gradual elimination of the unfolding agent and a search for the sequence of intermediate states. This is done first by dilution of the GuHCl while maintaining reducing conditions. Determination of intramolecular transfer efficiencies in reduced BPTI in a series of GuHCl concentrations, reveal a very broad transition. Yet the dependence of the distance between residues 1 and 26 on GuHCl concentration shows two distinct transitions, with, an intermediate state of reduced BPTI stable at 3.5 M GuHCl. Other derivatives do not reveal this intermediate state and hence it can be assigned to a local structure, probably associated with the β sheet and the N-terminal segment.

In very low CuHCl concentration, (.5 M GuHCl in 50 mM Bicine pH 7.5, 4mM DTT and 1mM EDTA), high excitation transfer efficiencies are observed. (Figure 5). Three features are shown in Figure 5: (a) In the low GuHCl concentration the excitation energy transfer efficiencies, E, are higher than those found for each derivative, in the reduced state in 6M GuHCl. (b) Except for R(1-15)BPTI, reduced BPTI has E values which are similar to those obtained in the native state. (c) Transfer effciency between residues 1 and 15 in the reduced state is higher than E found in this segment in both the native and the unfolded states.

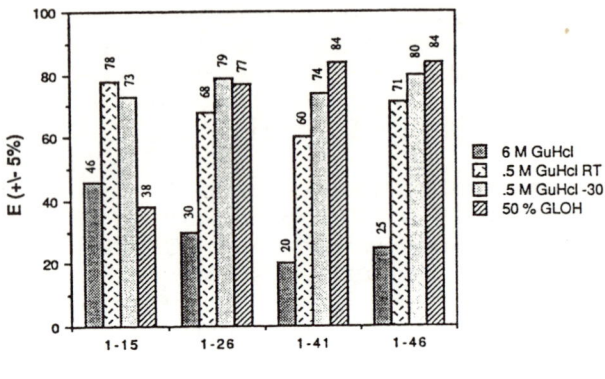

Figure 5. Transfer efficiencies, E, in BPTI derivatives in the native; reduced unfolded and reduced state. Four experiments are presented for each derivative. The conformational state and solvent in each column is as follows: first column (from left): 6.0 M GuHCl in 26% glycerol, .05 M bicine buffer pH 7.5, 10mM DTT, 1mM EDTA at -30°C. Second column: 0.5M GuHCl in 50% glycerol in .05M bicine buffer pH 7.5, 4mM DTT, 1mM EDTA at -30°C. Third column: 0.5M GuHCl in .05M bicine buffer pH 7.5, 4mM DTT, 1mM EDTA at room temperature. Fourth column: .05M bicine buffer pH 7.5 in 50% glycerol at -30°C. The percent transfer efficiencies calculated for each experiment are given on top of the corresponding column.

These changes in E are caused by changes of intramolecular distances in the labeled segments of the protein. These results show that even without formation of disulfide bonds, the average distances between residues separated by long segments of the backbone of BPTI approach those found in the native state; this indicating that the reduced molecule has a compact conformation. The efficiency of excitation energy transfer, E, is not correlated to the number of residues between the labeled sites; this indicates that in the reduced state BPTI is not in a random coil conformation. The high E observed for the pair of probes attached to residues 1 and 46 shows that in the reduced state the N-terminal segment of BPTI is close to the C-terminal segment. Figure 5 also shows that in the reduced state the distance between residues 1 and 15, is considerably shorter than in the native state, and this segment has a higher dynamic flexibility. Reduced BPTI thus seems to exist under folding conditions, in what has been characterized as a molten globule state (Dolgikh et al 1981) with some degree of local order. This conformational state has been observed in heat or acid unfolded α-lactalbumin (Dolgikh et al 1981) and cytochrome C (Ohgushi and Wada 1983). Here we find a similar state in the folding pathway of BPTI. This shows that this state may be a general functional intermediate in the folding pathway.

One consequence of the above conclusion is that the latter phases of the folding pathway are a sequence of rearrangements and formation of specific interactions taking place within a condensed volume element. This is probably an effective means of accelerating the rate of the folding as can also be found in Levitts' simulation studies (Levitt 1976).

The high flexibility of the 1-15 segment may contribute to the formation of the non-native intermediate disulfide bonds, 5-14 and 5-38 (Crieghton 1978). The flexibility of this segment may be instrumental for the folding pathway of BPTI. In the final phase of the pathway this segment is fixed in the native stretched conformation by the 14-38 disulfide bond.

Conclusions and Current Experiments

The localized intramolecular distances in reduced BPTI indicate some features of the initial phases of the pathway of folding of BPTI. At least three intermediate states in reduced BPTI, defined by structures confined to segments of the molecule are observed. A sequence of disulfide intermediates known from Creighton's work (Creighton 1978) constitute the next phase of the folding pathway. Long range interactions, (LRI), seem to be effective in early intermediates as indicated by the high transfer efficiencies between residues 1 and 46. The "message" coding for the interactions generating high probability of short distance between the two terminal segments are probably very effective in reduction of both the entropy of the unfolded state and its conformational space. Creighton has shown that in the first step of the pathway of disulfide bonds formation in BPTI, the 30-51 single disulfide intermediate is dominant (Creighton 1978). This can also be a result of other long range interactions between two secondary structures: α (47-56) and β (16-36). BPTI derivatives labeled at residues 26 and 58 or 46 are being prepared to probe this interaction. Short and medium range interactions stabilizing secondary structures like the β sheet, (residues 16 to 36), probably define the structural state of the intermediates and further reduce the complexity of the folding transition. The molten globule state could be another component in the coded pathway which facilitates the search for the native conformation by the combination of a condensed volume; partially stabilized, flexible local structures and fast dynamics. These may facilitate fast search for the folded conformation and rearrangements of abortive structures (local minima).

The above results and conclusions lead to the current and future experiments. New labeled derivatives targeted at subdomain structures are being prepared. One example is labeling of the C-terminal alanine-58, paired with labeling at lysine 46 which will probe the formation of the α helix. Combination of labeling either one of these residues with labeling at either residues 26 or 1, will let us probe the effect of long range interactions between the three main subdomain structures of BPTI at any phase of the pathway. Fragments of the protein are being labeled and the intrinsic stability of their structures and their dynamics are compared to the characteristics of the same segments labeled in the whole protein. Kinetic experiments in which the rate of change of each intramolecular distance is monitored also reveal the order of structure assembly in the folding protein. Fast mixing experiments show that the compact state of reduced BPTI is formed within less than 100 msec after dilution of the unfolding agent.

The experiments presented above show that our method is useful in addressing the challenge of determination of the structural characteristics of partially folded intermediate states of globular proteins. The method is characterized by time and distance resolutions, (10^{-10}-10^{-7} seconds and 10 to 80 Å respectively), most suitable to the framework of globular protein folding transitions. The possibilities for site directed labeling are increased by applications of the current technology of site directed mutagenesis for introducing new labeling sites. The use of external probes, site specifically attached to proteins, allows one to probe selected local structures in complex multicomponent systems, the rest of the complex being transparent to the fluorescence measurement.

The language in which the folding pathway is coded will eventually be deciphered by combination of many methods that will allow determination of the direct structural features instructed by native; modified or novel sequence messages. This will pave the way for design of proteins and their binding sites.

ACKNOWLEDGEMENTS

Supported in part by by US - ISRAEL binational science foundation (grant No. 212/86) and by National Institute of General Medical Sciences (grant No. GM39372-1).

ABBREVIATIONS

MNA: 2- methoxy-naphthyl-1- methyleneyl;
DA-coum: [7-(Dimethylamino-(-coumarin-4-yl]-acetyl; GuHCl: guanidinium hydrochloride; BPTI: Bovine pancreatic trypsin inhibitor; R-BPTI: reduced BPTI; (1-n)BPTI: N^{α}-MNA-arg^1-N^{ε}-DA-coum-lysn-BPTI; R(1-n)BPTI: reduced (1-n)BPTI; DTT: Dithiothriethol; DA-coum-BPTI: BPTI labelled by single DA-coum group; E: Transfer efficiency (percent); MD-lys: N^{α}-MNA-N^{ε}-DAcoum-lys.;

REFERENCES

Amir D, Haas E (1987). Biochemistry 26:2162-2175.
Amir, D, Haas E (1988). (Submitted).
Anfinsen CB, Haber E, Sela M, White EH Jr. (1961). Proc Natl Acad Sci USA 47:1309-1314.
Beechem J, Haas E (1988). (Submitted).
Creighton TE (1978) Prog Biophys Molec Biol 33:231-297.
Deisenhofer J, Steigmann W (1975). Acta Crystallogr Sect B: Struct Crystallogr Cryst Chem B31:238-250.
Dolgikh DA, Gilmanshin RI, Brazhnikov EV, Baychkova VE, Semisotnov GV, Venyaminov SYu, Ptitsyn OB (1981). FEBS Lett 136:311-315.
Forster Th (1948). Ann Physik 2:55-75.
Goldberg ME (1985). Trends in Biol Sci 388-391.
Haas E, Katchalski-Katzir E, Steinberg IZ (1978a). Biopolymers 17:
Haas E, Katchalski-Katzir E, Steinberg IZ (1978b). Biochemistry 17:5064-5070.
Haas E, Steinberg IZ (1984). Biophys J 46:429-437.

Haas E, McWherter CA, Scheraga HA (1988). Biopolymers 27:1-21.
Haas E (1986). In "Photophysical and Photochemical Tools in Polymer Science." (Winnik, M.A., Ed.) D. pp. 310-341, Reidel, Dordrecht.
Kim PS, Baldwin RL (1982). Ann Rev Biochem 51:459-489.
Levitt M (1978). Protein Folding as a Random Walk. (Personal communication).
Lotan N, Chen K, Roche RS (1974). Isr J Chem 12:207-218.
Matheson RR Jr., Scheraga HA (1978) Macromolecules 11:819.
Ohgushi M, Wada A (1983). FEBS Lett 164:21-24.
Ptitsyn OB (1973) Akad Nauk USSR Dokl Biophys 210:87.
Richardson JS (1985) Methods Enzymol 115:359-380.
Steinberg IZ (1971). Ann Rev Biochem 40:83-114.
Stryer L, Haugland RP (1967). Proc Natl Acad Sci USA 58:719-720.
Stryer L (1978). Ann Rev Biochem 47:819-846.
Wlodawer A, Walter J, Huber H, Sjolin L (1984). J Mol Biol 180:301-329.

Protein Adaptation to Extreme Salinity: The Crystal Structure of 2Fe-2S Ferredoxin from Halobacterium Marismortui

J. L. Sussman, M. Shoham and M. Harel

Department of Structural Chemistry
Weizmann Institute of Science, Rehovot,
76100 ISRAEL

INTRODUCTION

In the course of evolution different organisms have adapted to different environments, occasionally to very harsh ones. It is of great interest to explore the limits of biological adaptation to extreme temperature, pressure or salinity. Some understanding can be gained from the study of organisms which live in such inhospitable ecological niches where life seems almost impossible. These organisms have evolved special adaptation mechanisms to survive where other forms of life cannot exist. In an attempt to understand the requirements for maintaining life at high salinity on the molecular level, we undertook the study of proteins from halobacteria which live in the saltiest body of water found on earth, the Dead Sea (Steinhorn and Gat, 1983). Halophilic proteins are active at intracellular supersaturated salt concentrations (Eisenberg and Wachtel, 1987; Werber, Sussman, and Eisenberg, 1986), for example, the internal potassium concentration of Halobacterium marismortui is 4M (Werber, Mevarech, Leicht and Eisenberg, 1978). "Normal" proteins in non-halophilic cells would be precipitated out and cease to function under these conditions. What are the structural requirements for a protein to remain soluble and active at practically saturated salt solutions?

As a model for the adaptation process of a protein to an extremely saline environment, we have investigated the crystal structure of a 2Fe-2S ferredoxin from the halophilic archaebacterium Halobacterium marismortui (HmFd). Ferredoxin represents a good choice for studying haloadaptation because it is ubiquitous from prokaryotes to higher plants and because the three-dimensional structure of nonhalophilic ferredoxins with the same 2Fe-2S prosthetic group from the blue-green algae Spirulina platensis (SplFd) (K. Fukuyama et al., 1980; Tsukihara et al. 1981) and from Aphanothece sacrum (AsFd) (Tsukihara, et al., 1983) are known. Comparison of these ferredoxin structures highlights distinctive structural characteristics of the halophilic protein.

Some 50 2Fe-2S ferredoxins from different organisms have been sequenced (Tsukihara et al., 1982; Protein Identification Resource (PIR) 1986). They fall into two major categories, those from plants, algae and non-halophilic bacteria having about 98 residues, and those from halobacteria with 128 amino acids.

HmFd, a halobacterial 2Fe-2S ferredoxin, resembles the plant ferredoxins except that it is longer, with a molecular mass of 14,000 daltons, and richer in negatively charged residues (Hase, Wakabayashi, Matsubara, Mevarech, Werber, 1980; Kerscher, Oesterhelt, Cammack, Hall, 1976). HmFd has an extra 22-residue N-terminal and a 7-residue C-terminal region in comparison with its non-halophilic counterpart. Thirty four out of its 128 amino acid residues are either aspartic or glutamic acid, while on the other hand there are only six basic residues. The sequence alignment of HmFd (Hase, Wakabayashi, Matsubara, Mevarech, Werber, 1980 and SplFd (Matsubara, Wada, Masaki, 1976; Wada, Hase, Tokunaga, Matsubara, 1974) reveals extensive homology in the core parts, particularly in the vicinity of the four cysteine residues that comprise the ligands for the iron-sulfur cluster (Fig. 1). The homology gets progressively weaker farther from the active site in the direction of both termini (Sussman, Brown and Shoham, 1986).

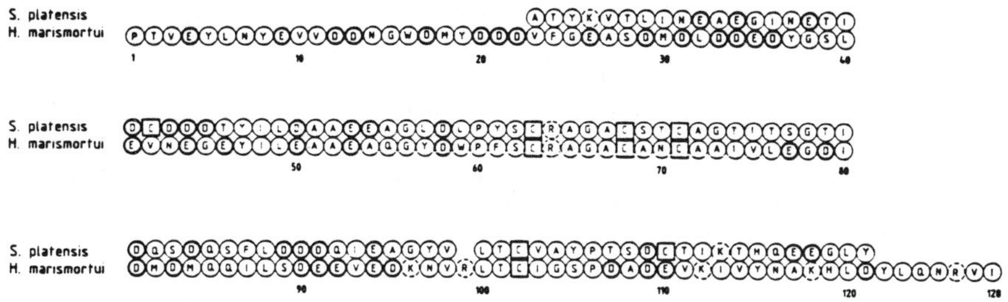

Figure 1. Amino acid sequence alignment of 2Fe-2S ferredoxins from a halophilic (Halobacterium marismortui) and a non-halophilic (Spirulina platensis) organism.

Here we report the crystal structure of a 2Fe-2S ferredoxin from Halobacterium marismortui at 2.5Å resolution. The most striking feature of the structure is the presence of a spatially distinct domain made up of regions from the amino- and carboxy termini. A possible function of this domain maybe to solvate the protein in the intracellular supersaturated salt solution.

STRUCTURE DETERMINATION

HmFd was isolated and purified according to Werber & Mevarech (Werber and Mevarech, 1978) and crystallized from 3.8 M phosphate buffer at pH 7.0 as reported previously (Sussman, Zipori, Harel, Yonath and Werber, 1979). Large hexagonal prisms, 1mm in cross section and 0.5mm in height, were grown by seeding over a period of several weeks. X-ray data were collected at room temperature on a

Nonius CAD4 diffractometer from native crystals and one suitable heavy-atom derivative, potassium tetracyanoplatinate to 3.3Å resolution. Recently another native dataset to 2.0Å resolution was measured at -150°C on a Rigaku AFC5-R rotating anode diffractometer, using a cryogenic technique developed by Hakon Hope (Hope, 1985). Additional phasing was provided by the anomalous scattering of the iron-sulfur cluster of native protein crystals. An anomalous difference Patterson and a difference Fourier synthesis using the Single Isomorphous Replacement (SIR) phases were used to locate the iron-sulfur cluster. SIR phases were combined with anomalous scattering phase information, using a method developed by Hendrickson et al. (Hendrickson, Smith and Sheriff, 1985) and yielded an electron-density map at 3.3Å resolution. The envelope of the protein could readily be outlined on this map. The polypeptide backbone was traced from stacked transparent sheets of this electron-density map. A protein model incorporating the known amino-acid sequence was fitted to the electron-density map using the computer program FRODO (Jones, 1978) on a Vector General computer graphics display. We have done a preliminary refinement of the model by the reciprocal space least-squares techniques CORELS (Sussman, Holbrook, Church and Kim, 1977) and PROFFT (Hendrickson and Konnert, 1980; Finzel, 1987) to an R-factor of 0.27 at 2.5Å resolution. Parameters of the crystallographic structure determination and refinement are summarized in Table 1. A representative portion of the resultant electron-density map is shown in Fig. 2. Other crystallographic computing was done with the software packages ROCKS (Reeke, 1984) and PROTEIN (Steigemann, 1974). The α-carbon coordinates have been deposited with the Protein Data Bank (Bernstein et al. 1977).

Molecular replacement attempts using the known structure of SplFd and AsFd in reciprocal space (Crowther, 1972) were unsuccessful. However, the similarity between these two homologous structures and HmFd could clearly be demonstrated **in real space** by an unequivocal positioning of the α-carbon skeleton of SplFd in the electron-density map at

3.3Å resolution. This was done by anchoring the 2Fe-2S cluster of the SplFd molecule onto the determined iron-sulfur position in the electron-density map and rotating the model about this position. The fact that a unique and clearly outstanding solution was found in this rotation search is a strong indication for the similarity of the two structures (Fig. 3).

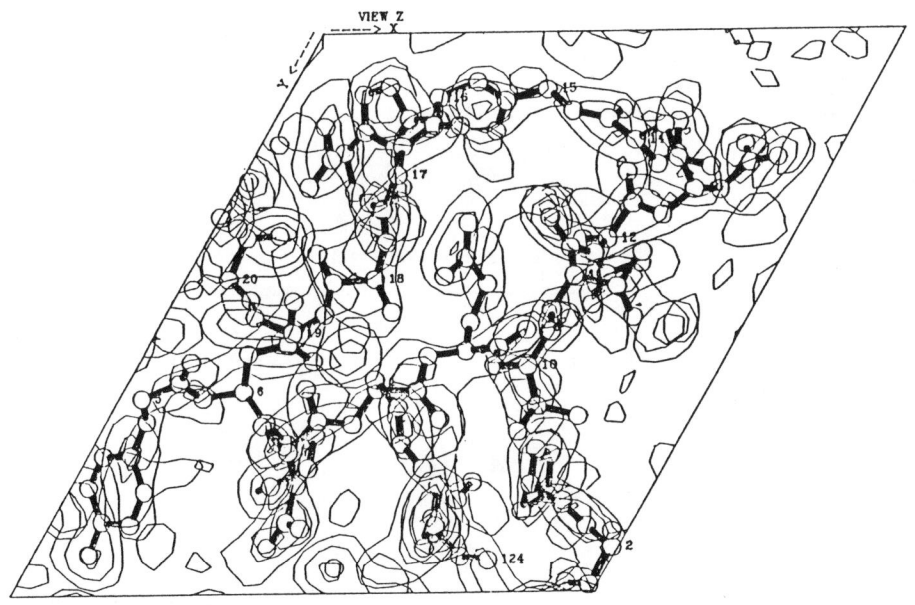

Figure 2. A 7Å thick portion of the 2Fo-Fc electron-density map at 2.5Å resolution of ferredoxin from Halobacterium marismortui as viewed down the unique z-axis. A ball and stick model of the corresponding part of the N-terminal solvation domain is superimposed on the map.

Table 1. Statistics for the structure determination of 2Fe-2S Ferredoxin from <u>Halobacterium</u> <u>marismortui</u>: phase determination and refinement.

			Single Isomorphous Replacement				
Heavy atom	Soaking time	Resolution	No. of unique reflections	R_{iso}	No. of sites		R_{cen}
$K_2Pt(CN)_4$	3 days	3.3Å	2127	.13	3		.46

Anomalous Scattering of native Crystals & Phase Combination

R_{ano}	number of acentric reflections	FOM_{ANO}	FOM_{SIR}	FOM_{COMB}
.04	1513	0.25	0.43	0.52

Refinement at 2.5Å

$R = \Sigma\|\|Fo\|-\|Fc\|\|/\Sigma\|Fo\|$	0.274
No. of reflections (6.0 to 2.5Å)	4483 (>3σ)
No. of atoms	1011
No. of Solvent atoms	32
No. of parameters (including individual B's)	4173
Average |Fo|	26.2 e
Average ||Fo|-|Fc||	6.5 e
Average B factor	11.9Å²
rms deviation bond length	0.04Å (0.030)*
rms deviation bond angle distance	0.09Å (0.040)
rms deviation from planar group	0.10Å (0.050)
rms deviation from trans peptide	5.0° (10.0)

Table 1. Statistics for the structure determination of 2Fe-2S Ferredoxin from Halobacterium marismortui: phase determination and refinement (continued).

FOM_{ANO} — Figure of merit due to anomalous scattering of native crystals and partial structure of the iron atoms.

FOM_{SIR} — Figure of merit due to single heavy atom derivative.

FOM_{COMB} — Combined figure of merit of ANO and SIR.

$$R_{ISO} = \frac{\sum_h ||F_{PH}obs| - |F_Pobs||}{\sum_h |F_Pobs|}$$

$$R_{CEN} = \frac{\sum_{h\ centric} ||F_{PH}obs| - |F_Pobs + F_Hcalc||}{\sum_{h\ centric} ||F_{PH}obs| - |F_Pobs||}$$

$$R_{ANO} = \frac{\sum_{h\ acentric} |F_{hkl} - F_{\bar{h}\bar{k}\bar{l}}|}{\sum_{h\ acentric} (F_{hkl} + F_{\bar{h}\bar{k}\bar{l}})/2}$$

F_P — Native protein structure factor.
F_{PH} — Heavy atom derivative structure factor.
obs — Observed.
calc — Calculated.
* — The numbers in parentheses represent the values to which these parameters were restrained.

MOLECULAR CONFORMATION OF 2Fe-2S FERREDOXIN

The core of the protein, residues 37-119, consists of a 10-stranded antiparallel β-barrel with a short α-helix at residues 48-55 (Fig. 4). The β-strands wrap around the barrel roughly perpendicular to the barrel axis. At the bottom of the barrel there is an external loop containing three of the four cysteine residues that serve as ligands to the iron sulfur cluster in the active site. The fold of the polypeptide chain around the iron-sulfur cluster is similar to that of SplFd and so is the tetrahedral coordination of the cluster. Cysteines 63 and 68 are bonded to one iron atom and cysteines 71 and 102 to the other.

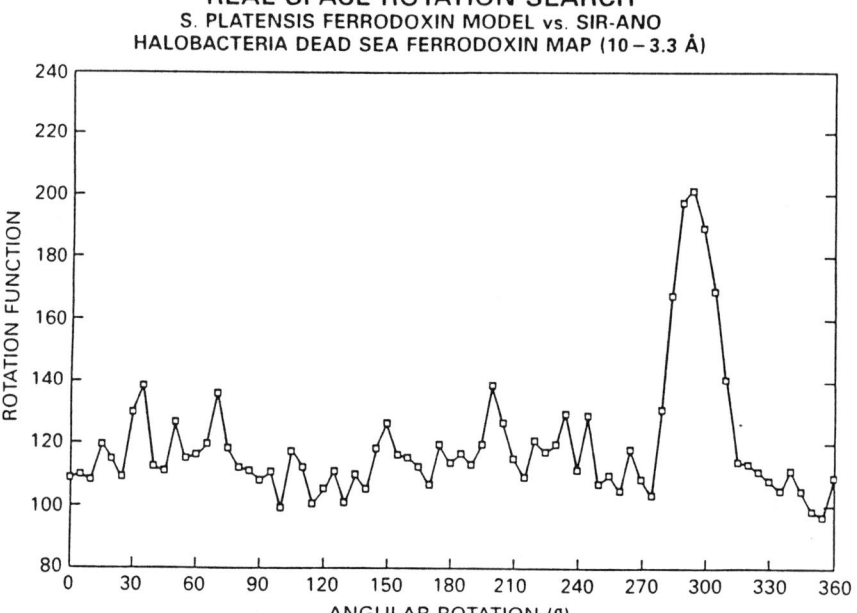

Figure 3. Real space rotation search (Steigemann, 1974) of the Spirulina platensis ferredoxin model on the Halobacterium marismortui 3.3Å resolution electron-density map. The map was calculated with phases based on a single isomorphous derivative (SIR), combined with anomalous scattering phases (ANO) inherent in the 2Fe-2S cluster. The highest peak in the rotation search in the $\theta 1, \theta 2$ plane is plotted as a function of the third rotation angle $\theta 3$. The single solution is an indication for the structural similarity of the two proteins.

A separate smaller domain, consisting of chain segments from the N- and C-termini of the molecule (residues 1-30 and 120-128) is attached to the β-barrel. The existence of such a domain extending out from the main body of the molecule is consistent with low angle X-ray scattering data of HmFd in solution (Eisenberg and Wachtel, 1987). It is rich in aspartic and glutamic acid residues and part of it forms an Ω-loop (Leszcynski and Rose, 1986). Residues 1-8 form an extended chain which packs against the wall of the β-barrel. Residues 8 through 18 form an Ω-loop with a distance of 4.5Å between α-carbon atoms 8 and 18 at the neck of the loop. In addition to hydrogen bonds at the neck and side chain interactions in its interior, extra stabilization of the Ω-loop is provided by stacking interactions between tyrosines 8 and 19 whose rings are roughly parallel at a distance of 3.4Å. Tryptophan 16 is part of this loop and is located on the surface of the protein and points towards the iron-sulfur cluster. The distance from the plane of the indole ring to the proximal iron atom is about 8Å. It is conceivable that Trp 16 serves as a mediator in the electron transport to and from the iron-sulfur cluster, possibly via Cys 68 which is located in between. Such a mechanism is consistent with the finding that the fluorescence of one tryptophan moiety in HmFd is quenched by the iron atoms (Gafni and Werber, 1979). This tryptophan is also present in 2Fe-2S ferredoxin from Halobacterium halobium (T. Hase et al. 1977). Aromatic residues have also been implicated in electron transfer of ferredoxin from Peptococcus aerogenes (Adman, Sieker and Jensen, 1972; Carter, Kraut, Freer and Alden, 1974).

SIMILARITY TO 2Fe-2S FERREDOXIN FROM SPIRULINA PLATENSIS

The folding of the core part of the protein is rather similar to that of the related SplFd, including the active site fold (Fig. 5). Conserved residues between these two proteins are mainly confined to the region in and around the active site (Fig. 1). Alignment of the sequence of 28 different 2Fe-2S ferredoxins indicates the same trend (Tsukihara et al., 1982). The common fold of HmFd and SplFd is confined to residues 37-119. In the remainder of the polypeptide chain, residues 1-36 and 120-128, there is no clear relationship between the structures of these two proteins.

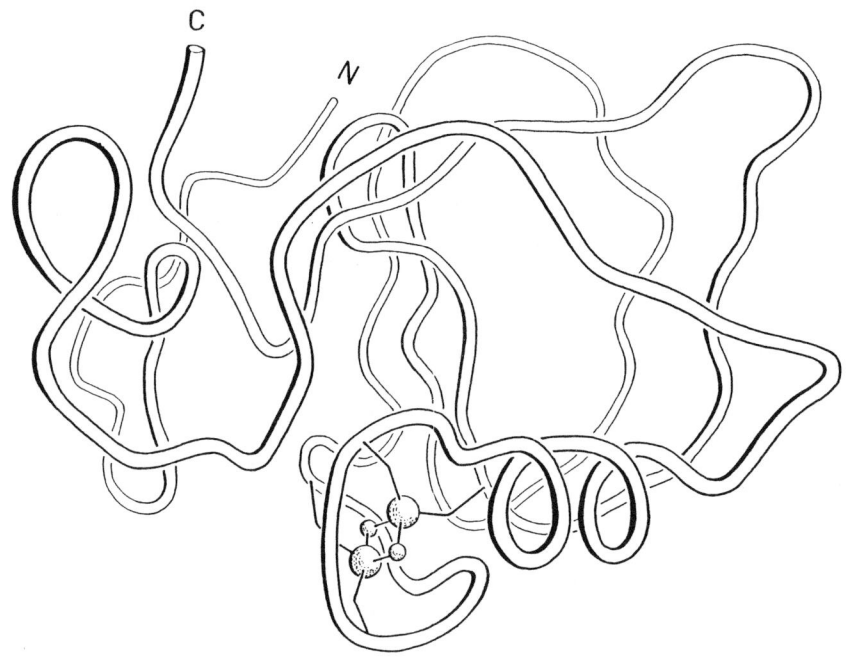

Figure 4. The fold of 2Fe-2S ferredoxin from Halobacterium marismortui. The 2Fe-2S cluster is shown as circles and its bonds to CYS 63, 68, 71 and 102 as solid lines. Residues 1-30 and 120-128, shown on the left-hand side of the diagram, form a distinct domain, rich in Asp and Glu, whose apparent function is to solvate the protein in concentrated salt solution.

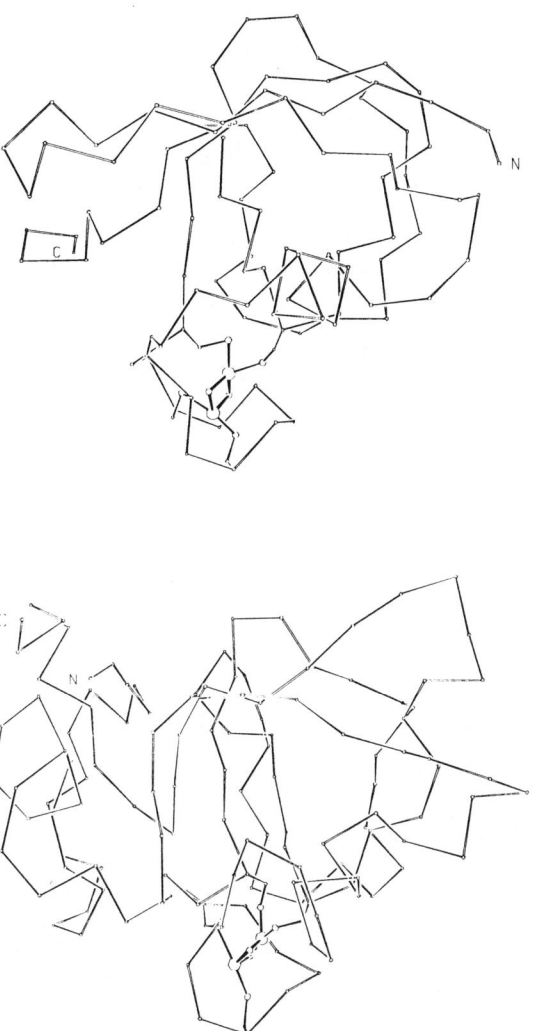

Figure 5. Schematic α-carbon drawings of 2Fe-2S ferredoxin from **(top)** Spirulina platensis (Tsukihara et al. 1981) and **(bottom)** Halobacterium marismortui in the same orientation. The depicted view yields emphasizes the unique handle-like structure in the halophilic ferredoxin.

CHARGE DISTRIBUTION ON THE SURFACE OF THE PROTEIN

The numerous carboxyl residues are distributed on the surface of the protein with concentrations in three regions: the extra domain, the hinge region in between the two domains (residues 31-36) and the solvent exposed wall of the β-barrel. The latter forms an amphipathic β structure in which some of the carboxyls appear at every other residue, i.e. Glu 77, Asp 79, Asp 81 and Asp 83 all pointing outward from the surface of the protein. Likewise, the single short α-helix has carboxyl groups at Glu 50 and Glu 53, both pointing roughly in the same direction. Interestingly, the active site region is entirely devoid of negative charges, although the iron-sulfur cluster is located close to the surface of the protein. Thus it would not be surprising if this region plays a role in a noncharged interaction with another protein.

ADAPTATION TO HIGH SALT

What does the adaptation to high salt environment involve? A possible mechanism of adaptation is the enhancement of the ability of the protein to bind sufficient amounts of water necessary for stability and function. Free water is scarce in the Dead Sea as well as inside the halobacterial cell. Therefore, halophilic proteins need residues capable of binding many water molecules. Consequently, a general characteristic of most of these proteins is the preponderance of aspartic and glutamic acid residues, which have the highest hydration capacity among the naturally occurring amino acids, with each carboxyl group capable of binding from 6-8 water molecules (Kuntz, 1971).

The adaptation of ferredoxin to high salinity might be achieved by the added extra domain. This domain, including the hinge region (1-36, 120-128), contains 36% acidic residues and may act as a water binding "handle". Indeed HmFd is soluble in very high ionic strength solutions. While the non-halophilic SplFd which is also quite acidic (~20%) was crystallized in 87% saturated ammonium sulfate

(Tsukihara et al. 1981) the HmFd crystals dissolve in 95% saturated solution.

We postulate that the function of the extra domain may be to solvate the protein in the intracellular supersaturated salt solution. This carboxyl-rich domain may provide the necessary water-binding capacity to allow for the solvation of the protein in an aqueous phase where free water is scarce in much the same way as a hydrophobic leader sequence causes solvation of a protein in a lipid bilayer.

Rao and Argos (Rao and Argos, 1981) made structural predictions about HmFd based on the sequence alignment with SplFd and its crystal structure. They assumed the two structures to have a common fold and correctly predicted most of the extra acidic residues in the core region to be located at the surface of the β-barrel or the α-helix. They also suggested that the additional N- and C-terminal segments form a structural fold protecting the barrel face which is uncharged in SplFd. However, rather than shielding parts of the protein, the small domain appears to provide extra solvation capacity for the entire molecule. Rao and Argos predicted an internal salt bridge between Asp 29 and Lys 112 in HmFd. These residues are in fact very far apart (~27Å), however there is an internal salt bridge between Asp 36 and Lys 118. NMR measurements of HmFd in solution indicate the single histidine in the molecule, His 119, to be located on the surface of the protein (Gochin and Degani, 1985). The crystal structure proves this to be correct. His 119 serves as ligand to the tetracyanoplatinate derivative together with Met 30, Asp 17 and Ser 28.

In view of the many carboxyl groups on the surface of the protein it is extremely interesting to explore the water structure around the protein. With the recently measured 2Å low temperature (-150°C) X-ray data set it is likely that we shall be able to observe more than one hydration layer. At the current stage of refinement, 2.5Å resolution, 32 solvent molecules have been located (Table

1), and further refinement is in progress.

Does the solvation domain represent a general phenomenon of haloadaptation? Some halophilic proteins are larger than their non-halophilic counterparts of known three-dimensional structure: malate dehydrogenase from Halobacterium marismortui (hMDH) and from porcine heart have molecular weights of 87,000 and 72,000 daltons, respectively (Mevarech, Eisenberg and Neumann, 1977; Pundak and Eisenberg 1981; Birktoft, Fernley, Bradshaw and Banaszak, 1982). Thus hMDH may possess an extra solvation domain while other halophilic proteins, which are no larger than their non-halophilic counterparts, like glutamate dehydrogenase or dihydrofolate reductase (Eisenberg and Tomkins, 1968; Moon and Smith, 1973), may lack it. The need for an extra solvation domain in HmFd may be a consequence of the covalently bound iron-sulfur cluster which confers extra rigidity on the molecule. In fact, removal of the iron-sulfur cluster causes the collapse of the structure into a random coil (Gochin and Degani, 1985). It is conceivable that in such a tightly restrained fold a more massive replacement of residues by aspartic or glutamic acid cannot be accommodated. Other proteins may be more flexible and may tolerate more replacements of nonacidic by acidic residues, thus having no need for an extra solvation domain. Nonetheless, the formation of a distinct solvation domain is apparently one way nature has chosen for the adaptation of a protein to extreme salinity.

ACKNOWLEDGEMENTS

We would like to thank Henryk Eisenberg, Ada Yonath, Hakon Hope, Felix Frolow & Jerry Brown for help during various stages of this work, Wolfie Traub, Jan Drenth, David Davies, Jim Mathew, Pat Weber, Barry Finzel, Mark Saper, Moshe Mevarech, Moshe Werber & Leemor Joshua-Tor for helpful and stimulating discussions, Janet Smith, Wayne Hendrickson, Wolfgang Steigemann, Gerson Cohen, George Reeke, Barry Finzel & Alwyn Jones for making their computer programs available to us, Yaacov Halfon,

Yaacov Shabtai & Gabriella Idan for skillful technical help and Haya Sussman for drawing the tube representation of the polypeptide fold (Figure 4). This work was supported by the U.S. Army Research Office - contract no. DAJA45-86-C-0016 (through its European Research Office) and the United States-Israel Binational Science Foundation (BSF), Jerusalem, Israel.

REFERENCES

Adman ET, Sieker LC, Jensen LH (1972). J. Biol. Chem. 248: 3987-3996.

Bernstein FC et al. (1977). J. Mol. Biol. 112: 535.

Crowther RA (1972). in The Molecular Replacement Method, (Int. Sci. Rev. Ser., 13:), M.G. Rossmann, Ed. (Gordon and Breach, New York), p.10.

Birktoft JJ, Fernley RT, Bradshaw RA, Banaszak LJ (1982). Proc. Natl. Acad. Sci. U.S.A. 79: 6166.

Carter CW Jr., Kraut J, Freer ST, Alden RA (1974). 249: 6339-6346.

Eisenberg H, Wachtel EJ (1987). Ann. Rev. Biophys. Biophys. Chem. 16: 69-92.

Eisenberg H, Tomkins GM (1968). J. Mol. Biol. 31: 37-49.

Finzel BC (1987). J. Appl. Cryst. 20: 53-55.

Fukuyama K et al. (1980). Nature 286: 522-524.

Tsukihara T et al. (1981). J. Biochem.(Tokyo) 90: 1763-1773.

Gochin M, Degani H (1985). J. Inorg. Biochem. 25: 151-161.

Hase T, et al. (1977). FEBS Letters 77: 308-310.

Hase T, Wakabayashi S, Matsubara H, Mevarech M,

Werber M, (1980). Biochem. Biophys. Acta 623: 139-145 (1980).

Hendrickson WA, Konnert JH (1980). in Computing in Crystallography, R. Diamond, S. Ramaseshan, K. Venkatesan, Eds. (Indian Academy of Sciences, Bangalore) pp. 13.01-13.25.

Hendrickson WA, Smith JL, Sheriff, S (1985). Methods in Enzymology 115: 41-55.

Hope H, (1985). in Amer. Cryst. Assoc., Abstracts Ser. 2, 13: abstract PA3.

Jones TA (1978). J. Appl. Cryst. 11: 268-272.

Kerscher L, Oesterhelt D, Cammack R, Hall, DO (1976). Eur. J. Biochem. 71: 101-108.

Kuntz ID (1971). J. Am. Chem. Soc. 93: 514-516.

Leszcynski JF, Rose GD (1986). Science 234: 849-855.

Matsubara H, Wada K, Masaki R, (1976). in Iron and Copper Proteins, K.T. Yasunobu, H.F. Mower, O. Hayaishi, eds. (Plenum Press, New York and London), pp. 1-15.

Mevarech M, Eisenberg H, Neumann E (1981). Biochemistry 16: 3781-3785

Pundak S, Eisenberg H (1981). Eur. J. Biochem. 118: 463-470.

Rao JKM, Argos P (1981). Biochemistry 20: 6536-6543.

Reeke GN Jr. (1984). J. Appl. Cryst. 17: 125-130.
Steigemann W (1974). Ph.D. thesis, Technische Universitaet Muenchen.

Steinhorn I, Gat JR (1983). Scientific American 249: 102-109.

Sussman JL, Holbrook SR, Church GM, Kim SH (1977). Acta Cryst. A33: 800-804.

Sussman JL, Brown J, Shoham M, (1986). in Iron-Sulfur Protein Research, H. Matsubara, Y. Katsube, K. Wade, eds. (Japan Sci. Soc. Press, Tokyo/Springer-Verlag, Berlin) pp. 69-82.

Sussman, JL, Zipori P, Harel, M, Yonath, A, Werber, MM (1979). J. Mol. Biol. 134: 375-377.

Tsukihara T, et al., (1982) BioSystems 15: 243-257.

Tsukihara T, Mizushima M, Fukuyama K, Katsube Y, (1983). in Int. Summer School on Crystallographic Computing II, Kyoto, Japan, pp. 63-64.

Protein Identification Resource (1986). (PIR) sequence data bank, The National Biomedical Research Foundation, Washington, D.C., release 10.0.

Wada K, Hase T, Tokunaga H, Matsubara H (1974). FEBS Lett. 55: 102-104.

Werber MM, Mevarech M (1978). Arch. Biochem. Biophys. 185: 447-456.

Werber MM, Mevarech M, Leicht W, Eisenberg H, (1978). Energetics and Structure of Halophilic Microorganisms, S.R. Caplan and M. Ginzburg, Eds. Elsevier/North-Holland Biomedical Press, 427-443.

Werber MM, Sussman JL, Eisenberg H (1986). FEMS Microbiol. Rev. 39: 129-135.

PARTIAL MOLAL VOLUME AND PHARMACODYNAMIC ACTIVITY

Sasson Cohen, Zipora Brif, *Frank Haberman, *Zvi Liron

Department of Physiology and Pharmacology,
Tel Aviv University, Tel Aviv 69978 and
*Institute for Biological Research, Ness Ziona
70450, Israel

INTRODUCTION

The use of the size and shape of drug molecules as a guide to the structure of their specific binding sites may have begun in 1857 when Claude Bernard, referring to curare, offered the following view: "...comme des espèces d'instruments physiologiques plus délicats que nos moyens mécaniques et destinés à disséquer, pour ainsi dire, une à une, les propriétés des éléments anatomiques de l'organisme vivant", (Holmstedt and Liljestrand, 1963). Ever since, this contention has evolved into a number of impressive methodologies, ranging from the familiar semi-empirical SAR, to the more sophisticated quantum-mechanical studies of conformation and charge distribution. In the long run, a given binding site has become to be characterized through its goodness-of-fit to a given set of ligands, expressed quantitatively as association or dissociation equilibrium constants. Molecular pharmacology may have come of age, fulfilling the anticipation of Gaddum (1942).

In a more critical view, the concept of molecular complementarity between binding site and drug molecule has its limitations. A persisting reminder is the disparity in structure between an agonistic molecule, say acetylcholine, and an antagonistic one, say hyoscine, both being assumed to bind to the same site, yet produce opposite effects. For such a pair, drug action cannot be measured by the same scale. Hence, the evidence of affinity becomes unconclusive, unless the binding site could alternate between two mutually-exclusive corresponding structures. Current theories of drug-receptor interaction do make a provision for such a model (Thron, 1973; Furchgott, 1978). But then admission of flexibility in the binding site implies fluidity to an extent that it may accomodate either of two extreme structures and perhaps intermediate ones as well. If such is indeed the case, drug-receptor interaction becomes analogous to a dissolution process, hence it may be approached as an extension of solution theory. The present article is concerned with the application of this rather uncommon approach, giving a brief overview of recent and ongoing work by the authors.

THEORETICAL CONSIDERATIONS

In a regular solution (Hildebrand, Pausnitz and Scott, 1970; Barton, 1983) of solute j at infinite dilution in solvent i, the excess free energy of the solute may be expressed as follows:

$$RT \ln{}^{j}f_{i}^{\infty} = {}^{j}V^{\circ}({}^{i}\delta - {}^{j}\delta)^{2} \qquad (1)$$

$$= {}^{j}V^{\circ}({}^{i}\delta^{2} + {}^{j}\delta^{2} - 2K{}^{i}\delta{}^{j}\delta) \qquad (2)$$

where δ represents the solubility parameter or square root of the cohesive energy density of i and j correspondingly, and ${}^{j}V^{\circ}$ is the molal volume of the solute. In regular solution, the interaction coefficient $K = 1$, hence the excess free energy term is always positive. In non-regular solution which is the more common case, $K \neq 1$. Application of equation (2) for complex molecules is not straightforward, because of the apparent inconstancy of either K or ${}^{j}\delta$ or both (Martin et al., 1985, Liron et al., 1986). The alternative course is to derive the excess free energy term from the difference between the partial molal volume of j in solvent i and its molal volume in a "standard" state:

$$RT \ln{}^{j}f_{i}^{\infty} = {}^{i}P_{i}({}^{j}V_{i} - {}^{j}V^{\circ}) \qquad (3)$$

where ${}^{i}P_{i}$ ($\approx {}^{i}\delta^{2}$) is the internal pressure of the solvent, ${}^{j}V_{i}$ and ${}^{j}V^{\circ}$ are the partial molal volume at infinite dilution and molal volume of j. Use of equation (3) does not necessarily imply strict regular behavior, since the value of the excess free energy term, positive or negative, will now depend on the measurable parameters ${}^{j}V_{i}$ and ${}^{j}V^{\circ}$. Partial molal volumes at infinite dilution of solute can be determined by means of high-precision densitometry, to within 0.1% or less (Liron and Cohen, 1983).Its major deficiency is that P_i values cannot be always approximated as $^i\delta^2$, especially in the case of highly polar solvents and that the value of $^jV°$ for most solids, and which is the molal volume of the pure solute as supercooled liquid, cannot be precisely determined. Still, these deficiencies may be of minor consequence as will be shown in the following sections. Equation (3) may be further refined by incorporating the Florey-Huggins correction for unequal molar size of solute and solvent:

$$RT\ln\,^jf_i^\infty = {}^iP_i(^jV_i - {}^jV°) + RT[(\ln\,^jV°/^iV) + (1 - {}^jV°/^iV)] \quad (4)$$

Various experimental findings relating to drug action will now be considered in the light of one or more of the above relationships.

DIFFERENTIAL SOLUBILITIES IN THE BIOPHASE

Some fluorinated ethers such as $(CF_3CH_2)_2O$ are potent convulsants; others such as $CHCl_2CF_2-O-CH_3$ are potent anesthetics. This apparently paradoxical effect could not be related to structure which lacks specificity, or rationalized in terms of the oil/water partition coefficient. When ten such agents were arranged in order of increasing δ over a range extending from 6.5 to 8.5 $(cal.cm^{-3})^{1/2}$, the most potent convulsant had the lowest δ value, the most potent anesthetic had the highest (Cohen et al, 1975). Equation (1) or its more evolved forms (Srebrenik and Cohen, 1976) predict that the differential distribution of an individual agent in a hypothetical biophase will be governed by the cohesive energy density of the biophase. Indeed, it could be shown that glutamate-induced conductance is preferentially depressed by high δ agents, while GABA-induced conductance is preferentially depressed by low δ agents (Richter et al.,1977). This finding led to the proposition that the specificity of drug action is also a consequence of the cohesive energy density of the drug and the holding phase (Landau et al., 1979). It has remained to be seen whether this approach could be applied to interpret the action of agents endowed with more specific structures.

PARTIAL MOLAL VOLUMES OF COMPLEX MOLECULES

Determination of the partial molal volumes at infinite dilution of a series of drugs known to interact at the muscarinic cholinergic receptor indicated that $^jV_i^\infty$ changed in accordance with some function of the cohesive energy density of the holding phase, fulfilling the anticipation of equation (3). Sample plots of $^jV_i^\infty$ against $^i\delta$ for a muscarinic agonist and a muscarinic antagonist, shown in fig. 1 (panels A, B), fail to reveal any particular trend, but do uphold the premise that $^jV^\circ$ is not necessarily the lowest recorded value. Thus, the molal volume of the "standard" state could be arbitrarily ascribed to any of the values found experimentally, the lowest volume being the one which is more convenient to use. In the present case, $^jV^\circ$ is represented by the partial molal volume in acetonitrile solution. Application of equation (4) permits now to calculate the excess free energy of a given molecule in a given solvent over its free energy in acetonitrile solution. A plot of RT ln $^jf_i^\infty$ against the molar attraction constant ($\delta V = (UV)^{1/2}$) of the respective solvent gives now a more rational trend (fig.1, panels C,D). The maximal difference in energy levels for 3-acetoxyquinuclidine is about 850 cal.mol^{-1}, a value consistent with the difference calculated between the energy levels corresponding to the two major conformers of this molecule (Weinstein et al., 1975). For the more rigid and bulkier 3-benziloxyquinuclidine (QNB), this difference is of the order of 2100 cal.mol^{-1}; for hyoscine, it is about 2000 cal.mol^{-1}. For these and each of twelve additional molecules, the dependence of RT ln $^jf_i^\infty$ on δV follows a complex cubic function with correlation coefficients of 0.99-0.95. The immediate implication of this finding is that the excess free energy of the solute over the "standard" state is a function of δV, rather than one of each of these parameters separately. For either 3-acetoxyquinuclidine and 3-benziloxyquinuclidine, RT ln f approaches zero at δV = 900-950 (cal.cm^3)$^{1/2}$, irrespective of their size difference. If a similar effect were to occur at the binding site as the holding phase of either one of two molecules of unequal size, then one can think of a situation where an increase in V will entail a decrease in δ, or a decrease in

Molal Volumes in Pharmacodynamics / 193

A - 3-acetoxyquinuclidine   B - 3-benziloxyquinuclidine

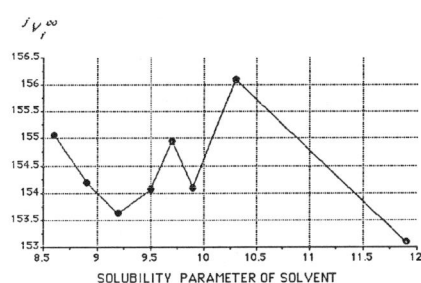 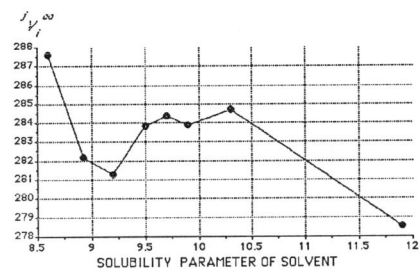

PARTIAL MOLAL VOLUME CHANGE AS A FUNCTION OF δ OF THE HOLDING SOLVENT

C - 3-acetoxyquinuclidine   B - 3-benziloxyquinuclidine

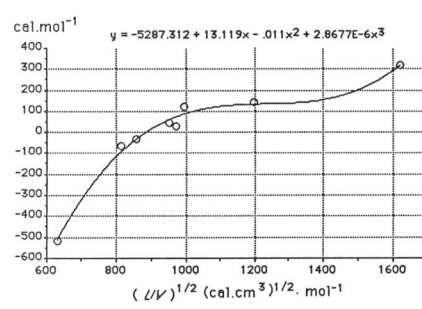 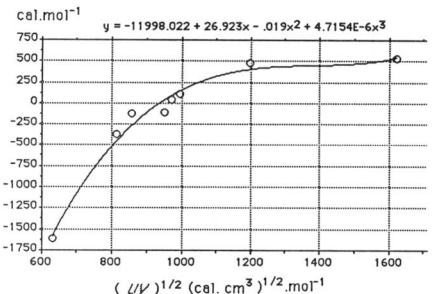

EXCESS FREE ENERGY (RELATIVE) AS A FUNCTION OF THE MOLAR ATTRACTION CONSTANT OF THE HOLDING SOLVENT

Figure 1. Interpretation of partial molal volume change (A,B) of two different solutes in terms of excess free energy (C,D), by application of equation (4).

V will entail an increase in δ, without much change in the overall value of their product. This is said with the reservation that such a process will occur only within certain permissible limits. What are these limits and in what way could the binding site be approached as a "solvent" for its ligand ?

ENTHALPY-ENTROPY RELATIONSHIP IN DRUG-RECEPTOR INTERACTION

An early clue to the current problem came with the finding that jV_i^∞ for atropine is significantly higher than the corresponding value for hyoscine. This finding, first reported by Barlow and Winter (1981), then confirmed by ourselves in eight different solvents (Cohen and Haberman, 1984), is not self-evident, because hyoscine has the greater molecular weight of the two. The two molecules share a common skeleton, interact with the same receptor site, with hyoscine showing a higher affinity. The following working hypothesis is now offered: (a) The molal volume of the binding site is equal to the the partial molal volume of hyoscine, determined in a suitable solvent, say acetonitrile; (b) ΔH in the binding of hyoscine is zero; ΔH in the binding of atropine must exceed ΔH for hyoscine (zero), in accordance with equation (3) above, as follows:

$$\Delta\Delta H = i\delta^2 (V^{atr} - V^{hyo}) \qquad (5)$$

For acetonitrile solution ($i\delta^2$ = 144 cal.mol^{-1}; V^{atr} = 247, V^{hyos} = 243 cm^3.mol^{-1}), $\Delta\Delta H$= 586 cal.mol^{-1}. From a knowledge of experimental ΔG values for atropine and hyoscine and assuming that the cohesive energy density of the receptor site is equal to that of acetonitrile, one may write:

for hyoscine, -13.5 (ΔG) = 0 (ΔH) - 13.5 (TΔS) kcal.mol^{-1}
for atropine, -12.8 (ΔG) =0.6(ΔH) - 13.4 (TΔS) kcal.mol^{-1}

Thus, the fit of the bulkier atropine to the receptor site requires investment of 0.6 kcal.mol^{-1} in excess of the process with hyoscine and which must be exacted from the free energy of binding.
Exactly the same procedure was applied to derive the

excess enthalpy (ΔΔH) over hyoscine (equation 5) for 13 additional drug molecules comprising agonists, partial agonists and antagonists of the the guinea-pig ileum muscarinic cholinoceptor (Cohen and Haberman, 1985). The list has been recently expanded to include acetylcholine chloride, methacholine chloride, carbachol, bethanechol, butyrylcholine chloride and McN-A343. From a knowledge of ΔG for each of these agents with respect to the common cholinoceptor, TΔS has been calculated (fig.2), then ΔS determined (T= 310°K). The data can be interpreted as follows:

For all antagonists, the increase in entropy is always greater than the increase in enthalpy. However, if the increase in enthalpy is not compensated by an equivalent increase in entropy, then the excess enthalpy change over entropy must come at the expense of the free energy of binding, hence relative affinity. The choice of hyoscine to represent the molal volume of the binding site is not arbitrary. Among all the antagonist known, hyoscine has the highest ratio of ΔG/molal volume, meaning the highest binding capacity per unit volume (Table 1).

Table 1. Partial molal volume ($cm^3.mol^{-1}$) of agonists and antagonists of the muscarinic acetylcholine receptor, at infinite dilution in acetonitrile

AGONISTS	jV_i^∞	$V/\Delta G$	ANTAGONISTS	jV_i^∞	$V/\Delta G$
Acetylcholine Cl	131.4	15.5	Benactyzine	293.9	26
Arecoline	147.4	16.0	Benztropine	279.2	23
Methacholine Cl	148.5	19.3	QNB	278.5	20,18
3-Ac-Q	153.1	19, 17	3-DiPh-Ac-Q	276.9	21
3-Pr-Q	171.4	30	Imipramine	266.8	28
Pilocarpine	172.4	24, 21	Nortriptyline	248.0	26
Pilocarpine HCl	175.7	23	Atropine	246.8	19
3-Bu-Q	188.9	24	Hyoscine	242.3	18
Oxotremorine	190.1	22, 17			
McN-A-343	221.3	33	AF-14	213.5	26

Abbreviations: Q=quinuclidine; Ac=acetoxy; Pr=propionoxy; Bu=butyroxy; QNB=3-benziloxyquinuclidine. $V/\Delta G$ is in units of cm^3/cal.

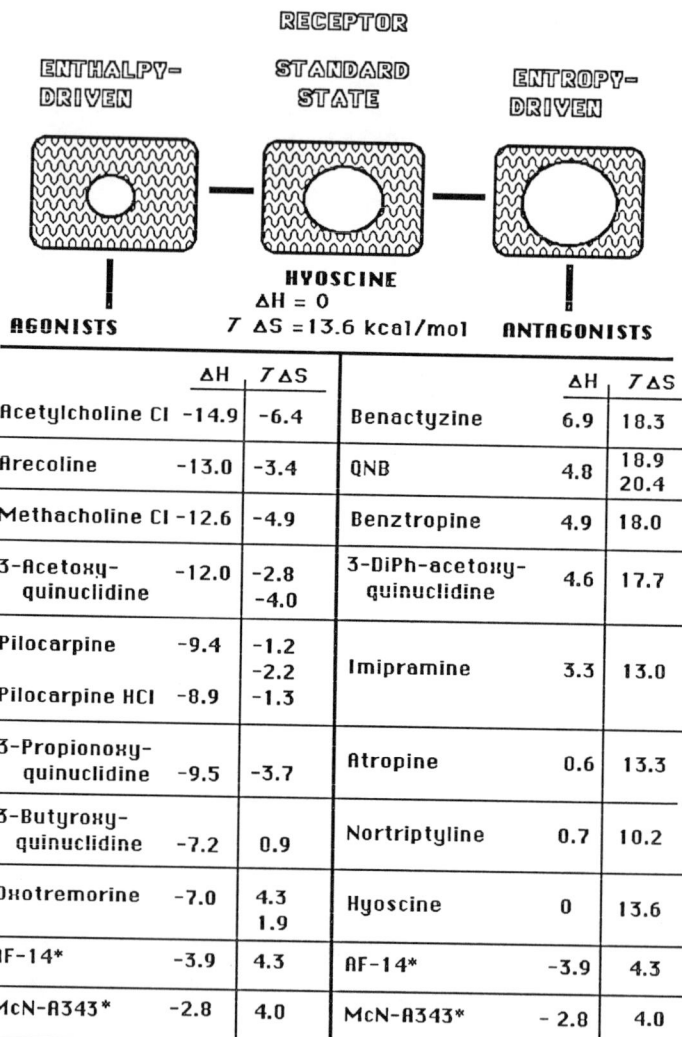

Figure 2. A physical interpretation of the thermodynamics of the binding of agonists and antagonists to a common cholinoceptor.

For all the full agonists, there is a decrease in entropy and in enthalpy, but the decrease in enthalpy outweighs the decrease in entropy. This difference between the two trends becomes gradually smaller as one reaches the partial agonists where the entropy change becomes positive, finally outweighing the small decrease in enthalpy found for this class of drugs.

The isokinetic plot of ΔH against ΔS (T=310 °K) (fig. 3) gives a more generalized view of the thermodynamics of the interaction of the cholinoceptor with the drug molecules under consideration. All antagonists fall in the region of positive ΔS and ΔH. All the agonists fall in the region of negative ΔS and ΔH. Those molecules with a mixed agonist-antagonist effect fall in the region of positive ΔS and negative ΔH. It is remarkable that oxotremorine, long considered to be a powerful agonist, is now relocated among the mixed agonists-antagonists. Recent evidence shows that this molecule may indeed exert a dual effect (Kilbinger, 1984).

The present findings are consistent with a physical model of fluidity in the structure of the binding site. Anatagonist molecules larger than hyoscine impose on this site an expansion in volume which must be coupled with the absorption of energy. The case of QNB is of particular interest since the entropy change in this case more than compensates for the enthalpy change, resulting in a permanent change in the receptor site, observed as isomerization to a higher affinity state (Galper et al., 1977). By the same token, agonistic molecules may produce contraction of the receptor site which must be coupled with the release of energy, hence their action is enthalpy driven rather than entropy driven as in the case of the antagonists. However, the enthalpy driven effect of the agonists may also have a different physical interpretation involving the participation of water molecules in the transition of the receptor to the active state.

A 'WATER-ICE' TRANSITION IN THE ACTION OF AGONISTS ?

The negative enthalpy change in the action of agonists could be ascribed entirely to a transition of water molecules associated with the receptor from the liquid to the crystal state:

$$\{R + n(H_2O)\} + \text{agonist} \rightleftharpoons \{R^*-(H_2O)_n\text{-agonist}\} \quad (6)$$

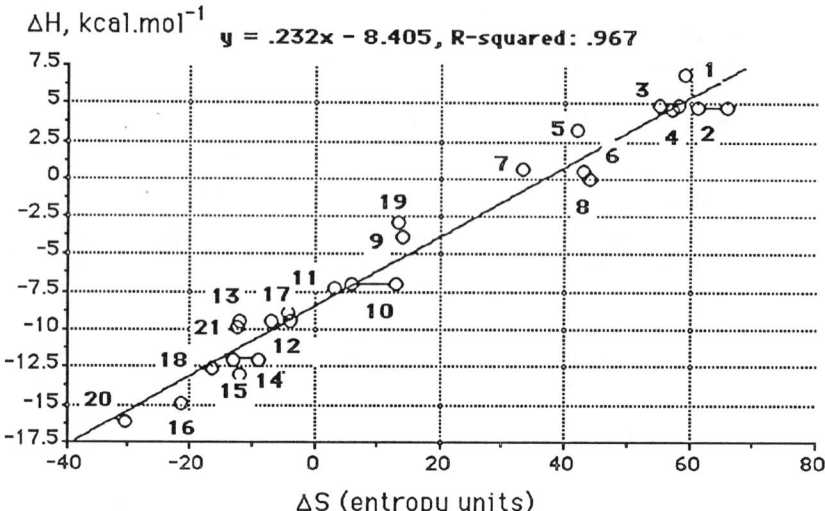

Figure 3. Isokinetic plot of ΔH against ΔS. Key: 1 Benactyzine; 2 3-benziloxyquinuclidine; 3 benztropine; 4 3-quinuclidyl-diphenylacetate; 5 imipramine; 6 atropine; 7 nortriptyline; 8 hyoscine; 9 AF-14; 10 oxotremorine; 11 3-butyroxyquinuclidine; 12 pilocarpine; 13 3-propionoxyquinuclidine; 14 3-acetoxyquinuclidine; 15 arecoline; 16 acetylcholine Cl; 17 pilocarpine HCl; 18 methacholine Cl; 19 McN-A343; 20 carbachol Cl; 21 butyrylcholine Cl

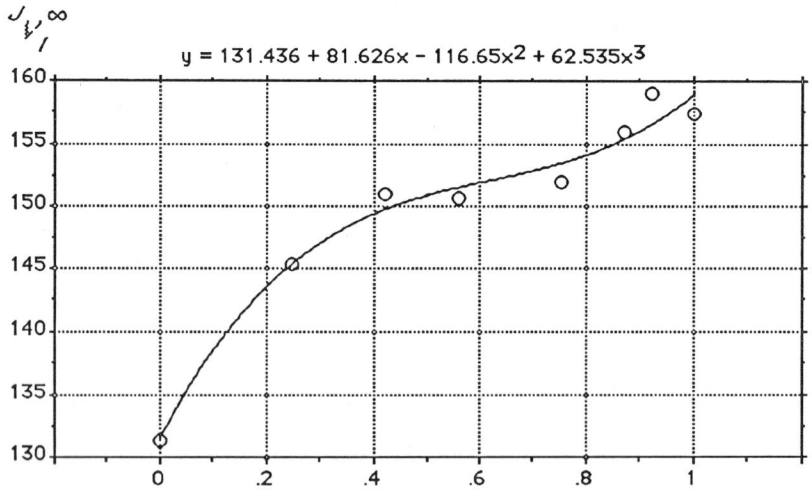

Figure 4. Partial molal volume of acetylcholine chloride at infinite dilution in mixed water-acetonitrile solvents

In the case of acetylcholine, ΔH is -14.9 kcal.mol-1, which if ascribed entirely to the transition of water molecules from the liquid to the solid state, must involve about 10 mols of water per mol acetylcholine. The volume change associated with such a transition is of the order of 2 $cm^3.mol^{-1}$ or 20 cm^3 per mol acetylcholine. In other words, the role of the agonist consists in shifting the equilibrium between the liquid-solid states of 'receptor' water in favor of the solid state.

If this contention has some truth in it, it must be shown that agonists do induce crystal formation in water solution and that their capacity to induce this effect is in direct relation with their corresponding ΔH values, or some parameter relating to their potency. The first evidence for such an effect comes from the measurement of the partial molal volume of acetylcholine chloride in mixed water-acetonitrile solutions (fig.4). At a volume fraction of water of 0.1, there is a rapid rise in the partial molal volume of acetylcholine chloride from 131.4 to 145.4 $cm^3.mol^{-1}$, $\Delta V=14$ $cm^3.mol^{-1}$. Since acetylcholine chloride is far more soluble in water than in acetonitrile, the observed expansion is ascribed to 'iceberg' formation involving about 7 mols of water per mol acetylcholine chloride at this particular composition of the mixture and $\Delta H \approx -11$ $kcal.mol^{-1}$.

The corresponding data for methacholine chloride, carbachol, bethanechol, pilocarpine hydrochloride and hyoscine hydrobromide are given in Table 2. It is remarkable that both methacholine chloride and bethanechol chloride which are methyl derivatives of acetylcholine and carbachol respectively seem to be more effective in this particular context than their non-methylated congeners. This finding suggests a role for conformation in crystal formation. Also, carbachol is less effective than acetylcholine, perhaps owing to its capacity for hydrogen-bond formation at the $-NH_2$ group. Pilocarpine hydrochloride is the least effective of the agonists investigated but surprisingly, McN-A343 seems to be as effective as bethanechol. Finally, the effect of hyoscine hydrobromide is negligible.

In general terms, the above data are consistent with a hypothesis ascribing to water an integral role in the agonist-receptor interaction. Considered at the two extremes of the full agonist (acetylcholine) and the full antagonist (hyoscine), the case for 'iceberg' formation seems to be well borne out. Considered as indicative of the rank order of activity of the individual members, the data are less satisfactory.

Table 2. Partial molal volume of some molecules related to acetylcholine

Compound	$^jV_i^\infty$, cm^3.mol^{-1}		ΔV	n(H$_2$O)	$-\Delta H$ kcal.mol^{-1}
	CH$_3$CN	CH$_3$CN/H$_2$O 90/10			
Acetylcholine chloride	131.4	145.4	14	7	11
Methacholine chloride	148.5	164.2	16	8	11.5
Carbachol chloride	122*	129.3	7	3-4	4.3-5.8
Bethanechol chloride	137*	150.5	13	6-7	8.6-10
Pilocarpine hydrochloride	175.7	179.9	7	3-4	4.3-5.8
McN-A343 chloride	221.3	234.3	13	6-7	8.6-10
Hyoscine hydrobromide	290.0	290.2	0	0	0

* estimated by extrapolating to water mass fraction=0

One reason for this apparent deficiency is the failure to account for the entropy change associated with the enthalpy change presently calculated from water crystallization alone. This is particularly true for McN-A343 which appears to be more active than anticipated from its thermodynamic profile, or carbachol which appears less active than anticipated from its pharmacological profile. Altogether, one is led to the conclusion that the major difference between the two states of the receptor site, one accomodating the antagonist and being inactive, the other accomodating the agonist and becoming active may be sought not only in the dimensions of the respective states, but also in their content of structured water. One is tempted to view the activating process simply in the terms conveyed by equation (6), with the agonist mediating the process between receptor and water as a catalyst would do.

REFERENCES

Barlow RB, Winter EA (1981). Affinities of the protonated and non-protonated forms of hyoscine and hyoscine N-oxide for muscarinic receptors of the guinea-pig ileum and a comparison of their size in solution with that of atopine. Br J Pharmacol 72: 657-664.
Barton AFM (1983). Handbook of Solubility Parameters and Other Cohesion Parameters. CRC Press, Boca Raton, Fl.
Cohen S, Goldschmid A, Shtacher G, Srebrenik S, Gitter S (1975). The inhalation convulsants: A pharmacodynamic approach. Mol Pharmacol 11: 379-385.
Furchgott RF (1978). Pharmacological characterization of receptors: its relation to radio-ligand binding studies. Fed Proc 37: 115-120.
Gaddum JH (1942). Edinburgh Med J 49: 731.
Hildebrand JH, Pausnitz JM, Scott RL (1970). Regular and Related Solutions. Van Nostrand-Reinhold, New York.
Holmstedt B, Liljestrand G (1963). Readings in Pharmacology. Pergamon, London.
Kilbinger H (1984). Presynaptic muscarine acetylcholine receptors modulating acetylcholine release. TIPS 5: 103-105.
Landau EM, Richter J, Cohen S (1979). Differential solubilities in subregions of the membrane: a non-steric mechanism of drug specificity. J Med Chem 22: 325-327.
Liron Z, Cohen S (1983). Densitometric determination of the solubility parameter and molal volume of compounds of medicinal relevance. J Pharm Sci 72: 499-504.
Liron Z, Srebrenik S, Martin A, Cohen S (1986). Theoretical derivation of solute-solvent interaction parameter in binary solution: Case of the deviation from Raoult's law. J Pharm Sci 75: 463-468.
Martin A, Wu PL, Liron Z, Cohen S (1985). Dependence of solute solubility parameters on solvent polarity. J Pharm Sci 74: 638-642.
Richter J, Landau EM, Cohen S (1977). Anaesthetic and convulsant ethers act on different sites at the crab neuromuscular junction in vitro. Nature 266: 70-71.
Srebrenik S, Cohen S (1976). Theoretical derivation of partition coefficient from solubility parameters. J Phys Chem 80: 996-999.
Thron CD (1973). On the analysis of pharmacological experiments in term of an allosteric receptor model. Mol Pharmacol 9: 1-19.

POLYMORPHISM IN DRUG DESIGN AND DELIVERY

Joel Bernstein

Department of Chemistry, Ben-Gurion University of the Negev, Beer Sheva 84105, Israel

INTRODUCTION

The complete development of a new drug, from conception of the medical need to a commercially marketed substance, is a process which involves a very substantial investment in time and money. Current estimates suggest that about ten years and $100 million may be required for any totally new material. In the drug development process, an enormous variety of chemical, biological, physiological and mechanical techniques and technologies are called upon for the thorough analysis of the structure and properties of both the molecular moiety in question and its activity in various formulations. Clearly, any means of streamlining this long and expensive process can represent a considerable saving in time and money.

In this volume the emphasis is on one of the most exciting developments in the area of drug design: interfacing computers and humans, with the ultimate aim of understanding the molecular nature of drug action by studying receptor-ligand interactions. Progress in this area will most certainly lead more quickly to the design of target molecules for clinical testing and thus considerably shorten the time and reduce the guesswork involved in the early stages of the development process.

In view of the interests of the participants in the conference which formed the basis of this volume, it is not surprising that considerably less emphasis is placed on the more technological, but no less important aspects of drug formulation and development of delivery systems.

Having thus divided the drug development process into two major stages we will want to show below that there is at least one common connection which may be utilized to facilitate progress in both stages; namely crystalline polymorphism in substances with potential pharmaceutical applications. We will start out with some definitions, followed by a summary of the importance of polymorphism in the formulation stage, which in turn will be followed by a review of the relevance of polymorphism to progress of drug design on the molecular level.

Polymorphism is defined as the ability of an element or compound to crystallize in more than one distinct crystalline species. Inclusion of solvent in the crystalline solid leads to *solvates*, and in the special case of water as a solvent, *hydrates*. Solvates may lose solvent by heating, or even simply upon standing. If the original crystal structure is maintained then the relationship between the solvate and the desolvated material is denoted as *pseudopolymorphism*; if desolvation is accompanied by a transformation to another crystalline form then the relationship is one of *polymorphic solvates*. These have been discussed in greater detail elsewhere, including a number of examples (Byrn, 1982).

The above definitions relate to differences in the spacial arrangement of molecules between different crystal forms. Additional of types of polymorphism relate to variations in molecular structure among polymorphs. *Conformational polymorphism* refers to the existence of molelcules of differing molecular conformation in the different crystal structures (Corradini, 1973; Panagiotoupoulis *et al*, 1974; Bernstein and Hagler, 1978; Bernstein, 1987). When different molecular configurations exist in different crystal structures, the phenomenon is described as *configurational polymorphism* (Byrn, 1982).

Since the properties of any solid depend upon its structure, it is clear that the properties polymorphic forms of a single drug material may be expected to vary from one form to another. In terms of the physical properties of the material, variations may be found in the habit (external crystal shape), in the melting point, solubility and dissolution behavior, in the tabletting behavior, in the thermal and/or photostability and shelf-life, suspension properties, and even in the physiological absorption rate. In general, these properties relate to the whole process of drug delivery to a receptor site, rather than the ligand-receptor interaction. However, since the vast majority of the detailed molecular geometric information used to study and model these systems is derived from x-ray crystal structure analyses, the existence of conformational polymorphs indicates that basing such studies on a single molecular conformation derived from a particular crystal structure involves an approximation which may not always be valid. Hence, there is clearly a need for an awareness of the possible existence of polymorphism and its implications in the drug development process.

At this point it is probably natural to inquire as to how widespread the phenomenon of polymorphism really is. In one of the early studies of the occurrence of polymorphism, Kuhnert-Brandstatter (1965) surveyed the available data on three widely used classes of drugs: the steroids, sulfonamides and barbituates. Of 126 samples 57% were found to be polymorphic; moreover, of the marketed samples 17% were found to be unstable polymorphic forms. In spite of this and increasing evidence for the widespread nature of polymorphism in pharmaceuticals, the bodies entrusted with creating and maintaining standards for drug preparations have been somewhat slow in recognizing its importance for inclusions in manuals and pharmacopoeia. For instance, of all the substances in the Pharmacopoeia Europa of 1980, only four substances were specifically noted as being polymorphic, while nearly a third of the materials were polymorphic and another third exhibited solvated crystalline forms.

In fact, in what has almost become a classic declaration of the ubiquity of polymorphic materials, Walter McCrone wrote a quarter of a century ago, "...every compound has different polymorphic forms and the number of forms known for a given compound is proportional to the time and energy spent in research on that compound." (McCrone, 1963).

A few examples are worthy of note here. The listing of the coenzyme of vitamin B_{12} in the Tenth Edition of the Merck Index indicates that it is marketed under eighteen different trade names as a hematopoietic vitamin. A 1972 patent issued to the Yamanouchi Pharmaceutical Company describes this material as existing as 'conventional crystals" and as two hydrates, specified A and B. In separate experiments the three forms were exposed direct sunlight for three hours and to a temperature of 85°C for one hour. The conventional crystals and Form B both showed approximately 25% and 70% degradation in the two tests, while Form A showed virtually no degradation in sunlight and less than 30% degradation upon heating, and would clearly be the preferred form in a commercial preparation (See also Haleblian, 1975). The entry in the Merck Index does make note of the sensitivity of the substance to light, but does not specify whether the crystalline material is a hydrate nor is there any mention of the existence of more than one hydrated form.

Another example is that of sulfapyridine, one of the early antibacterial sulfonamide drugs. The polymorphic behavior of the material has been investigated periodically for forty years by a number of groups, with accumulated evidence for the existence of at least five, and perhaps six polymorphic forms (Castle and Witt, 1946; Mesley and Houghton, 1967; Kuhnert-Brandstatter and Wunsch, 1969; Yang and Guillory, 1972; Gouda *et al*, 1977; Burger *et al*, 1980). The melting points of the various forms have been reported (see Bar and Bernstein, 1985, for a summary of these data) and they fall in the range 167-196°C (including solvates). Yet, somewhat surprisingly, the material is still a melting point standard in the United States Pharmacopoeia.

Gouda *et al* (1967) have measured the dissolution rates of three of the crystalline forms of sulfapyridine, as well as the amorphous form (Figure 1). The rates vary over a considerable range (approximately a factor of two), and such a variation could considerably influence the therapeutic effect of the material in its end-use form.

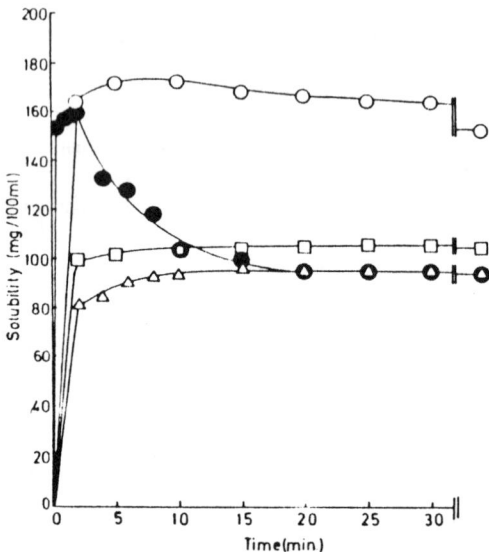

Figure 1. Dissolution rates of sulfapyridine crystal forms. △ Form I, • Form II, □ Form V, ○ Amorphous Form (reprinted, with permission, from Gouda, et al, 1977).

The dissolution of the solid form of a pharmaceutical preparation is usually considered to be the important initial step in the often complex process of physiological absorption. Hence the choice of the most suitable polymorph for effective drug administration can often be made on the bases of *in vitro* studies such as that summarized in Figure 1. However, there are a number of documented cases in which the evidence strongly suggests that a direct reaction between the solid drug preparation and, say, an enzyme is the initial step in physiological absorption. Perhaps the classic case in this regard is that of chloramphenicol palmitate, CAPP.

Chloramphenicol is a broad spectrum antibiotic and antirickettsial which is marketed as an oral suspension of the tasteless 3-palmitate due to the exceedingly bitter taste of the parent material. The early physical and physiological studies on this material were summarized by Aguiar *et al* (1967). There are three polymorphic crystalline forms (A, B, C) in addition to an amorphous form. The characterization of the various forms by the standard melting point and IR analyses was less than conclusive, and in fact, often problematic, due to polymorphic transitions during grinding for sample preparation (Borka and Backe-Hansen, 1968). In their IR studies Aguiar *et al* found only two different spectra, one of which was specific for polymorph A, while the other shows minor variation for any of the "non-A" phases, i.e. the spectra of phases B and C are indistinguishable. Since only the B and amorphous forms are biologically active (*vide infra*) several pharmacopeias required revisions in the IR analyses of these materials, generally allowing a maximum of only 10% of the more stable, but less active A form in the final preparation.

Aguiar *et al* also determined the physiological absorption rate as a function of the ratio of A and B polymorphs, Figure 2. The suspesion containing only the metastable B form gives higher blood levels following oral doses than thsoe containing only Form A, by nearly an order of magnitude. Since particle size was shown to have little effect on blood levels, it was concluded that the structure of the solid plays an intimate role in determining the physiological absorption rate.

The mechanism of this absorption was also investigated in considerable detail. CAPP is nearly insoluble in water; hence it must by hydrolyzed by enzymes in the small intestine before absoprtion can take place. According to one possible proposed mechanism (Aguiar *et al*, 1967) the first and rate deter-

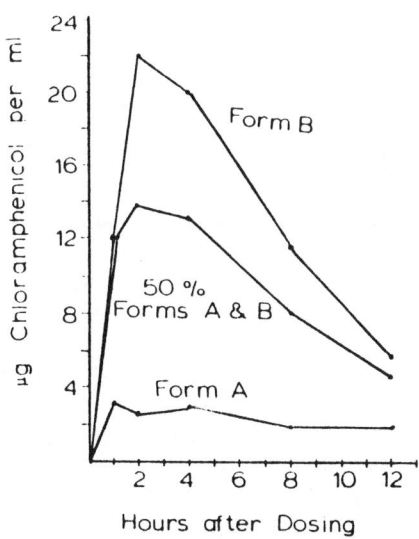

Figure 2. Peak blood serum levels of chloramphenicol two hours following dosing for pure polymorphs A and B and a 50:50 mixture. After Aguiar *et al*, 1967, with permission).

mining step in the total process is a dissolution of the ester followed by enzymatic hydrolysis of CAPP. However, a second mechanism, proposed by Andesgaard *et al* (1974) is that *solid* CAPP is enzymatically attacked in the small intestine. If dissolution is the first and rate determining step of the total process, then there should be a close relationship between the rates of dissolution and the rates of enzymatic hydrolysis of the polymorphs A and B. On the other hand, no relationship of this sort is expected if CAPP is attacked in the undissolved state.

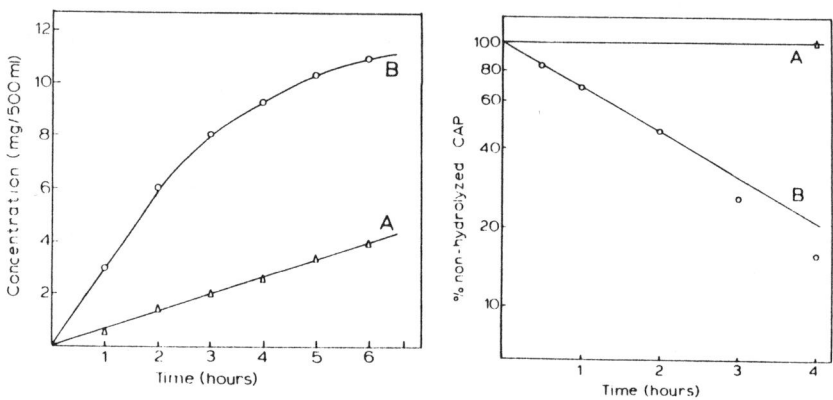

Figure 3. (a) Rates of dissolution of polymorphs A and B of CAPP. (b) Rates of *in vitro* enzymatic hydrolysis by pacreatic lipase of polymorphs A and B of CAPP (After Andesgaard *et al*, 1965, with permission)

The rates of dissolution on the one hand and *in vitro* hydrolysis of the solid by the enzyme pancreatin on the other hand, are given in Figure 3. If dissolution is the first step in the total hydrolysis process, the reaction scheme may be written as

undissolved CAPP ⇒ dissolved CAPP ⇒ hydrolyzed CAPP

Since the rate of the second step of this process must be the same for forms A and B of CAPP, this hypothesis leads to the conclusion that any differences in the rate of formation of hydrolyzed CAPP must be due to a difference in the rate of dissolution of the two polymorphs. The results presented in Figure 3 are not compatible with the assumption that dissolution is the first and rate determining step, since the slopes at time zero are significantly different from those expected for such a mechanism. Rather, Andesgaard *et al* claim that it is more reasonable to assume that CAPP is attacked in an undissolved state, probably by pancreatic lipase, which is known to act on substances insoluble in water (Waki, 1970). Further studies on CAPP and the analogous stearate by Cameroni *et al* (1976) apparently corroborate this assumption.

While this apparently enzymatic hydrolysis of a solid may not specifically be a receptor-ligand interaction in the purest sense, it represents the same kind of challenge for computer-assisted modelling that characterizes many of the other subjects touched upon in this volume. Some of the information required for such a study is avaliable from the crystal structure of the A form (Eguchi and Iitaka, 1974; Szulzewsky *et al,* 1981), and a rather detailed characterization of the polymorphic behavior of the material, including a structure analysis of Form C (Szulzewsky *et al*, 1982). Both these latter authors and Cameroni *et al* (1976) suggest that the difference in crystal structures is accompanied by differences in the molecular conformation within the crystal (i.e. conformational polymorphism), and this may be an important factor in the relative physiological activity of the various polymorphs. Attempts to probe these questions would involve, among other things, modelling studies of the enzyme interactions at the surface of the crystal, with of course variations in parts of the molecule accessible to the enzyme depending on the packing, the crystal face in question, and the conformation of the molecule. While this represents a considerable challenge for such studies, the presence of known or in principle determinable substrate geometric arrangements with respect to an enzyme do restrict some of the degrees of freedom and may lead to more detailed understanding of such processes.

This brings us naturally back to the subject of conformational polymorphism itself. In the case of CAPP it was suggested as a factor in determining the bioavailability of the drug. In the framework of receptor-ligand interactions and the general scheme of drug design the phenomenon can be utilized in a much broader way.

The vast majority of the currently available geometric information which is employed in setting up models to study, for instance, receptor-ligand interactions, is based on data taken from X-ray crystal structure analyses. The extracted data may involve some simple characteristic parameters, such as bond lengths and angles, 'building block' components such as amino acids or nucleotides, or whole molecules, from small substrates or ligands to proteins or viruses. Whatever the particular use or even degree of precision of the structure, one basic assumption is virtually always implied: namely that the geometric information is transferable from the crystalline state in which is was determined to whatever state is being assumed for the system under study. For the most part such an assumption has proven correct; without it, virtually no progress could have been made, and the progress to date has indeed been impressive.

However, the proliferation of crystal structure analyses in the past two decades has provided the opportunity to pursue the study of the constancy and/or variablity of various geometric parameters (see, for instance, Allen *et al*, 1983; Taylor *et al, 1983;* Ashida *et al*, 1987; Elder *et al,* 1984). The increased availability of structural data has in some ways been a double-edged sword. On the one hand, it has led to greater confidence in the crystallographically derived geometric data, but has also led to a greater awareness of the statistical nature of that data and the possibility of variations from 'normal' or expected values. When the latter are encountered they have often been attributed to the influence of 'crystal forces' with scant additional discussion. In fact, those 'crystal forces' are the same types of intermolecular interactions which are of crucial importance in determining the ways in which receptors and ligands interact. In a crystal a particular molecule occupies a site whose geometry and energetics are determined by the collection of molecules surrounding that site. For a ligand the geometry and energetics are determined in general by a specific part of a much larger molecule. To understand the interaction between the individual molecule and its surroundings in a crystal or a ligand and a receptor, what we want to understand is then, the geometry of the site and the energetics connected with it. Our general lack of information about the structure of receptor sites, of course complicates this aspect of the task. With crystals, we are at a distinct advantage since we have detailed geometric information about the surroundings of any particular molecule, and the aim of the remainder of this chapter is to outline how this information may be put to use in investigating and understanding intermolecular interactions by studying conformational polymorphs.

It is almost axiomatic that in comparing any number of systems, the more features the systems have in common the easier it is to extract those which distinguish among them. As noted above, the two main considerations in attempting to understand the environment of a molecule are geometry and energy. The existence of conformational polymorphism by its very nature provides two or more different, but very well defined, geometric and energetic environments. Yet simply because the crystal structures are built up from the same chemical moiety they have a great deal in common, and a considerable amount of information may be obtained from studies of such systems. We will examine briefly the energetic aspects of conformational polymorphism, follow that with a few examples relevant to the subject of this volume, and then conclude with an outline of the steps and procedures for utilizing the phenomenon to obtain information on molecule...molelcule interactions which could prove useful in the study of ligand...receptor interactions. A considerably more detailed description of conformational polymorphism may be found in a recent review (Bernstein, 1987).

What features of the molecular geometry may be expected to vary among polymorphs? Among the most difficult to perturb are bond lengths, followed by bond angles, and depending on the nature of the particular bond in question, torsion angles about bonds. As a rule of thumb, in the same order these differences correspond to a range of approximately two orders of magnitude with torsion angles on the low end. For instance, the barrier to rotation about the single bond in ethane is *ca* 2.5 kcal/mole, while the barrier to free rotations of methyl groups in dimethylacetylene is *ca* 0.5 kcal/mole (Brand and Speakman, 1960). The differences in lattice energy among different crystal forms of an organic compound generally do not exceed 1-2 kcal/mole (McCrone, 1963), putting us on the same energy scale as the energies involved in torsional rotations. Hence, among polymorphs we can expect fairly large variations in torsional conformations, but considerably less variation among bond angles and bond lengths.

A number of examples serve to illustrate the variety and magnitude of the conformational differences which may be encountered in the types of molecules which might be considered in studies of ligand-receptor interactions. A compound which has received a great deal of attention recently is the antiviral agent ribavarin (virazole), a potential AIDS-retarding drug (Chemical & Engineering News, 1987). Considerable differences in the conformations found in the two known crystal structures (Prusiner and Sundaralingham, 1976) are readily apparent from Figure 4. Adenosine-5'-monophosphate is also dimorphic (Kraut and Jensen, 1963; Neidle *et al*, 1976), with considerable variation in the side-chain conformation (Figure 5). Finally, L-glutamic acid, which might be used as a model for studies concerned with the action of monosodium glutamate, is at least dimorphic (Lehmann *et al*, 1972; Hirayam *et al*, 1980) with significant conformational differences (Figure 6). (Interestingly, the Merck Index makes no mention of the polymorphism of any of these three compounds.)

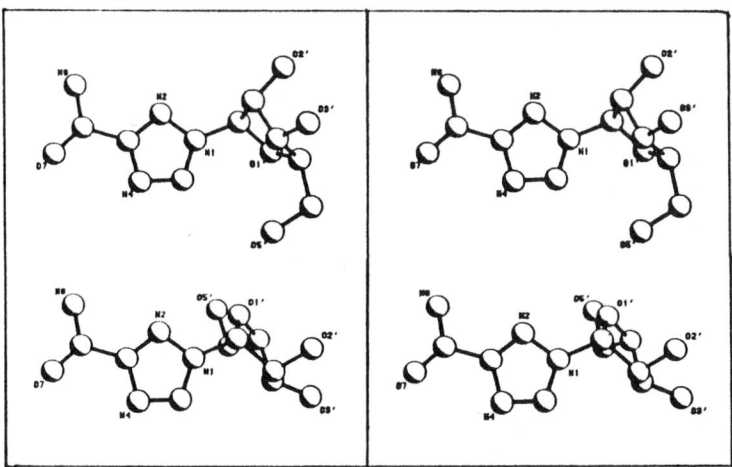

Figure 4. Stereviews of the two forms of virazole. In both cases the view is on the best plane of the triazole ring, noted by N1, N2 and N4. For clarity, the carbons have been left unlabelled.

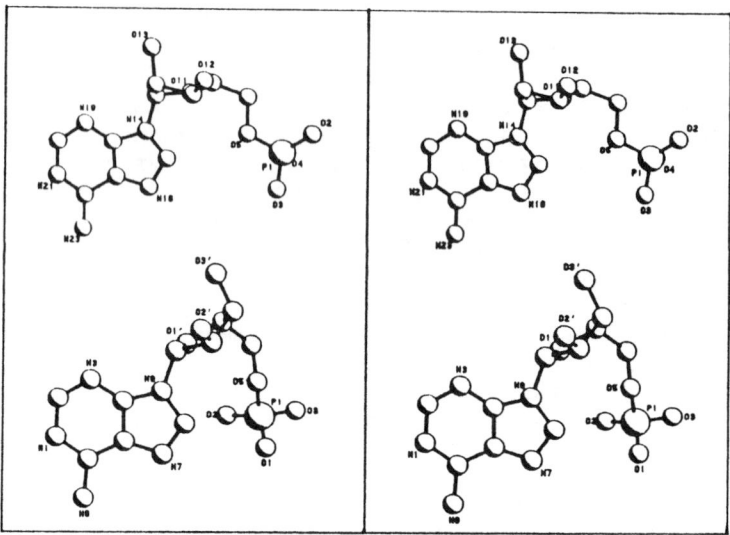

Figure 5. Stereoviews of the two forms of adenosine-5'-monophosphate. In both cases the view is on the best plane of the 6-membered ring of the base. For clarity carbons have been left unlabelled.

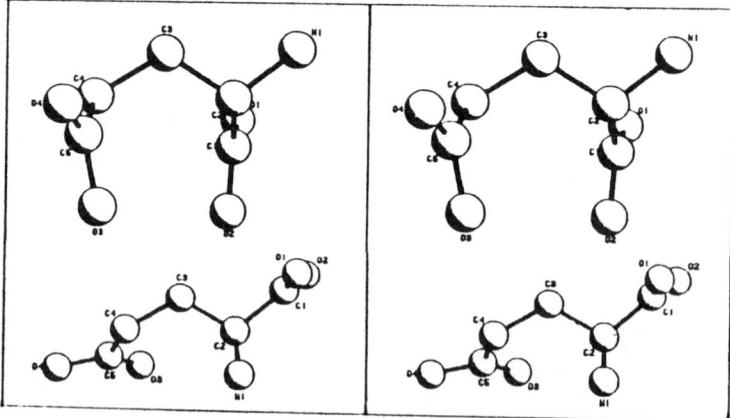

Figure 6. Stereoviews of the two forms of L-glutamic acid. In both cases the view is on the plane of C2-C3-C4, which particularly highlights the conformational differences about C2-C3 and C3-C4.

A typical investigation of a system exhibiting conformational polymorphism involves the application of a number of now fairly standard techniques for studying geometry and energetics. Polymorphism may be detected and studied by a variety of methods (Bernstein, 1987; Haleblian, 1975) and physical evidence for the existence of conformational polymorphs may be detected (Bernstein, 1987). X-ray crystal structure analyses are then carried out to verify the presence of conformational polymorphs, and to obtain the geometric details regarding both the molecule and the crystal packing required for further analysis. Observed differences in molecular conformation may be evaluated energetically by employing molecular orbital calculations at various levels of approximation or, particularly for larger molecules, molecular mechanics. A considerable advantage in computing time is gained in anaylzing these polymorphic systems, since the points of real interest on the multidimensional energy surface are the ones observed in the crystal structures and a large number of conformers may, at least initially, be eliminated from consideration.

To this point one has obtained the necessary geometric and energetic information on the molecular level. How does one examine the environment of the molecule? The geometric information, namely interatomic distances and the orientation of the molecule (the "ligand") with respect to its neighbors (the "receptor") is contained in the crystal structure analysis. The energy information is most directly obtained via computation of the lattice energy, employing a now quite well-developed technique involving the use of atom...atom potential functions (Kitaigorodsky, 1973; Williams, 1981; Simonetta, 1984). The technique is based on the assumption that the total lattice energy may be considered as a sum of individual atomic contributions, which themselves may be approximated by, usually, a Lennard-Jones or Buckingham potential function for the interaction between two atoms. The total energy, which should be compared to the sublimation energy, may be readily partitioned into its individual atomic contributors for a rather detailed analysis of the specific atomic interactions. These may be used to understand, for instance, the stabilization of a higher energy molecular conformation in one crystal structure compared to other polymorphs, or to examine the differences in energetic environments among polymorphs (Bernstein and Hagler, 1978).

The number of conformational polymorphs treated in this manner is still somewhat limited, in spite of the availability many potential candidate systems. Clearly, there is still much to be learned about the general potential of the technique in obtaining information about the interactions between a molecule and its environment. In the meantime, additional tools are constantly being added to the computational armory to aid the analysis of these and other systems of interacting molecules. For instance, an approach based on deformation forces rather than on energy can be used to determine which atoms are moved from the positions expected in the absence of intermolecular interactions (Krygowski and Turowska-Tyrk, 1987). In another development, the role of molecular shape and molecular free surface in determining molecular conformation and molecular motion in organic crystals have been investigated (Gavezzotti, 1982, 1983, 1985). The experience gained from these well defined systems of conformational polymorphs can be expected to contribute as well to the more complex systems of ligand-receptor interactions.

Just as the techniques of molecular mechanics and molecular dynamics were developed on small model systems and extended to increasingly larger systems, we are confident that the experience gained from studying conformational polymorph of relatively small molecules will be eventually applied to interactions involving ligands and receptors.

REFERENCES

Aguiar AI, Krc J, Kinkel AW and Samyn JC (1967). J Pharm Sci 56: 847-853.
Allen FH, Kennard O and Taylor R (1983). Accts Chem Res 16: 146-153.
Andersgaard H, Finholt P, Gjermundsen R and Hoyland T (1974). Acta Pharm Suecica 11: 239-248.
Ashida T, Tsunogae Y, Tanaka I, and Yamane T (1987). Acta Cryst B43: 212-218.
Bar I and Bernstein J (1985). J Pharm Sci 74: 255-263
Bernstein J (1987). Conformational Polymorphism. In Desiraju GR (ed): "Organic Solid State Chemistry," Amsterdam: Elsevier, pp 471-518.
Bernstein J, Hagler, AT (1978). J. Am Chem Soc 100: 673-681.
Borka L and Backe-Hansen K (1968). Acta Pharm Suecica 5: 271-278.
Brand JCD and Speakman JC (1960). "Molecular Structure, The Physical Approach." London: E. Arnold, p. 248.
Byrn SR (1982). "Solid State Chemistry of Drugs." New York: Academic Press.
Cameroni R, Coppi G, Gamberini G and Forni F (1976). Farmaco Ed Prat 31: 615-624.
Castle RN and Witt NF (1946). J Am Chem Soc 68: 64-66.
Chemical & Engineering News (1987): Jan. 26, p. 18.
Corradini P (1973). Chim Ind (Milan) 55: 122-129.
Elder M, Machin P and Hull SE (1984). J Molec Graphics 2: 70-78.
Gavezzotti A (1982). Nouv J Chem 6: 443-450.
Gavezzotti A (1983). J Am Chem Soc 105: 5220-5225.
Gavezzotti A (1985). J Am Chem Soc 107: 962-969.
Gouda MW, Ebian AR, Moustafa MA and Khalil SA (1977). Drug Dev Ind Pharm 3: 273-290.
Haleblian JK (1975). J Pharm Sci 64: 1269-1288.
Hirayam N, Shirahata Y, Ohashi Y and Sasada Y (1980). Bull Chem Soc Jpn

53: 30-37.
Kitaigorodsky AI (1973). "Molecular Crystals and Liquid Crystals." New York: Academic Press, pp. 134-190.
Kraut J and Jensen LH (1963). Acta Crystallogr 16: 79-86.
Krygowski ETM and Turowska-Tyrk I (1987). Chem Phys Lett 138: 90-96.
Kuhnert-Brandstatter M (1965). Pure Appl Chem 10: 136.
Kuhnert-Brandstatter M and Wunsch S (1969). Mikrochim Acta: 1297-1308.
Lehmann MS, Koetzle TF and Hamilton WC (1972) J Cryst Mol Struct 2: 225-236.
McCrone WC (1963). Polymorphism. In Fox D, Labes MM, Weissberger A (eds): "Physics and Chemistry of the Organic Solid State, Vol. I," New York: Interscience, p. 725.
Merck Index (Tenth Ed.) (1983). Rahway, New Jersey: Merck & Co.
Mesley RJ and Houghton EE (1967). J Pharm Pharmacol 19: 295-304.
Neidle S, Kuhlbrandt W and Achari A (1976). Acta Crystallogr B32: 1850-1854.
Panagiotoupoulis NC, Jeffrey GA, LaPlaca, SJ and Hamilton, WC (1974). Acta Crystallogr. Sect. B30: 1421.
Prusiner P and Sundaralingham M (1976). Acta Crystallogr B32: 419-424.
Simonetta M (1984). Int Revs Phys Chem 4: 39.
Taylor R, Kennard O, and Versichel W (1983). J Am Chem Soc 105: 5761-5766.
Waki SJ (1970). "Lipid Metablism." New York: Academic Press.
Williams DE (1981). Topics in Appl Phys 26: 3-40.
Yamanouchi Pharmaceutical Company (1972). Japanese Patent 7,222,716 (June 24,1972).
Yang SS and Guillory JK (1972). J Pharm Sci 61: 26-40.

THE SOLUTION CONFORMATION OF THE ANTIBIOTIC ANTICANCER CHROMOMYCIN A_3 BY TWO-DIMENSIONAL NMR SPECTROSCOPY

Elisha Berman and Michal Kam

Organic Chemistry Department, Weizmann Institute of Science, Rehovot 76100, Israel

ABSTRACT: The solution conformations of chromomycin A_3 (CRA) and dechromose-A chromomycin A_3 (CRA-B) in dichloromethane and methanol were studied by two-dimensional (2D) NMR techniques. In dichloromethane, the drugs are found in a compact wedged-like conformation, with the phenolic hydroxyls at the tip and the side chains folded back to one side of the aglycon plane, oriented parallel to each other. The overall structure is stabilised by intramolecular hydrogen bonds and by the formation of a hydrophobic pocket, enclosed by the three side chains. In methanol, the drugs have the expected open conformation with extended side chains.

Like its parent drug, CRA-B binds to d(ATGCAT)$_2$ duplex with a major groove orientation, a 2:1 drug/duplex ratio and a two-fold symmetry of the resultant complex. The drug molecule is suggested to reside diagonally across the two strands of the duplex and to span 3 base pairs, while all three side chains of the drug are folded away from the major groove of the DNA. The observed nonequivalent positions of CRA and CRA-B within the major groove of the duplex results from the different conformation adopted by the sugar side-chains in the two complexes.

INTRODUCTION

Chromomycin A_3 is an antitumour, antiviral and antibiotic agent, produced by *Streptomyces griseus* No. 7, and belongs to the aureolic group of antibiotics (Remers, 1979). It is composed of the chromomycinone aglycon and of five sugar units arranged in two chains (units A,B,C,D and E in Fig. 1)(Miyamoto et al., 1967; Thiem and Meyer, 1979). Although CRA is active against a variety of experimental and human tumours, its clinical use is restricted because of high toxicity and limited antitumour spectrum (Remers, 1979). The binding properties of CRA have been extensively studied in the past, though the position and the exact mode of binding remains a matter of controversy.

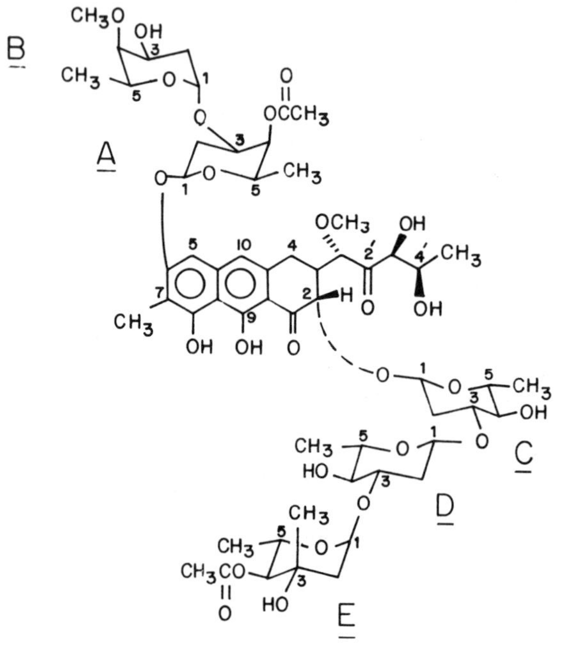

Fig. 1: The structure of chromomycin A_3.

Recently, 2D NMR studies have been performed to elucidate the three-dimensional structure of a CRA/d(ATGCAT)$_2$ complex (Keniry et al., 1987). It was suggested that the aromatic chromophore of CRA lies in the major groove of the duplex, with its hydrophilic side proximal to the helix centre. The aliphatic side chain is oriented such that it partially protects the aromatic protons from the external solvent. The study did not point out to the location of the sugar units of CRA. As the sugar side chain were reported to be tightly bound in the DNA complex (Berman et al., 1985). The positions of the sugar units relative to the DNA are of interest, since the biological activity of the drug and its rate of dissociation from the DNA are strongly correlated with the number of the sugar units present on the molecule (Remers, 1979).

EXPERIMENTAL

CRA (98% purity) was purchased from Sigma and used without further purification. The d(ATGCAT) was prepared by the standard solid-phase triester techniques in an applied Biosystems 380B DNA synthesiser, desalted and purified by gel-filtration. CRA-B was prepared from CRA by acid hydrolysis (Miyamoto et al., 1967).

The DNA sample was prepared by dissolving 7 mg of d(ATGCAT) in 0.5 ml borate buffer (D_2O, 100 mM, 180 mM NaCl, 0.2 mM EDTA at pH 8.2) containing 20 mM $MgCl_2$. After recording the spectrum of the duplex alone, dry CRA-B (2 mg) was added to the sample. More CRA-B and $MgCl_2$ were added stepwise to the CRA-B/duplex sample terminating at $\sim$2:1 drug/duplex ratio. The final sample contained 3.9 mM d(ATGCAT)$_2$, 6.6 mM CRA-B and 60 mM $MgCl_2$.

Spectra were recorded using a Bruker WH-270 spectrometer operating at 270 MHz for protons and a Bruker AM-500 spectrometer operating at 500 MHz for protons. Two-dimensional experiments were performed using standard pulse sequences issued by Bruker. Spectra of the CRA-B/duplex were recorded at 14°C with the COSY spectrum recorded using spectral window of 4.4 KHz, and a data matrix of 480x2048, which was zero-filled to yield a final matrix of 1024x2048 and processed with a squared sine bell function ($\pi/16$). NOESY spectra were recorded with varying mixing-time (0.05, 0.1, 0.2, 0.3 and 0.4 s) using the same spectral parameters.

RESULTS AND DISCUSSION

A. Solution Conformation of CRA and CRA-B:

Careful analysis of the proton spectra yielded complete assignment for the proton resonances of CRA and CRA-B. The three-dimensional solution conformation of these drugs in organic solvents was determined from the coupling constants information (Thiem and Meyer, 1979) and the firmly assigned strong NOE contacts (Table). The analysis yielded a unique, well-defined conformation for CRA (Kam at al., 1988).

Interproton distances (0.3-0.4 nm) across the two sugar side chains (entries 4 and 6 in Table) were estimated by comparison with the observed intensity for the fixed distance NOE contacts (entries 1, 5 and 7). Similar interproton distances were assumed for the NOE contacts which help fix the side chains orientation (entries 2 and 3). A molecular model representing the conformation of the CRA molecule in solution was constructed and optimised to satisfy as many of the observed low intensity NOE contacts as possible. At the same time, we avoided short distances between protons that did not show any NOE contact in the spectrum.

Table: NOE Contacts Observed for CRA and for CRA-B

NOE contacts	CRA	CRA-B
(1) 1' → 1'-OMe	++	++
(2) 2 → 1'-OMe	++	++
(3) 5 → 1A	+	++
(4) 4A → 4D	++	
(5) 4B → 4B-OMe	++	
(6) 4B → 1E	++	
(7) 1E → 2E	++	+

The predominant three-dimensional conformation of CRA in dichloromethane is unexpectedly compact. The two sugar side chains are extended parallel to each other, on the same side of the aglycon, with sugar units B and E and units A and D proximate to each other respectively (Fig. 2). Together with the aliphatic side chain and the aglycon they form hydrophobic pocket. A sideways projection of the molecule resembles a wedged-slab shape with one relatively thin face (1.0-1.2 nm across) and with the phenolic hydroxyls at its tip (Fig. 3). The overall structure is held by intramolecular hydrogen bonds and an additional stabilisation provided by the hydrophobic interactions within the pocket.

A similar conformation was constructed for CRA-B, though the two important inter-residue NOE contacts between sugars B and E and units A and D were missing (entries 4 and 6 in Table). Thus, the two oligosaccharide side chains of CRA-B must be further apart from each other relative to CRA.

The conformation of CRA in methanol, determined from the NOE data (Kam et al., 1988), clearly indicated that the drug is found in a fully extended conformation (Fig. 4) as predicted from the energy-minimisation calculation performed for CRA in vacuum (B. Meyer, private communication). Thus, in protic solvents the open conformation is preferred, though, the observed structural flexibility of the drug is probably reflected in its conformational adaptability for binding with the DNA.

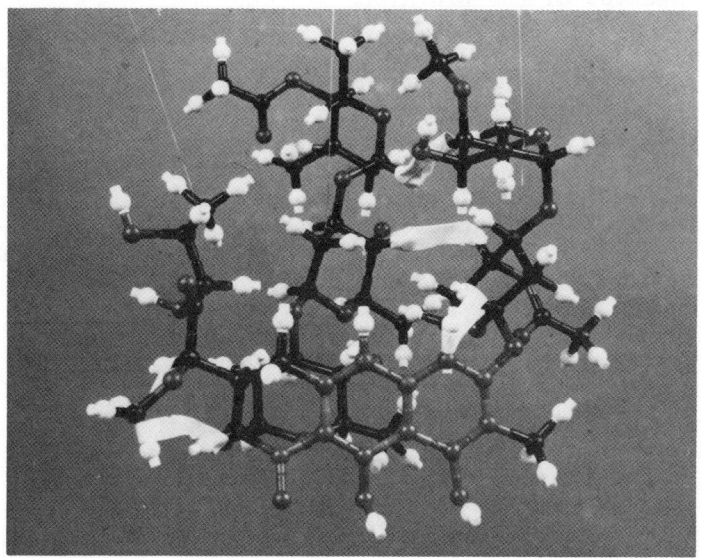

Fig. 2: Model showing the conformation of CRA. The strips symbolise observed NOE contacts (see Table).

Fig. 3: Projection of the CRA molecule displaying its wedged-like shape with the hydrophilic part of the aglycon at its tip (bottom left).

B. NMR Studies of CRA-B/d(ATGCAT)$_2$ Complex:

Following the study of the structure of the CRA/d(ATGCAT)$_2$ complex (Keniry et al., 1987), we have investigated the binding of CRA-B to the same duplex at 14°C. Complete binding of the drug was ensured by monitoring changes in the intensities of the aromatic proton resonances of the free drug (Weinberger et al., 1988) resulting from a titration with the duplex. At the final point of the titration, 85% of the duplex molecules were bound by the drug as determined from the relative intensities of corresponding free and bound duplex resonances.

Resonance Assignments: The protons are denoted by their DNA-base identification letter (Fig. 4, bottom), followed by their position number. The sugar protons are identified by the symbol of their base followed by their position number (primed). Only the C5 and C6 proton resonances are expected to give a COSY cross peak in the aromatic region (6.5-8.5 ppm) as well as corresponding strong NOE cross peaks (Fig. 4). These were used as a starting point for the sequential assignment of other duplex resonances. For DNA in the B-conformation, NOE contacts between the base protons and their 1' protons (on the deoxyribose moiety) are clearly observed. Together with the somewhat weaker NOE contacts to the 1' protons of the neighbouring nucleotide (on the 5' side), it is possible to perform a sequential "walk" through the cross peaks which will correspond to "sequencing" the individual DNA strands. Another independent "walk" was done using the NOE contacts between the DNA bases and the 2',2" sugar protons. The NOE contacts observed between the 1' protons and their own 2 and 2' protons have further supported our assignments.

The ability to "walk" through the NOE cross peaks, and the presence of all but one of the expected inter-base NOE contacts (i.e. At8/Ti5-CH$_3$, At8/Ti6, Ti6/G8, G8/C6, C6/Ai8, Ai8/Tt6, Ai8/Tt5-CH$_3$, At2/Ai2) is a clear indication that the B-conformation of the duplex is retained in the bound state and that intercalation of the drug between the bases is excluded. The absence of the C5/G8 NOE contact indicates that these two protons are further apart from each other relative to normal B form of DNA. Hence, the binding of CRA-B produces a minor distortion only in the B-conformation of the duplex, a result which is consistent with the findings for the binding of CRA to the same duplex (Keniry et al., 1987). Half as many of the expected duplex proton resonances were observed for the 2:1 CRA-B/d(ATGCAT)$_2$ complex indicating a 2-fold symmetry.

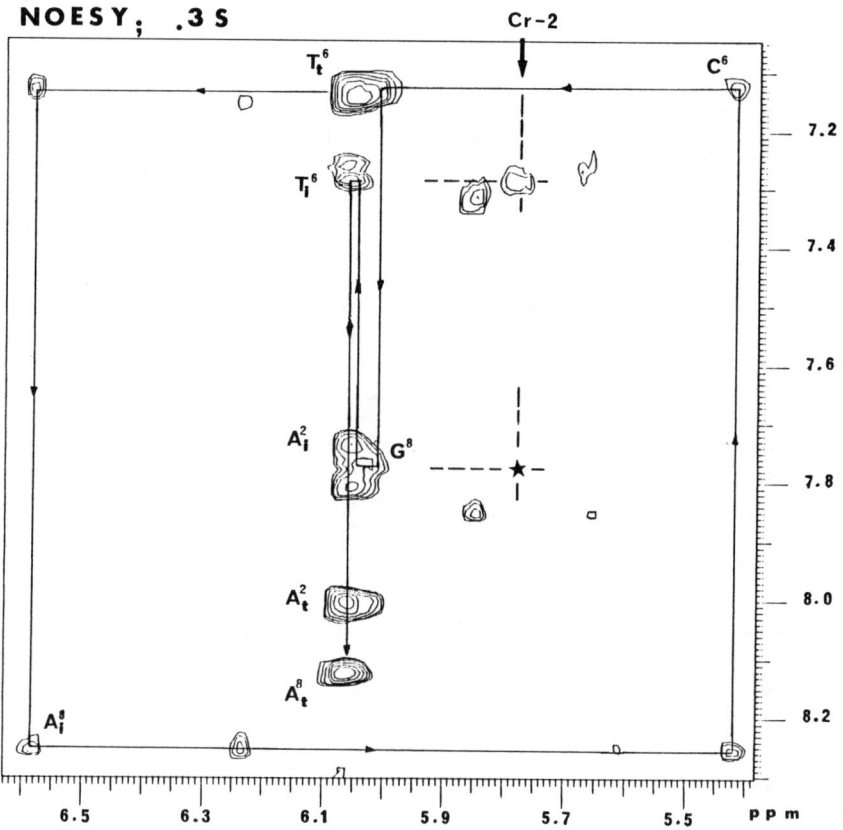

(5') A_t T_i G C A_i T_t (3')

Fig. 4: Part of the NOESY spectrum of CRA-B/duplex complex at $14^\circ C$, showing contacts between the base and the sugar anomeric protons. A line is drawn through the cross peaks corresponding to a sequential "walk" through the DNA. The arrow indicates the position of proton 2 of CRA-B that has NOE contact to the Ti6 and to the G8 protons. The latter contact (*), was clearly observed at a longer mixing time (0.4 s). Unmarked cross peaks arise from the free duplex present in solution.

Chemical shift assignments obtained for CRA-B in dichloromethane and in methanol (Kam et al., 1988) have provided the initial basis for the assignment of CRA-B resonances when bound to the d(ATGCAT)$_2$ duplex. Some 22 protons of the drug were assigned with identified NOE connectivities.

Duplex-Drug Contacts: Few NOE contacts between the drug and the duplex have been identified in the spectra, but these were sufficient to indicate the location of the drug within the duplex. Two such contacts were observed for proton 2 of CRA-B (Fig. 4), which was thus placed close to both the Ti6 and the G8 protons of the duplex. NOE contacts between the Tt5-methyl group and protons 5 and 10 of the aglycon were also observed (Fig. 5). From the time dependence course of the cross peak intensities it was concluded that proton 5 is closer to the duplex relative to proton 10. In fact, the Tt5-methyl resonance is splited up into two components, one is closer to proton 5 and the other resonance is closer to proton 10 of CRA-B (Fig. 5). As the only other splited resonance was assigned to the Tt6 proton, we concluded the Tt base may adopts two distinct conformations as a consequence of fraying at the duplex ends. Thus, the disaccharide side chain (sugar units A and B in Fig. 1) must be close to the 3' end of the duplex.

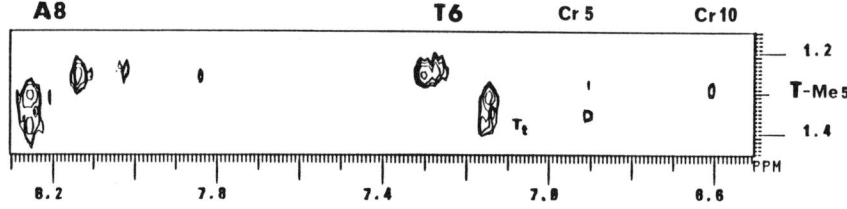

Fig. 5: Part of the NOESY spectrum (mixing time: 0.2 s) of the CRA-B/d(ATGCAT)$_2$ complex at 14°C showing NOE contacts of the T5-methyl groups to the A and T base protons and to the aglycon 5 and 10 protons.

To satisfy both the symmetry of the complex and the observed contacts between the duplex and the drug, the aglycon of the bound drug was located in the major groove of the duplex such that it is aligned diagonally across the helix between the Tt and the Ai bases of one strand and the Ti and G bases of the other strand, (Fig. 6). Each aglycon spans nearly 3 base-pairs in excellent agreement with the DNA footprinting results (Van Dyke and Dervan, 1983). Observed intramolecular NOE contacts in CRA-B indicated that the two sugar chains of CRA-B are folded toward each other, which suggests that the two side chains are residing on the same side of the aglycon. The accommodation of two CRA-B molecules within the short nucleotide helix supports such folding which may resemble the compact conformation adopted by the drug in dichloromethane.

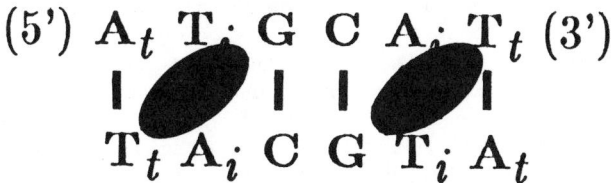

Fig. 6: The orientation of CRA-B in the major groove of the DNA.

C. Comparison between the CRA and CRA-B Complexes

The removal of sugar unit B from CRA, does not alter the number of drug molecules which are bound to the DNA. The aglycons of both CRA and CRA-B reside in the major groove, close to the guanine base, with the hydrophilic part of the aglycon close to the helix centre. The chemical shift values of many of the corresponding duplex protons resonances in the two respective drug/DNA complexes was different and interpreted as an indication of nonequivalent positions for the two drugs in major groove. It is further supported by observation of NOE contacts between a duplex proton (resonating at 4.40 ppm) and both the 1A and 4A protons of CRA (at 5.36 and 5.23 ppm) which were not in the case of CRA-B. Also none of the CRA-B/duplex NOE contacts (i.e., 2/Ti6, 2/G8, 5/Tt5-CH_3, 10/Tt5-CH_3) are observed for the CRA/duplex. Only one type of Tt5-methyl resonance is observed in the spectrum of the CRA/duplex complex, while two such resonances were observed for the same group in the CRA-B/duplex complex.

CONCLUSIONS

Free CRA adopts two predominant conformations, one which is dominant in methanol and has an elongated shape with extended sides chains conformation (Kam et al., 1988); while the other conformation, which is predominant in dichloromethane, has compact wedged-like appearance with all three side chains folded. The side chains are folded parallel to each other on the same side of the aglycon and are held together by intramolecular hydrogen bonds. In this compact conformation, the drug molecule has the appropriate shape and dimensions needed to penetrate deeply into the major groove of d(ATGCAT)$_2$. Analysis of the NOE data obtained for the CRA-B/DNA complex has suggested that the drug adopts a folded conformation when bound to the duplex. In addition, two drug molecules are accommodated by the short duplex which also implies a folded conformation for the bound drug.

The relative position of CRA and CRA-B in the major groove are not the same which lead to the suggestion that the one function of the sugar moieties in CRA, is to enable the compact conformation needed for the drug to be recognised at the binding site. Following the penetration of the drug into the major groove, these sugar side sugar chains may undergo some conformational change that locks the drug within the binding site. On that basis, structural modifications of the side chains are expected result in the unfolding of conformation of the drug making it less recognisable for the duplex. Alternatively, the modified drug will be unable to lock itself within the binding site.

ACKNOWLEDGMENTS

A full account of this work is described in the thesis for the Degree of Master of Science presented to the Feinberg Graduate School at the Weizmann Institute by Miss Kam on February 1988. The financial support of the Israel-USA Binational Foundation is gratefully acknowledged.

REFERENCES

Berman E, Brown SC, James TL, Shafer RH (1985). NMR studies of chromomycin A_3 interaction with DNA. Biochemistry 24:6887-6893.

Kam M, Shafer RH, Berman E (1988). Solution conformation of the antitumour antibiotic chromomycin A_3 determined by two-dimensional NMR spectroscopy. Biochemistry 27: May issue.

Keniry MA, Brown SC, Berman E, Shafer RH (1987). NMR studies of the interaction of chromomycin A_3 with small DNA duplexes. Biochemistry 26:1058-1067.

Miyamoto N, Kawashima K, Shinohara M, Tanaka K, Tatsuoka S, Nakanishi K (1967). Chromomycin A_2, A_3 and A_4. Tetrahedron 23:421-437.

Remers WA (1979). "The Chemistry of Antitumour Antibiotics", Vol. 1, pp. 135-175, Wiley, New York.

Thiem Y, Meyer B (1979). Studies on the structure of chromomycin A_3 by proton and carbon NMR spectroscopy. J.

Chem. Soc. Perkin Trans 2:1331–1336.

Van Dyke MW, Dervan PB (1983). Chromomycin, Mithramycin, and Olivomycin binding sites on Heterogeneous DNA. Footprinting with (Methidiumpropyl-EDTA)iron(II). Biochemistry 22:2373–2377.

Weinberger S, Itzhaki L, Livnah O, Livnah N, Berman E. (1988). Unique binding cavity for divalent cations in the DNA-metal-chromomycin A_3 complex. Biochemistry: submitted.

SECTION III. RECEPTORS AND TRANSMEMBRANE SIGNALING

STRUCTURAL MODELS OF ALPHA HELICAL MEMBRANE PEPTIDES AND THE GABA RECEPTOR CHANNEL

H. Robert Guy and G. Raghunathan

Laboratory of Mathematical Biology, National Cancer Institute, National Institutes of Health, Bethesda, Maryland 20892

INTRODUCTION

The transmembrane region of many membrane proteins appear to be comprised primarily of α helices. Several models for the transmembrane portions of transmitter- and voltage-activated channels have been proposed; however, these models are typically too imprecise for detailed analysis of possible drug-receptor interactions. To develop precise models of these proteins one must be able to accurately predict the conformation of α helical segments including their side chain conformations and accurately pack α helices next to each other. In developing methods to achieve this, it is important to begin with fairly simple proteins and to be able to experimentally test the results.

We have been using energy calculations with computer graphics to predict structures of several peptides that lyse membranes and/or form membrane channels: melittin, δ lysin, magainin, PGLa, XPF, and pardaxin. All of these peptides have sequences consistent with their forming amphipathic α helices. We have modeled these systems to study how these monomers aggregate on one surface of a membrane, how they insert across the membrane, and how they form channels. Melittin and δ lysin are lytic toxins from bee venom and staphylococcus. The crystal structure of melittin has been determined Terwilliger and Eisenberg, 1982). δ Lysin has been crystallized (Thomas, et al., 1986) and an α helical backbone structure has been determined by NMR (Tappin, et al., 1988). Magainin, PGLa, and XPF are potent antimicrobial agents secreted form Xeneopus granular glands (Zasloff, 1987). They resemble melittin in that they can form positively charged amphipathic α helices; however, they are not hemolytic at concentrations that kill microbes. Magainin induces anion selective channels in artificial membranes (Cruciani, et al., 1988). PGLa and XPF are homologous peptides that have no homology to Magainin but have very similar antimicrobial effects. Pardaxin is a potent shark repellant made by the Red Sea Moses Sole (Lazarovici et al., 1988). It forms channels in artificial membranes that, unlike most other α helical peptide channels, appears to have only one size. Pardaxin channels allow all inorganic monovalent cations and guanidine but not Tris base to pass. This suggests that the smallest region of the channel is about five to six Å in diameter. Pardaxin channels are more permeant to cations than anions but anions can pass.

We postulate that all of these proteins form antiparallel dimers that remain relatively unaltered in the different types of oligomers. Adjacent α helices in these dimers typically cross at a small positive angle typical of '3-4 ridges-into-grooves' helix packing (Chothia, et.al., 1981). All of these dimers can be used to construct membrane surface 'raft' type structures in which one side of the 'raft' is comprised of hydrophobic side chains and the other side is comprised of hydrophilic side chains (see Fig. 1) and cylindrical transmembrane channels with hydrophobic side chains on the exterior and hydrophilic side chains lining the channel. The helical interactions in the 'rafts' between the initial dimers tend to have negative crossing angles and could be classified and '3X4' or '4X4 crossed ridges' packing. These interactions usually change to '3-4 or 4-4 ridges-into-grooves' packing in the channel structures.

This paper deals with a description of how we modeled the different oligomers of pardaxin, how some of the methods used to model the peptide channels may be extended to the transmembrane region of the GABA receptor, and how some of the features of the GABA receptor channel may be similar to those of the peptide channels.

α Helical Membrane Peptides and GABA Receptor Channel / 233

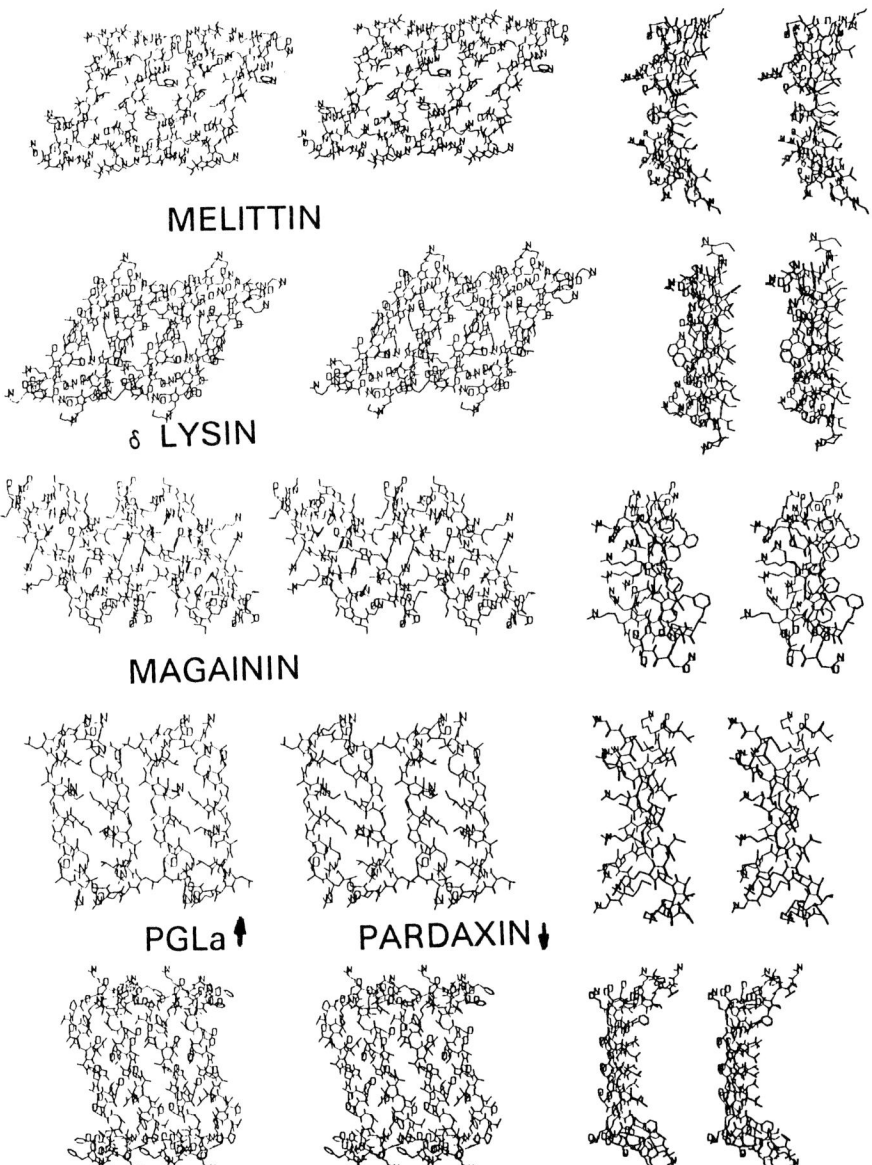

Figure 1. Stereo views of 'Raft' conformations of membrane peptides. Side chain and termini oxygens and nitrogens are labeled O and N. Left view has the hydrophobic surface up and the hydrophobic surface down. Right view shows the principle dimers with the darker monomer nearer the viewer.

PARDAXIN

We have modeled pardaxin as a tetramer in water, as a 'raft' that that displaces a lipid monolayer on the surface of the membrane, as an aggregate of two 'rafts' as it folds into the membrane, and as a transmembrane channel.

Our pardaxin monomer structures are more complex than those for the other channel-forming peptides. We have divided the peptide into four segments: the N-terminus segment (NS) for residues 1-10 (NH_2-G-F-F-A-L-P-K-I-I) which is helical in the channel and 'raft' models and extended in the solution model, the bend segment (BS) for residues 11 and 12 (S-S), the amphipathic α helix segment (HS) for residues 13-26 (P-L-F-K-T-L-L-S-A-V-G-S-A-L), and a flexible C-terminus segment (CS) for residues 27-33 (S-S-S-G-G-Q-E-COOH). Possible secondary structures of pardaxin were analyzed by the Delphi program designed to predict secondary structure and by the Amphi program (Guy, unpublished) that identifies possible amphipathic α helices and β strands. Delphi predicted that BS and CS were not α helical or extended structures and that NS and HS would be either α helices of β strands depending upon the decision constants used. Amphi predicts that SN and SH could both form amphipathic α helices. Our final selection of secondary structure was based primarily on which of these possibilities could be used to construct apparently stable oligomers for different environments. The initial α helix for HS was created by a program that generates α helices with the backbone structure and side chain conformations most commonly observed for α helices in crystal structures (Cornette, Guy, and Margalit, unpublished). NS was modeled as either an α helix (using the backbone structure of a helix in lactose dehydrogenase with a similar sequence) or as a β strand. BS and CS were assigned phi and psi backbone torsion angles commonly observed in random coil segments and that allowed the protein segments to interact well.

To simplify the analysis we usually assume that all monomers in an aggregate have identical conformations and that each monomer has identical interactions with surrounding monomers, solvent, and lipids. These assumptions reduce the number of conceivable packing arrangements greatly and allow the aggregate to be generated by a series of translations and/or rotations of the original monomer so that the energies of only one monomer and surrounding atom within a certain distance of the monomer need be calculated. Connolly (1983) surfaces were added to the structures and monomers were manually docked using the Mogli program on an Evans and Sutherland computer graphics monitor. Energies for these models then were refined with the CHARMM program (Brooks, et al., 1983) using adopted basis Newton-Raphson method. Connolly surfaces were used to determine that the final structure is tightly packed and has no internal cavities large enough to hold a water molecule.

The channel model we favor is comprised of twelve antiparallel monomers (see Fig. 2); however, similar structures comprised of eight or ten monomers cannot be excluded. Antiparallel models appear favorable because they allow: (i) negative carboxyls of the C-terminus glutamate to interact with positive charges of Lys-16 and Lys-8; (ii) backbone dipoles of adjacent HS's to interact favorably; (iii) the only negatively charged portion of the molecule, the C-terminus, to be near the radial center of the pore where it could make the channel selective to cations, and (iv) large hydrophobic side chains on HS to pack tightly next to each other. Parallel models appear unlikely because: (i) all models with parallel α helices have highly positively charged regions that should impede cation permeability and (ii) we could find no parallel models in which side chains pack next to each other as well as in antiparallel models. HS's pack very closely in an antiparallel manner to form a dimer with the following features: all the large alkyl side chains and the phenlyalanine side chain form a two-residue-wide hydrophobic column on one side of the dimer, several serines on the opposite side of the dimer form hydrogen bonds with each other, there are no cavities or overlapping regions between helices, Lys-16 extends over the C-terminus of the adjacent α helix where it interacts favorably with the negative end of the helix dipole and carboxyls of the terminus Glu-33, the helices cross each other at an angle of $-15°$ predicted by '3-4 ridges into grooves' helix packing theory (Chothia, et.al., 1981) (See Fig. 2). This dimer for the HS's also was used to construct models of pardaxin in solution as a tetramer and on the membrane surface as a 'raft'-like structure. In constructing the channel, HS's of the initial dimers were packed next to each other so that Ala-21 and Ala-25 pack between each other and a hydrogen bond forms between Thr-17 and Ser-29. The helices cross at an angle of $25°$.

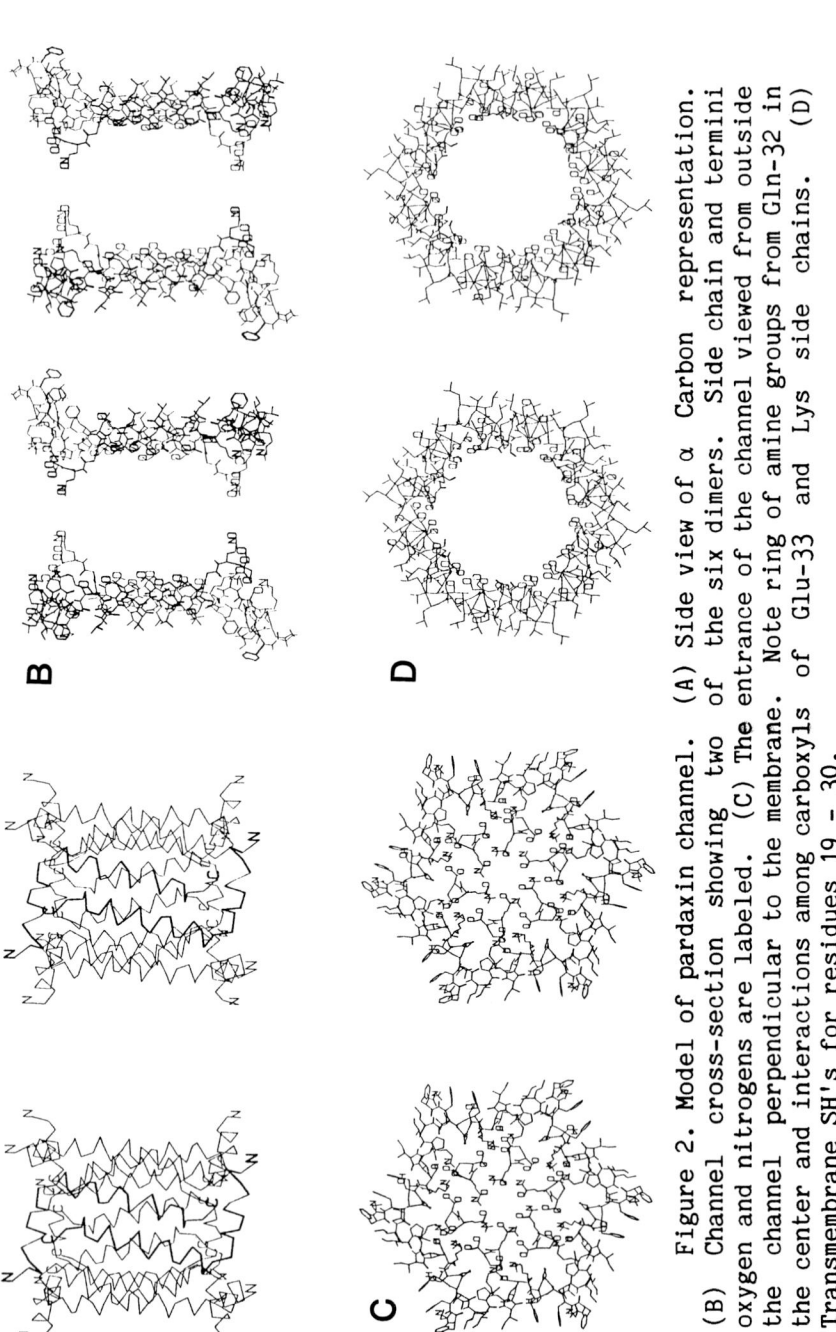

Figure 2. Model of pardaxin channel. (A) Side view of α Carbon representation. (B) Channel cross-section showing two of the six dimers. Side chain and termini oxygen and nitrogens are labeled. (C) The entrance of the channel viewed from outside the channel perpendicular to the membrane. Note ring of amine groups from Gln-32 in the center and interactions among carboxyls of Glu-33 and Lys side chains. (D) Transmembrane SH's for residues 19 - 30.

Fig 2c shows our model of the channel's entrances formed by residues 1-16 and 31-33. NS's form a hydrophobic and positively charged outer ring that surrounds the negatively charged C-termini. The narrowest portion of the channel is formed by a ring of six amide groups from the Gln-32 side chains. A 4.8 angstrom diameter sphere could just pass through this ring. This is large enough to pass all known permeant cations but small enough to exclude Tris base, which is impermeant. These form a network of hydrogen bonds with each other and with the carboxy terminus. Lys-8 and Lys-16 form salt brides to the two carboxyl groups of Glu-33. The channel should be cation selective because: the amine groups are farther from the center of the channel than the carboxyl groups, the C-terminus of HS is nearer the channel entrance than is its N-terminus, the narrow opening is lined by oxygens from the Gln-32 amide groups, and the positive charges of the N-termini may interact with negatively charged lipid head groups.

Fig 2D shows the non-charged central region of the channel formed by residues 17-30. Only serine and threonine side chains line the hexagonally-shaped pore. Only alkyl side chains are on the exterior of the structure where they would be exposed to lipid. This region of the pore is about twenty angstroms wide.

The 'raft' model in Fig. 1 is similar to the channel model except that adjacent dimers are related to each other by a linear translation instead of a 60° rotation about a channel axis. We postulate that these 'rafts' displace the lipid molecules on one side of the bilayer. When two or more 'rafts' meet they can insert across the membrane to form a channel in a way that never exposes the hydrophilic side chains to the lipid alkyl chains. Fig. 3 illustrates a transition 'boat' conformation in which each 'raft'is tilted from the plane of the membrane by 40°. Formation of this state could explain the voltage dependence of channel formation since two positive charges on NS cross the membrane for half of the monomers. The conformational change from the 'boat' to the channel structure involves primarily a pivoting motion about the 'ridge' of side chains formed by Thr-17, Ala-21, Ala-25, and Ser-29. These small side chains present few steric barriers for the postulated conformational change.

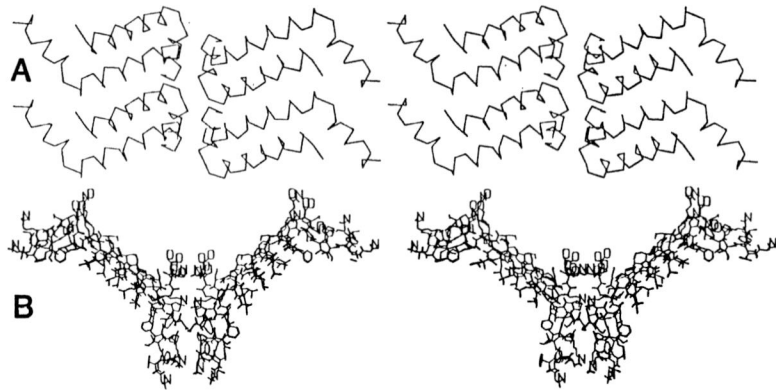

Figure 3. Stereo views of the 'boat' conformation of pardaxin. (A) View perpendicular to the membrane surface of α carbons representation. (B) View along plane of the membrane. Note negatively charged C-termini (O's) exposed to the outer surface (up) and positively charged N-termini (N's) exposed to the inner surface in the center.

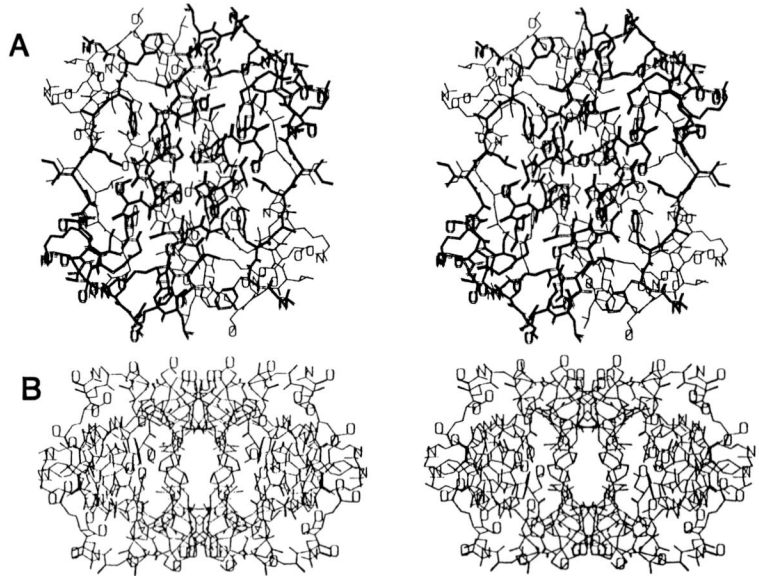

Figure 4. Stereo views of soluble tetramer. (A) Note α helices in the center and β strands on the sides. (B) Top view of 4A. Note two rows of hydrophobic side chains from SH's in the center and β strands fitting into hydrophobic grooves on the sides.

Pardaxin appears to exist primarily as a tetramer in solution. Circular dichroism studies indicate that pardaxin in water is about 22 % α and 27 % β sheet and that the α helix content increased when it associates with membranes (Lazarovici et al., 1988). In our tetramer model two HS helices form a dimer similar to that proposed for the channel (see Fig. 4). Two of these dimers associate with two-fold symmetry so that the two-reside-wide hydrophobic strip of these dimers pack next to each other. The helices between dimers cross at an angle of 40° consistent with '4-4 ridges-into-grooves' packing. This packing arrangement is similar to the crystal structure of melittin (Terwilliger and Eisenberg, 1982). The structure formed by the four HS helices leaves a hydrophobic groove on each side of the tetramer. In the model shown in Fig. 4 two NS segments form a two-stranded antiparallel β sheet in the grooves on each side of the tetramer.

GABA AND GLYCINE RECEPTORS

In developing a GABA receptor channel model we have assumed that each subunit has four transmembrane helices (M1 - M4) and that the receptor is comprised of two α and two β subunits, as postulated by Schofield, et al. (1987) when they published the sequence. We postulate that polar faces of M1 and/or M2 line the pore, that each tetramer has two channels and that the tetramers aggregate.

The first step in the modeling was to pack M1 next to M2 so that primarily polar side chains were on one side of the complex and primarily hydrophobic side chains were on the other side. Only the portion of M1 following a highly conserved proline was used at this stage. Both M1 and M2 can form amphipathic α helices in which the polar faces are comprised primarily of serine and threonine side chains that are seven residues apart. This type of seven residue repeating pattern is seen often in coiled-coils in which α helices cross each other at the angle predicted by '3-4 ridges-into-grooves packing.

M3 and M4 are more hydrophobic and less well conserved than M1 and M2. Comparison of M3 and M4 sequences among the GABA receptor α and β subunits and the Glycine receptor subunit indicates that one face of each putative α helix is conserved much less well than the opposite face. We have observed this type of 'unilateral conservation' pattern in other membrane proteins and have postulated that poorly conserved hydrophobic α helix faces are probably exposed to

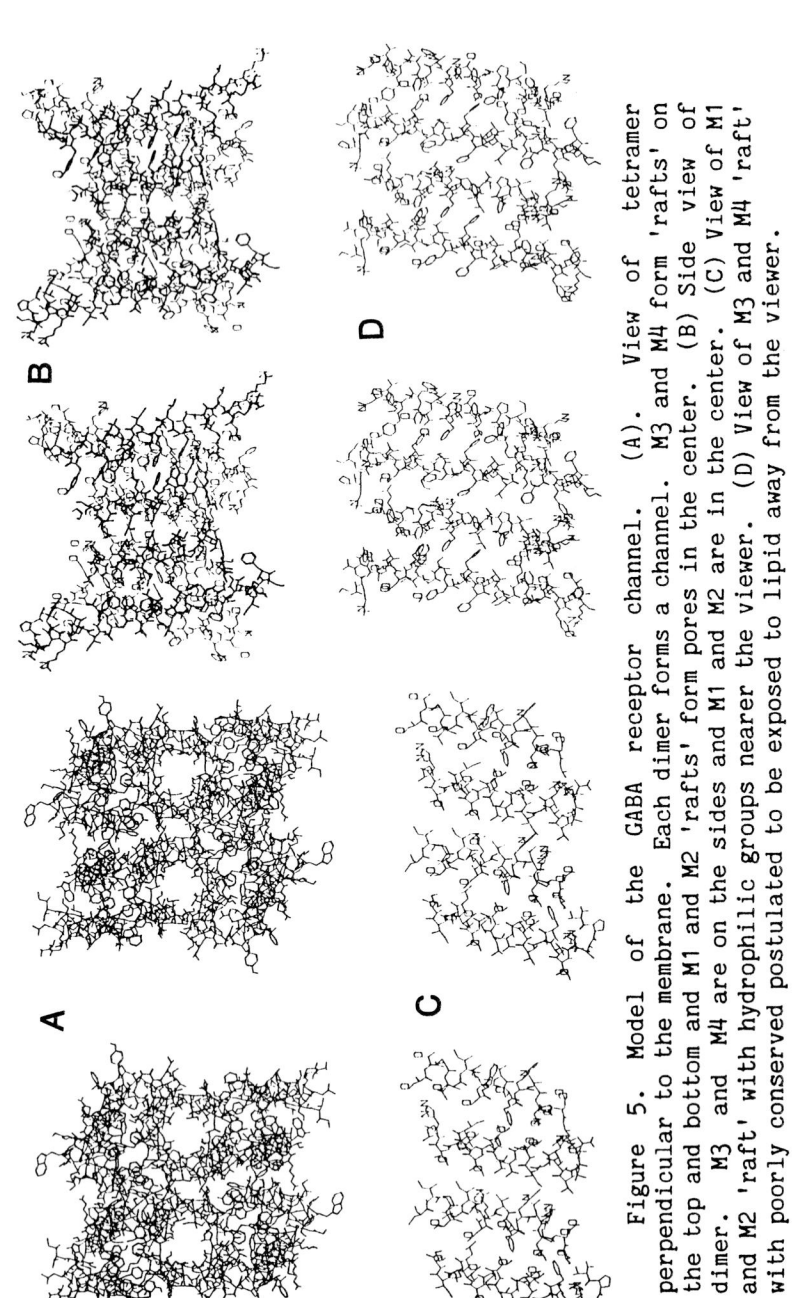

Figure 5. Model of the GABA receptor channel. (A). View of tetramer perpendicular to the membrane. Each dimer forms a channel. M3 and M4 form 'rafts' on the top and bottom and M1 and M2 'rafts' form pores in the center. (B) Side view of dimer. M3 and M4 are on the sides and M1 and M2 are in the center. (C) View of M1 and M2 'raft' with hydrophilic groups nearer the viewer. (D) View of M3 and M4 'raft' with poorly conserved postulated to be exposed to lipid away from the viewer.

lipid. M3 was packed next to M4 so that poorly conserved hydrophobic faces were on one side of the complex. The conserved faces of M3 and M4 both contain phenylalanine and tyrosine side chains that are seven residues apart. '3-4 ridges-into-grooves' packing was used to obtain an arrangement in which phenyl rings stack next to each other and a salt bridge occurs between a glutamate near the C-terminus of M3 and an arginine near the N-terminus of M4.

The two pairs of helices were combined to form models of a single subunit. The model that appears to satisfy the side chain packing criteria best is a bundle of four antiparallel α helice that has '3-4 ridges-into-grooves' packing between adjacent helices. This helix motif occurs in several soluble proteins.

The subunit models were then used to construct tetramers in which the polar groups of M1 and M2 lined the channel and the poorly conserved faces of M3 and M4 could be in contact with lipid. We were not able to construct a model that satisfied all our criteria in which the channel is formed between four subunits. We could however construct a model in which a channel is formed between two subunits if we assumed that the dimers aggregate to form larger aggregates (see Fig. 5). In this model the pore is formed between a bundle of four antiparallel helices, M1 and M2 from the α and β subunits, that have '3-4 ridges-into-grooves' packing. The pore, which is lined primarily with hydroxyl groups from the threonine and serine side chains, is just large enough to allow passage of the largest known permeant organic anions. Positively charged arginines at each end of M2 and a histidine near the C-terminus of the β subunit M2 should be sufficient to make the channel selective for anions. Only a few side chains postulated to be buried between dimers are not conserved among both GABA subunits and the Glycine subunit. Most of these side chains are hydrophobic. The few buried side chains that can form hydrogen bonds are positioned where they can form hydrogen bonds with hydrogen bond donors and acceptors on other side chains. The helix interactions between dimers have '4-4 ridges-into-grooves packing'. Several charged side chains near the ends of the helices or in the loops connecting helices can form salt bridges. The side chains on the exterior of the aggregate are hydrophobic and poorly conserved. We consider this model as very tentative because the number of subunits per channel is not known.

The GABA receptor model has several features similar to the peptide models. The channel lining is comprised primarily of serine and threonine side chains like the pardaxin model. In the postulated aggregates 'rafts' are formed by M3 and M4 and by M1 and M2 (see Fig. B and C). The interactions in the aggregates of M1 and M2 are very similar to those of the HS's in the pardaxin tetramer model. Interactions of the phenylalanine side chains on M3 and M4 are similar to those of the magainin model. These similarities suggest information obtained by determining structures of membrane peptide aggregates may be useful in understanding structures of larger membrane channel proteins.

REFERENCES

Brooks, B.R., Bruccoleri, R.E., Olafson, B.D., States, D.J., Swaminathan, S., and Karplus, M. (1983). CHARMM: A program for macromolecular energy, minimization and dynamics calculations. J. Compt. Chem. 4, 187-217.

Chothia, C., Levitt, M., Richardson, D. (1981) Helix to helix packing in proteins. J. Mol. Biol. 145, 215-250.

Connolly, M.L. (1983) Analytical molecular surface calculation. J. Appl. Cryst. 16, 548-558.

Garnier, J., Osguthorpe, D.J., and Robson, B., (1978). Analysis of the accuracy and implications of simple methods for predicting the secondary structure of globular proteins. J. Mol. Biol. 120, 97-120.

Lazarovici, P., Primor, N., Lelkes, P.I., Fox, J., Shai, Y., Raghunthan, G., Guy, H.R., Shih, Y.L., Edwards, C. (1988) Pardaxin interaction with artifical membranes: Sequence determination, toxin conformational changes and phosphatidylserine vesicles aggregation. Biochemistry (submitted for publication)

Schofield, P.R., Darlison, M.G., Fujita, N., Burt, D.R., Stephenson, F.A., Rodriguez, H., Rhee, L.M., Ramachandran, J., Reale, V., Glencorse, T.A., Seeburg, P.H., and Barnard, E.A. (1987) Sequence and functional expression of the $GABA_a$ receptor shows a ligand-gated receptor super-family. Nature 328:221-227.

Tappin, M.J., Pastore, A., Norton, R.S., Freer, J.H., and Campbell, I.D. (1988) High-resolution ^{1}H NMR study of the solution structure of δ-Hemolysin. Biochemistry 27:1643-1647.

Terwilliger,T.C.,Eisenberg, D. (1982) The structure of melittin. I. Structure 1mlta 2 determination and partial refinement. J.Biol.Chem. 257:6010

Thomas, D.H., Rice, D.W., and Fitton, J.E. (1986) Crystallization of the delta toxin of Staphylococcus aures. J. Mol. Biol. 192: 675.

Zasloff, M., Martin, B., and Chen, H-C. (1987) Antimicrobial activity of synthetic magainin peptides and several analogues. Proc. Natl. Acad. Sci. (USA). *4:

STRUCTURE OF NICOTINIC ACETYLCHOLINE RECEPTORS FROM MUSCLE AND NEURONS

Jon Lindstrom, Paul Whiting, Ralf Schoepfer, Michael Luther, and Manoj Das

The Salk Institute for Biological Studies
P.O. Box 85800, San Diego, California 92138 USA

Nicotinic acetylcholine receptors (AChRs) from muscles and nerves evolved as part of a gene family composed of ligand-gated ion channels formed by several homologous subunits (reviewed in Lindstrom et al., 1987b). The structure of muscle AChRs has evolved conservatively, as indicated by the 80% sequence identity between the ACh-binding subunits of AChRs from an elasmobranch electric organ and human skeletal muscle (Noda et al., 1983a). By contrast, the structure of neuronal nicotinic receptors has evolved rapidly, perhaps in response to various functional roles in the nervous system, resulting in multiple neuronal AChR subtypes, which differ significantly within and between species (Lindstrom et al., 1987b). Whereas muscle AChRs have four kinds of subunits in a pentameric arrangement (Kubalek et al., 1987), neuronal AChRs are composed of only two kinds of subunits in an as yet uncertain arrangement (Lindstrom et al., 1987b).

In the hopes of clarity, we will review our structural studies starting from the best understood AChR, the Torpedo electric organ, AChR and proceed through human muscle AChR to the least well-characterized AChRs, those of neurons.

AChRs from Torpedo californica electric organ are composed of four kinds of subunits termed α (molecular weight calculated from cDNA 50,116), β (mw 53,681), γ (mw 56,279), and δ (mw 57,565) (Noda et al., 1983b,c). Their sequence homology (19% of the aligned amino acids are identical in all four subunits) suggests evolution by gene duplication (Raftery et al., 1980; Noda et al., 1983b). The α subunits contain the ACh binding site as shown by affinity labeling of a unique disulfide-linked pair of cysteines at positions α192,193 by the antagonist MBTA (Kao et al., 1984; Kao and Karlin, 1986). Subunit stoichiometry is $\alpha_2\beta\gamma\delta$ (Reynolds and Karlin, 1978; Lindstrom et al., 1979; Raftery et al., 1980). By optical diffraction analysis of electronmicrographs of 2D crystalline arrays of these AChRs, Nigel Unwin and a coworker (Brisson and Unwin, 1985) have shown that the subunits are organized like barrel staves around a central channel, presumed to be the cation channel whose opening is triggered by the binding of ACh. We have provided subunit-specific monoclonal antibody (mAb) probes which helped to determine that the subunits are organized in the order $\alpha\beta\alpha\gamma\delta$ around the channel as shown in Figure 1 (Kubalek et al., 1987). This order of subunits is supported by the results of some (Hamilton et al., 1985) but not others (Karlin et al, 1983).

αBungarotoxin (αBgt) binding localizes the ACh binding site to the top of α subunits, whereas binding of mAb35 localizes the main immunogenic region (MIR) to the side of the extracellular surface of α subunits (Kubalek et al., 1987).

The ACh binding site is formed at least in part by sequences near α192,193, as shown by several different affinity labeling experiments and experiments with αBgt binding to synthetic peptides (Kao et al., 1984; Wilson et al., 1985; Neumann et al., 1986; Dennis et al., 1986; Ralston et al, 1987).

The MIR, like the ACh binding site, depends on the native conformation of the AChR, but there are no affinity labeling reagents for this site; thus it is difficult to map the MIR precisely on the primary structure. By peptide mapping we localized the MIR to within α46-127 (Ratnam et al., 1986b), but using small synthetic peptides we could not map it more precisely (Ralston et al., 1987). Mapping of the MIR at higher resolution within this sequence has been reported by others using synthetic peptides (Tzartos et al., 1988) or bacterially expressed fragments of α subunits (Barkas et al., 1987). AChRs from Xenopus muscle do not react with mAbs to the MIR, unlike AChRs from muscles of all other species tested (Sargent et al., 1983); yet in the sequence α74-78, which Tzartos et al. (1988) report to form part of the MIR, α subunits from Xenopus do not differ from all others (Baldwin et al., 1988), which argues against this high resolution mapping of the MIR.

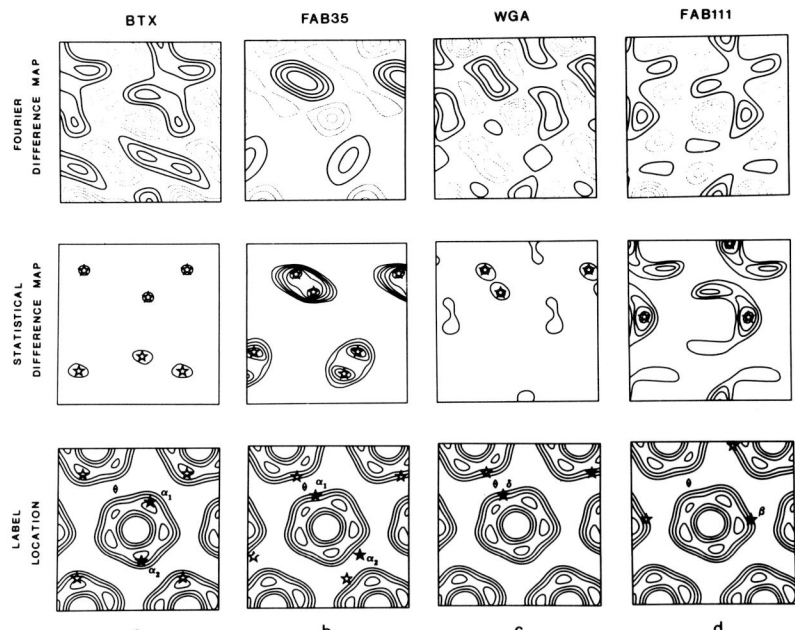

Figure 1. Location of subunits, ACh binding sites, and the MIR on AChRs in 2D crystalline arrays. α subunits were localized with αBgt (to the ACh binding site) and Fab fragments of mAb35 (to the MIR). β subunits were localized with Fab fragments of mAb111. δ subunits were localized with wheat germ agglutinin. γ subunits were localized by difference from the others. Tubular crystalline arrays of <u>Torpedo marmorata</u> AChRs were labeled with each ligand, then electron micrographs were analyzed by optical diffraction. The electron density maps were compared with unlabeled AChRs to map the sites of ligand binding. Reproduced from Kubalek et al. (1987).

The transmembrane orientation of the subunit polypeptide chain has been investigated by several methods. Sequence homologies among subunits suggest that they should all have basically the same transmembrane orientation (Noda et al., 1983b). In this paragraph we consider the transmembrane orientation starting at the N-terminus and concentrate primarily on α subunits. The presence of leader sequences (Noda et al., 1983b) and other evidence (Anderson et al., 1983) suggests that the N-termini may be on the extracellular surface. The N-termini are not accessible to antibodies in the native AChR (Neumann et al., 1985; Lindstrom et al.,

1984). Both in vitro mutagenesis experiments (Mishina et al., 1985) and competition experiments using anti-peptide antibodies and lectins (Criado et al., 1986) suggest that α subunits are glycosylated at asparagine α141 on the extracellular surface. There is a major epitope around α160 (see Fig. 2). This is interesting because all antibodies to denatured subunits are thought to bind to the cytoplasmic surface (Froehner, 1981). mAb236 to this epitope binds to only about 20% of AChRs in solid phase assays or in detergent solution (unpublished). Some evidence suggests that this mAb binds to the cytoplasmic surface (Criado et al., 1985a). This would imply the existence of transmembrane domains linking this sequence to the surface. However, by colloidal gold electron microscopy, mAb236 does not bind unless the membranes are treated with 3M KSCN, and then binding is observed on the extracellular surface (Peter Sargent et al., unpublished). Antibodies to this epitope which have been affinity purified from antiserum to α subunits are not efficiently bound by native AChR (Fig. 3). Thus, this epitope is not readily accessable in the native AChR, and its transmembrane orientation is still uncertain. A problem hampering studies of the transmembrane orientation of the polypeptide chain has been that anti-peptide antibodies frequently do not bind to the native AChR. Evidence that sequences around α192,193 are near the ACh binding site places this region on the extracellular surface. A hydrophobic sequence between α210 and α236 is the first of four hydrophobic sequences, termed M1-M4, observed in each subunit. It has been suggested that M1-M4 form hydrophobic transmembrane α helices (Claudio et al., 1983; Noda et al., 1983b; Devillers-Thiery et al., 1983). M1-M3 are highly conserved between species and subunit types. The requirement to simply pass through the hydrophobic core of the lipid bilayer would not seem to require great conservation of sequence, whereas the requirement to specifically interface with sequences of the same subunit or other subunits might well require great conservation of sequence. Some evidence suggests that mAbs to α235-242 bind to the cytoplasmic surface, which is consistent with M1 being a transmembrane domain (Criado et al., 1985b). M2 corresponds to α244-277. M3 corresponds to the α sequence 273-296. Two types of experiments suggest that M2 may compose or be near to the 'barrel stave' contributed by each subunit to the lining of the cation channel: 1) noncompetitive channel blocking inhibitors affinity label serines homologous to α248 in α, β, and δ subunits (Hucho et al., 1986; Giraudat

Nicotinic AChR Structure / 249

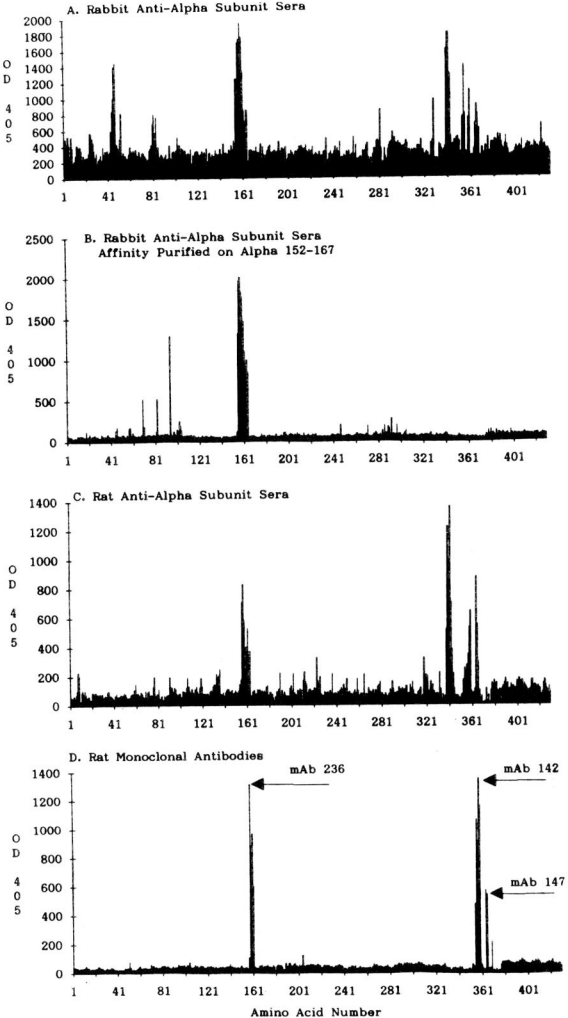

Figure 2. Mapping epitopes on AChR α subunits by the technique of Geysen et al. (1987). In these experiments overlapping synthetic peptide octomers 1-8, 2-9, etc. for the complete sequence of α subunits of AChRs from Torpedo californica were synthesized on plastic pegs in a format that fits 96-well microliter dishes used for ELISA assays. Antibodies were allowed to bind to the pegs, and, after rinsing, bound antibodies were detected using peroxidase-labeled anti-antibody. After the assay the antibodies were eluted from the pegs with SDS and the solid phase assays were repeated with other antibodies. In A notice that there are

(Fig. 2 legend continued)
several epitopes on α detectable in this way. In B antibodies to one of these epitopes were affinity purified and retested, validating the technique. In C note that rat antisera to α subunits detect many of the epitopes detected by rabbit antisera. In D note that rat mAbs are available for several of these epitopes.

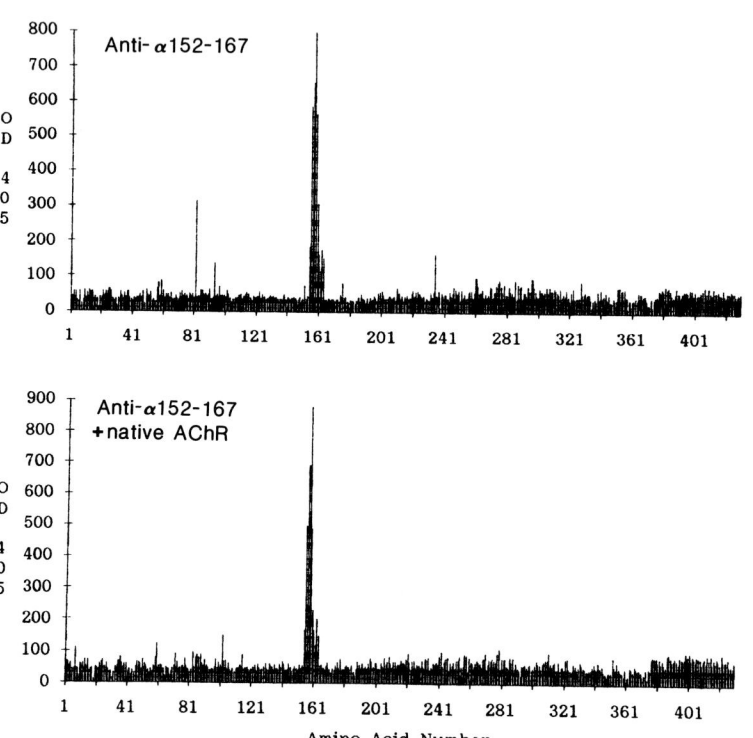

Figure 3. Affinity-purified antibodies to the epitope around α160 are not efficiently adsorbed by native AChR. Rabbit antibodies to α subunits affinity purified on α152-167 were assayed by the Geysen et al. (1987) method as shown in Figure 2. The affinity-purified antibodies had a titer of only ~2nM against ^{125}IαBgt-labeled AChR, due either to weak affinity for the native AChR or to presence of a small fraction of partially denatured AChRs. A 10-fold excess of native AChR was added while the affinity-purified antibody was again exposed to the solid phase synthetic peptides. Note that antibody binding to the peptides was not greatly inhibited.

et al., 1987), and 2) expression in Xenopus oocytes of chimeric cDNAs for Torpedo and bovine AChR δ subunits indicate that M2 and the sequence between M2 and M3 are involved in determining the rate of ion transport through the open channel (Imoto et al., 1986). The sequence α339-378 is not highly conserved between subunits or species. Perhaps as a consequence, all subunits seem to have several epitopes in this region (Ratnam et al., 1986b). Antibodies to these epitopes bind well to both native and denatured AChRs (Kordossi and Tzartos, 1987), suggesting that these sequences are not highly structurally constrained in the native molecule. mAbs and anti-peptide sera specific for several epitopes in several subunit sequences corresponding to this region have been mapped to the cytoplasmic surface (Ratnam et al., 1986a, b; Young et al., 1985; LaRochelle et al., 1985). Amphipathic transmembrane α helices from each subunit corresponding to α373-α397 have been proposed to form the 'barrel staves' which line the central cation channel (Guy, 1983; Finer-Moore and Stroud, 1984). This sequence is not as conserved between subunits and species as are M1-M3. Antisera to α389-396 bind to the cytoplasmic surface, suggesting that the amphipathic sequence α373-397 is not a transmembrane domain, but exposed on the cytoplasmic surface (Ratnam et al., 1986a,b). Deletion or replacement of the entire amphipathic sequence of α subunits still yields partially functional AChRs in oocyte expression studies (Mishina et al., 1985; Tobimatsu, 1987). Trypsinization can remove the amphipathic sequence (Roth et al., 1987) yet trypsinized AChRs retain channel function (Lindstrom et al., 1980). All of these results argue that this amphipathic sequence is not the barrel stave contributed by each subunit to the lining of the channel. M4, between α409 and α428 is less well conserved than M1-M3. Antibodies to the C-termini are found to bind to the cytoplasmic surface (Lindstrom et al., 1984; Ratnam et al., 1986a, b; Young et al., 1985). M4 of α subunits can be replaced by foreign sequences without loss of channel activity (Tobimatsu et al., 1987). Thus evidence suggests that much or all of the sequence C-terminal of M3 is on the cytoplasmic surface, and that M4 is not an important transmembrane domain. However, some evidence may conflict with immunological localization of the C-termini on the cytoplasmic surface. There is a cysteine penultimate to the C-terminus of Torpedo AChR δ subunits that may be the cysteine through which δ subunits of AChR dimers are disulfide linked. There is evidence that the disulfide bond between δ subunits is accessible to reduction from the extracellular surface of reconstituted vesicles (McCrea et al., 1987).

Theoretical models have been proposed with four (Claudio et al., 1983; Devillers-Thiery et al., 1983; Noda et al., 1983b) or five (Finer-Moore and Stroud, 1984; Guy 1983) transmembrane domains in each AChR subunit. Clearly, unequivocal experimental data for the transmembrane orientation of many parts of the subunit polypeptide chains has been difficult to obtain. There is data in conflict with both models. In vitro mutagenesis does not readily give information about transmembrane orientation. Antibodies can be made to many parts of the sequence, but often these do not bind to the native AChR. Thus the transmembrane orientation of many parts of the sequence cannot be determined using antibodies. Ultimately, X-ray crystallography should determine the transmembrane orientation of the polypeptide chains, but this may still take a while.

AChRs from muscle are presumed to have the same basic structure as AChRs from electric organs, as a result of the extensive sequence homologies between their subunits. We have recently been studying human muscle AChRs.

We have made the curious, but useful, observation that the human neuromedulloblastoma cell line TE671 expresses muscle AChRs (Lindstrom et al., 1987a; Schoepfer et al., 1988a; Luther et al., submitted). This result is curious because muscle AChRs are not expressed in adult human brain (Whiting et al., 1987a) and because the αBgt binding sites produced by this cell line had been thought to correspond to those found in human brain (Syapin et al., 1982; Lukas, 1986). This result is useful because no cell line which produces human muscle AChRs had existed (the only source for purification had been amputated legs) and because human AChRs are needed for studies of myasthenia gravis (MG), a disease in which muscle weakness is caused by autoantibodies to AChRs (Lindstrom et al., 1988).

Muscle AChRs from TE671 cells are found to have four kinds of subunits which correspond antigenically to those of Torpedo AChR by Western blot analysis (Luther et al., submitted). The sequence of the α subunits of AChRs determined by cDNA analysis is ≥95% identical to α subunits of AChRs from muscles of other mammals (Schoepfer et al., 1988a). The sequence is 80% identical to Torpedo α subunits and preserves cysteines at α128,142,192, and 193, a putative N-glycosylation site at α141, and many other features. The δ subunit shows ≥90% sequence identity with δ subunits from AChRs of other mammalian muscles (Luther et al., submitted). There is 30% sequence identity between α and δ subunits, revealing their evolution from a common ancestor. For example, δ has cysteines corresponding to α128,142; however, δ subunits lack cysteines corresponding to α192,193 that are unique to ACh- binding subunits. They also lack a cysteine at δ500 which is unique to Torpedo (Noda et al., 1983c). This is interesting because Torpedo AChRs exist as dimers linked by disulfide bonds between their δ subunits (Chang and Bock, 1977) and this cysteine may be the site of this bond (Noda et al., 1983c). Muscle AChRs are usually found as monomers. However, AChRs in TE671 cells are found as noncovalently associated dimers (Luther et al., submitted). This may be a peculiarity of these AChRs, or it may be true of other muscle AChRs and not usually detected due to proteolysis or some other effect.

There are nicotinic AChRs in neurons, but until recently they have not been studied at the molecular level for lack of suitable probes. However, in the past few years rapid progress has been made using mAb and cDNA probes initially based on Torpedo AChR (reviewed in Lindstrom et al., 1987b). Our approach has been to use crossreacting mAbs to identify and purify neuronal AChR proteins, then use mAbs to these neuronal AChRs to immunoaffinity purify others. Despite the presence of significant overall sequence homologies, the great specificities of mAbs for small surface features and their frequent conformation dependence results in very little immunological crossreaction between AChRs from muscle and nerves, or between neuronal AChRs of one species and another.

Figure 4 summarizes the results of our immunoaffinity purification studies. AChRs from eel electric organ and muscle have four kinds of subunits, one of which binds ACh as shown by affinity labeling with MBTA, and three of which serve 'structural' roles (Lindstrom et al., 1980, 1983; Conti-Tronconi et al., 1982). mAb35 raised against eel AChRs (Tzartos et al., 1981) crossreacts with muscle AChRs from many species including human (Tzartos et al., 1982) and chicken (Role et al., 1985). It also crossreacts with both ganglionic (Smith et al., 1986) and brain AChRs from chicken (Swanson et al., 1983; Whiting and Lindstrom, 1986a). mAb35

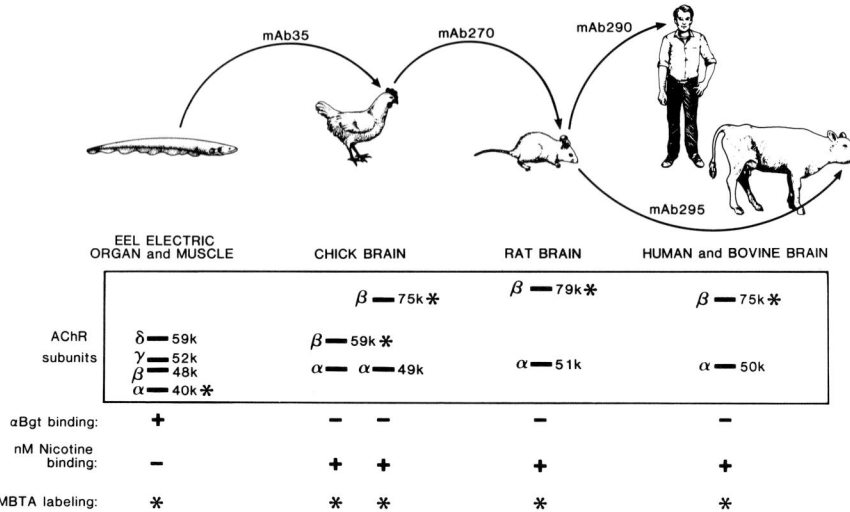

Figure 4. Summary of immunoaffinity purification of neuronal AChRs. The ACh-binding subunits were identified by affinity labeling with MBTA and are indicated by *. The other subunits are referred to as 'structural.' The α, β... terminology is in order of increasing molecular weight according to the convention used for electric organ AChRs. Reproduced from Lindstrom et al. (1987b).

binds to the main immunogenic region, a highly conserved determinant of unknown function which is very conformation-dependent; consequently, we have detected no reaction of mAb35 with synthetic peptides (Ralston et al., 1987). mAb35 immunoaffinity purifies AChRs from chicken brains which are composed of two kinds of subunits (Whiting and Lindstrom, 1986a). mAb270 raised to these purified AChRs binds to the structural subunit. Using mAb270 a second AChR subtype was purified from chicken brain (Whiting et al., 1987b). It has a structural subunit identical in molecular weight, N-terminal amino acid sequence, and peptide map to the structural subunit of the AChR subtype recognized by mAb35 (Whiting et al., 1987b; Schoepfer et al., 1988b), but the ACh-binding subunit of this AChR has a higher molecular weight. Both subtypes are present in approximately equal amounts in adult chicken brain, have similarly high affinity for nicotine, and do not bind αBgt. Both brain AChR subtypes differ from the AChR in chicken ganglia, which has much lower affinity for ACh and nicotine, indicating that at least its ACh-binding subunit differs (Halvorsen and Berg, 1986). mAb270 crossreacts with both the ganglionic AChR in rat PC12 cells (Whiting et al., 1987c) and with AChRs with high affinity for nicotine in rat brain (Whiting and Lindstrom, 1986b, 1987a; Swanson et al., 1987). Only one AChR subtype comprising >90% of the high affinity nicotine binding sites in rat brain is immunoaffinity purified on mAb270 (Whiting and Lindstrom, 1987a). Its two subunits correspond immunologically and by N-terminal protein sequence to those of the AChR subtype from chicken brain which has an ACh-binding subunit of similar molecular weight (Whiting et al., 1987b). In both species, AChRs of this subtype exist at least in part in presynaptic or extrasynaptic locations (Swanson et al., 1987) by contrast with the postsynaptic functional role of AChRs from muscle or ganglia. Using mAbs to AChRs from rat brain, similar AChRs have been identified in bovine and human brains, and have been subsequently purified from bovine brains (Whiting and Lindstrom, 1988).

cDNAs corresponding to some of the subunits of neuronal AChRs have been identified, including cDNAs for both of the subunits of the major subtype of mammalian brains and their chicken counterpart. By low stringency hybridization several cDNAs with homologies to muscle AChR ACh-binding subunits have been identified (Boulter et al., 1986, 1987; Goldman et al., 1987; Nef et al., 1988), as have cDNAs corresponding to a structural subunit (Deneris et al., 1988; Nef et al., 1988; Schoepfer et al., 1988b). Coexpression in Xenopus oocytes of cDNAs for the structural subunit and any of these putative ACh-binding subunits yields functional AChRs (Boulter et al., 1987). N-terminal amino acid sequence analysis of the ACh-binding subunit of AChRs purified from rat brain shows that they correspond to the cDNA α4 (Whiting et al., 1987d). N-terminal amino acid sequence analysis of the structural subunits (Schoepfer et al., 1988b) shows that they correspond to the rat cDNA β_2 or the chicken cDNA nα (Deneris et al., 1988; Nef et al., 1988). Immunoaffinity purification on mAb270 to the structural subunit reveals only one AChR subtype from rat brain (Whiting and Lindstrom, 1987a). If there are other subtypes that share this common structural subunit, they must be present in small amounts, or their ACh-binding subunits must obscure the mAb270 epitope. The cDNA for the structural subunit shared by the two AChR subtypes in chicken brain has been identified by

N-terminal amino acid sequence analysis (Schoepfer et al., 1988b). It is 84% identical in sequence to the rat cDNA β_2 and 100% identical to the chicken cDNA nα. The cDNA for the larger apparent molecular weight ACh-binding subunit from chicken brain has also been identified by N-terminal amino acid sequence analysis (Whiting et al., unpublished). It is 71% identical in sequence to the rat cDNA α4 (Goldman et al., 1987) and 100% identical to the chicken cDNA α4 (Nef et al., 1988).

Figure 5. Comparison of amino acid sequence of some AChR subunits deduced from cDNAs. The chicken brain AChR structural subunit sequence is from Schoepfer et al. (1988b). The other sequences are from Boulter et al. (1987). This figure is reproduced from Schoepfer et al. (1988b).

Analysis of the sequences of the cDNAs for subunits of neuronal nicotinic AChRs reveals the presence of four hydrophobic domains corresponding to those of muscle AChR. Some of these sequences are shown in Figure 5. The second hydrophobic segment is especially well conserved. All of the identified neuronal AChR subunits have cysteines corresponding to muscle α128 and 142, and an N-glycosylation site corresponding to α141. The neuronal ACh-binding subunits have cysteines corresponding to muscle α192,193 which accounts for their MBTA reactivity. There are additional unique potential sites for glycosylation. All of the subunits have long sequences in the region corresponding to the nonhomologous, highly immunogenic, loosely conformed sequences on the cytoplasmic surface of muscle AChRs. This would seem to allow the other more rigidly conformed homologous regions to stay in register. In the case of the larger molecular weight ACh-binding subunits this domain is especially protracted. Muscarinic AChR subtypes show a similar segmental homology, with each subtype having a unique putative cytoplasmic domain (Peralta et al., 1988).

The cytoplasmic domains of neuronal AChRs are especially interesting for several reasons. These domains would be exposed on the surface of vesicles containing newly synthesized AChRs, and presumably would be the surface by which they would be recognized for the axonal transport to distant presynaptic or extrasynaptic locations. The cytoplasmic surface would presumably interact with extrinsic membrane proteins and cytoskeletal elements to anchor these AChRs at their functional locations. Finally, this surface might be subject to regulatory actions like phosphorylation, which could affect AChR function, desensitization, or turnover.

Subunit stoichiometries of neuronal AChRs composed of only two kinds of subunits are interesting to consider. On sucrose gradients these AChRs sediment as monomers, slightly larger than electric organ AChR monomers (Whiting and Lindstrom, 1986a, 1987a, 1988; Whiting et al, 1987b, d). AChRs immobilized on agarose beads via mAb35 can still bind ^{125}ImAb35. Experiments of this type using mAbs specific for both subunits show that there must be at least two subunits of each type per AChR monomer (Whiting and Lindstrom, 1986a, 1987a; Whiting et al., 1987b, d). Their sedimentation properties could accommodate no more than three of one type of subunit (Whiting et al., 1987d). If they preserve the pentagonal symmetry of AChRs from Torpedo, there would be two ACh-binding subunits and three structural subunits (as shown in Fig. 6). The problem with this arrangement is that a structural subunit would have to specifically associate at the same site with both an ACh-binding subunit and another structural subunit. A simple tetrameric arrangement would avoid this complication. GABA$_A$ receptors, which are members of the same gene superfamily are thought to have this subunit arrangement (Barnard et al., 1987). The characteristic feature of this gene family seems to be the formation of a

Figure 6. Subunit stoichiometries of AChRs from muscle and electric organs. Reproduced from Lindstrom et al. (1987b).

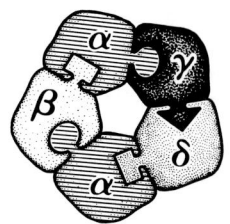

Muscle-type AChR
$\alpha_2\beta\gamma\delta$

ligand-gated ion channel from multiple homologous subunits. This requires highly specific association between subunits, but may not require the conservation of a pentameric symmetry any more than the conservation of the ligand-binding site, or cation, or anion specificity of the channel. However the subunits associate to form the channels of neuronal AChRs, the electrophysiological properties of the channel (Lipton et al., 1986) are not much different from those of muscle AChR (Sakmann et al., 1983).

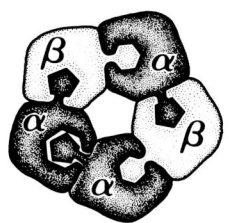

Neuronal AChR
$\alpha_3\beta_2$

In conclusion, AChRs of muscle and neurons are members of a gene family of ACh-gated cation channels formed from multiple homologous subunits. The properties of the ion channel and the sequences of the subunits responsible for specific association of the subunits within the membrane to form this channel appear to be highly conserved. Unique putative cytoplasmic domains distinguish the subunits of members of the AChR family. The neuronal cousins in the family all appear similar in having only two kinds of subunits, rather than the four which characterize muscle AChRs. The ACh-binding

Neuronal AChR
$\alpha_2\beta_2$

subunits in all members of the family appear to be characterized by the presence of cysteines near the ACh binding site which are homologous to cysteines α192,193 of muscle AChR. These cysteines are present despite differences in pharmacological properties between these AChRs. The nicotinic AChR gene family is part of a gene superfamily of receptors with ligand-gated ion channels (reviewed in Barnard et al, 1987), including the receptors for GABA and glycine which not only bind different ligands and lack cysteines corresponding to α192,193 in their ligand-binding subunits, but also form anion channels. Yet aspects of their sequences which may correspond to areas involved in specific association of the subunits to form channels are the most conserved. These other receptors most closely resemble the neuronal lineage of the AChR gene family, especially in having only two kinds of subunits.

ACKNOWLEDGMENTS

Research in the laboratory of J.L. is supported by grants from the NIH (NS11323), the Muscular Dystrophy Association, the U.S. Army (DAMD17-86-C-6148), the Alexander Onassis Public Benefit Foundation, the Council for Tobacco Research-USA, Inc., and the California chapter of the Myasthenia Gravis Foundation. R.S. was supported by a fellowship from the Deutsche Forschungsgemeinschaft. M.L. was supported by a fellowship from the Muscular Dystrophy Association. We thank Lisa Churchill-Roth for helping to prepare the manuscript.

REFERENCES

Anderson D, Blobel G, Tzartos S, Gullick W, Lindstrom J (1983). Transmembrane orientation of an early biosynthetic form of acetylcholine receptor delta subunit determined by proteolytic dissection in conjunction with monoclonal antibodies. J Neurosci 3:1773-1784.

Baldwin T, Yoshihara C, Blackmer K, Kintner C, Burden S (1988). Regulation of acetylcholine receptor transcript expression during development in Xenopus laevis. J Cell Biol 106:469-478.

Barkas T, Mauron A, Roth B, Alliod C, Tzartos S, Ballivet M (1987). Mapping the main immunogenic region and toxin binding site of the nicotinic acetylcholine receptor. Science 235:77-80.

Barnard E, Darlison M, Seeburg P (1987). Molecular biology of the $GABA_A$ receptor: The receptor/channel superfamily. Trends in Neurosci 10:502-509.

Boulter J, Connolly J, Deneris E, Goldman D, Heinemann S, Patrick J (1987). Functional expression of two neuronal nicotinic acetylcholine receptors from cDNA clones identifies a gene family. Proc Natl Acad Sci USA 84:7763-7767.

Boulter J, Evans K, Goldman D, Martin G, Treco D, Heinemann S, Patrick J (1986). Isolation of a cDNA clone coding for a possible neural nicotinic acetylcholine receptor alpha subunit. Nature 319:368-374.

Brisson A, Unwin P (1985). Quaternary structure of the acetylcholine receptor. Nature 315:474-477.
Chang H, Bock E (1977). Molecular forms of the acetylcholine receptor: Effects of calcium ions and a sulfhydryl reagent on the occurence of oligomers. Biochemistry 16:4513-4520.
Claudio T, Ballivet M, Patrick J, Heinemann S (1983). Torpedo californica acetylcholine receptor 60,000 dalton subunit: Nucleotide sequence of cloned cDNA deduced amino acid sequence, subunit structural predictions. Proc Natl Acad Sci USA 80:1111-1115.
Conti-Tronconi B, Hunkapillar M, Lindstrom J, Raftery M (1982). Subunit structure of the acetylcholine receptor from Electrophorus electricus. Proc Natl Acad Sci USA 79:6489-6493.
Criado M, Hochschwender S, Sarin V, Fox JL, Lindstrom J (1985a). Evidence for unpredicted transmembrane domains in acetylcholine receptor subunits. Proc Natl Acad Sci USA 82:2004-2008.
Criado M, Sarin V, Fox JL, Lindstrom J (1985b). Structural localization of the sequence α235-242 of the nicotinic acetylcholine receptor. Biochem Biophys Res Commun 128:864-871.
Criado M, Sarin V, Fox JL, Lindstrom J (1986). Evidence that the acetylcholine binding site is not formed by the sequence α127-143 of the acetylcholine receptor. Biochemistry 25:2839-2846.
Deneris E, Connolly J, Boulter J, Wada E, Wada K, Swanson L, Patrick J, Heinemann S (1988). Primary structure and expression of β2: A novel subunit of neuronal nicotinic acetylcholine receptors. Neuron 1:45-54.
Dennis M, Giraudaut J, Kotzba-Hibert F, Goeldner M, Hirth C, Chang J-Y, Changeux J-P (1986). A photo affinity ligand of the acetylcholine binding site predominantly labels the region 179-207 of the alpha subunit on native acetylcholine receptor from Torpedo marmorata. FEBS Lett 207:243-249.
Devillers-Thiery A, Giraudat J, Bentaboulet M, Changeux J-P (1983). Complete mRNA coding sequence of the acetylcholine-binding alpha subunit of Torpedo marmorata acetylcholine receptor: A model for the transmembrane organization of the polypeptide chain. Proc Natl Acad Sci USA 80:2067-2071.
Finer-Moore J, Stroud R (1984). Amphipathic analysis and possible formation of the ion channel in an acetylcholine receptor. Proc Natl Acad Sci USA 81:155-159.

Froehner S (1981). Identification of exposed and buried determinants of the membrane-bound acetylcholine receptor from **Torpedo californica**. Biochemistry 20:4905-4915.

Geysen H, Rodda S, Mason T, Tribbick G, Schoofs P (1987). Strategies for epitope analysis using peptide synthesis. J Immunol Meth 102:259-274.

Giraudat J, Dennis M, Heidmann T, Hanmont P-Y, Lederer F, Changeux J-P (1987). Structure of the high-affinity binding site for noncompetitive blockers of the acetylcholine receptor: [^{3}H] Chlorpromazine labels homologous residues in the β and δ chains. Biochemistry 25:2410-2418.

Goldman D, Deneris E, Luyten W, Kochhar A, Patrick J, Heinemann S (1987). Members of a nicotinic acetylcholine receptor gene family are expressed in different regions of the mammalian central nervous system. Cell 48:965-973.

Guy R (1983). A structural model of the acetylcholine receptor channel based on partition energy and helix packing calculations. Biophys J 45:249-261.

Halvorsen S, Berg D (1986). Identification of a nicotinic acetylcholine receptor on neurons using an alpha neurotoxin that blocks receptor function. J Neurosci 6:3405-3412.

Hamilton S, Pratt D, Eaton D (1985). Arrangement of the subunits of the nicotinic acetylcholine receptor of **Torpedo californica** as determined by alpha neurotoxin cross-linking. Biochemistry 24:2210-2219.

Hucho F, Oberthur W, Lottspeich F (1986). The ion channel of the nicotinic acetylcholine receptor is formed by the homologous helices MII of the receptor subunits. FEBS Lett 205:137-142.

Imoto K, Methfessel C, Sakmann B, Mishina M, Mori Y, Konno T, Fukuda K, Kurasaki M, Bujo H, Fujita Y, Numa S (1986). Location of a delta subunit region determining ion transport through the acetylcholine recptor channel. Nature 324:670-674.

Kao P, Dwork A, Kaldany R, Silver M, Wideman J, Stein S, Karlin A (1984). Identification of the alpha subunit half cystine specifically labeled by an affinity reagent for the acetylcholine receptor binding site. J Biol Chem 259:11622-11665.

Kao P, Karlin A (1986). Acetylcholine receptor binding site contains a disulfide crosslink between adjacent half-cystinyl residues. J Biol Chem 261:8085-8088.

Karlin A, Holtzman E, Yodh N, Label P, Wall J, Hainfeld J (1983). The arrangement of the subunits of the acetylcholine receptor of **Torpedo californica**. J Biol Chem 258:6678-6681.

Kordossi A, Tzartos S (1987). Conformation of cytoplasmic segments of acetylcholine receptor alpha and beta subunits by monoclonal antibodies: Sensitivity of the antibody competition approach. EMBO J 6:1605-1610.

Kubalek E, Ralston S, Lindstrom J, Unwin N (1987). Location of subunits within the acetylcholine receptor: Analysis of tubular crystals from Torpedo marmorata. J Cell Biol 105:9-18.

LaRochelle W, Wray B, Sealock R, Froehner S (1985). Immunochemical demonstratioon that amino acids 360-377 of the acetylcholine receptor gamma subunit are cytoplasmic. J Cell Biol 100:684-691.

Lindstrom J, Cooper J, Swanson L (1983). Purification of acetylcholine receptors from the muscles of Electrophorus electricus. Biochemistry 22:3796-3800.

Lindstrom J, Cooper J, Tzartos S (1980). Acetylcholine receptors from Torpedo and Electrophorus have similar subunit structures. Biochemistry 19:1454-1458.

Lindstrom J, Criado M, Hochschwender S, Fox JL, Sarin V (1984). Immunochemical tests of acetylcholine receptor subunit models. Nature 311:573-575.

Lindstrom J, Criado M, Ratnam M, Whiting P, Ralston S, Rivier J, Sarin V, Sargent P (1987a). Using monoclonal antibodies to determine the structures of acetylcholine receptors from electric organs, muscles, and neurons. Ann NY Acad Sci 505:208-225.

Lindstrom J, Gullick W, Conti-Tronconi B, Ellisman M (1980). Proteolytic nicking of the acetylcholine receptor. Biochemistry 19:4791-4795.

Lindstrom J, Merlie J, Yogeeswaran B (1979). Biochemical properties of acetylcholine receptor subunits from Torpedo californica. Biochemistry 18:4465-4470.

Lindstrom J, Schoepfer R, Whiting P (1987b). Molecular studies of the neuronal nicotinic acetylcholine receptor family. Molec Neurobiol 1(4):218-337.

Lindstrom J, Shelton GD, Fujii Y (1988). Myasthenia gravis. Adv Immunol, in press.

Lipton S, Aizeman E, Tauck D (1986). Patch clamp recordings of nicotinic cholinergic responses in solitary rat retinal ganglion cells in culture. Neurosci Soc Meet Abstr 175.7.

Lukas R (1986). Characterization of curaremimetic neurotoxin binding sites on membrane fractions derived from the human medulloblastoma clonal line TE671. J Neurochem 46:1936-1941.

Luther M, Schoepfer R, Whiting P, Blatt Y, Montal MS, Montal M, Lindstrom J (submitted). The human medulloblastoma cell

line TE671 expresses a muscle-like acetylcholine receptor.
McCrea PD, Popot J-L, Engelman DM (1987). Transmembrane topography of the nicotinic acetylcholine receptor delta subunit. EMBO J 6(12):3619-3626.
Mishina M, Tobimatsu T, Imoto K, Tanaka K, Fujita Y, Fukuda K, Kurasaki M, Takahashi H, Morimoto Y, Hirose T, Inayama S, Takahashi T, Kuno M, Numa S (1985). Location of functional regions of acetylcholine receptor alpha subunit by site-directed mutagenesis. Nature 313:364-369.
Nef P, Oneyser C, Alliod C, Couturier S, Ballivet M (1988). Genes expressed in the brain define three distinct neuronal nicotinic acetylcholine receptors. EMBO J 7:595-601.
Neumann D, Barchan D, Fridkin M, Fuchs S (1986). Analysis of ligand binding to the synthetic dodecapeptide 185-186 of the acetylcholine receptor alpha subunit. Proc Natl Acad Sci USA 83:9250-9253.
Neumann D, Gershoni J, Fredkin M, Fuchs S (1985). Antibodies to synthetic peptides as probes for the binding site on the alpha subunit of the acetylcholine receptor. Proc Natl Acad Sci USA 82:3490-3493.
Noda M, Furutani Y, Takahashi H, Toyosato M, Tanabe T, Shimizu S, Kikyotani S, Kayano T, Hirose T, Inayama S, Numa S (1983a). Cloning and sequence analysis of calf cDNA and human genomic DNA encoding alpha subunit precursor of muscle acetylcholine receptor. Nature 305:818-823.
Noda M, Takahashi H, Tanabe T, Toyosato M, Kikyotani S, Furutani Y, Hirose T, Takashima H, Inayama S, Miyata T, Numa S (1983b). Structural homology of Torpedo californica acetylcholine receptor subunits. Nature 302:528-532.
Noda M, Takahashi H, Tanabe T, Toyosato M, Kikyotani S, Hirose T, Asai M, Takashima H, Inayama S, Miyata T, Numa S (1983c). Primary structures of β- and δ-subunit precursors of Torpedo californica acetylcholine receptor deduced from cDNA sequences. Nature 301:251-255.
Peralta E, Winslow J, Ashkenazi A, Smith D, Ramachandran J, Capon D (1988). Structural basis of muscarinic acetylcholine receptor subtype diversity. TIPS Feb suppl 6-11.
Raftery M, Hunkapillar M, Strader C, Hood L (1980). Acetylcholine receptor: Complex of homologous subunits. Science 208:1454-1457.
Ralston S, Sarin V, Thanh H, Rivier J, Fox JL, Lindstrom J (1987). Synthetic peptides used to locate the alpha-bungarotoxin binding site and immunogenic regions on alpha subunits of the nicotinic acetylcholine receptor. Biochemistry 26:3261-3266.
Ratnam M, LeNguyen D, Rivier J, Sargent PB, Lindstrom J

(1986a). Transmembrane topography of nicotinic acetylcholine receptor: Immunochemical tests contradict theoretical predictions based on hydrophobicity profiles. Biochemistry 25:2633-2643.

Ratnam M, Sargent PB, Sarin V, Fox JL, LeNguyen D, Rivier J, Criado M, Lindstrom J (1986b). Location of antigenic determinants on primary sequences of subunits of nicotinic acetylcholine receptor by peptide mapping. Biochemistry 25:2621-2632.

Reynolds J, Karlin A (1978). Molecular weight in detergent solution of acetylcholine receptor from Torpedo californica. Biochemistry 17:2035-2038.

Role L, Matossian V, O'Brien R, Fischbach G (1985). On the mechanism of acetylcholine receptor accumulation of newly formed synapses on chick myotubes. J Neurosci 5:2197-2204.

Roth B, Schwendimann B, Hughes G, Tzartos S, Barkas T (1987). A modified nicotinic acetylcholine receptor lacking the 'ion channel amphipathic helices.' FEBS Lett 221:172-178.

Sakmann B, Bormann J, Hamill O (1983). Ion transport by single receptor channels. Cold Spring Harbor Symposia on Quantitative Biology 48:247-257.

Sargent P, Hedges B, Tsavaler L, Clemmons L, Tzartos S, Lindstrom J (1983). The structure and transmembrane nature of the acetylcholine receptor in amphibian skeletal muscles revealed by crossreacting monoclonal antibodies. J Cell Biol 98:609-618.

Schoepfer R, Luther M, Lindstrom J (1988a). The human medulloblastoma cell line TE671 expresses a muscle-like acetylcholine receptor: Cloning of the alpha subunit cDNA. FEBS Lett 226(2):235-240.

Schoepfer R, Whiting P, Esch F, Blacher R, Shimasaki S, Lindstrom J (1988b). cDNA clones coding for the structural subunit of a chicken brain nicotinic acetylcholine receptor. Neuron, in press.

Smith M, Margiotta J, Franco A, Lindstrom J, Berg D (1986). Cholinergic modulation of an acetylcholine receptor-like antigen on the surface of chick ciliary ganglion neurons in cell culture. J Neurosci 6:946-953.

Swanson L, Lindstrom J, Tzartos S, Schmued L, O'Leary D, Cowan W (1983). Immunohistochemical localization of monoclonal antibodies to the nicotinic acetylcholine receptor in the midbrain of the chick. Proc Natl Acad Sci USA 80:4532-4536.

Swanson L, Simmons D, Whiting P, Lindstrom J (1987). Immunohistochemical localization of neuronal nicotinic recep-

tors in the rodent central nervous system. J Neurosci 7:3334-3342.
Syapin P, Salvaterra P, Engelhardt J (1982). Neuronal-like features of TE671 cells: Presence of a functioning nicotinic cholinergic receptor. Brain Res 231:365-377.
Tobimatsu T, Fujita Y, Fukuda K, Tanaka K, Mori Y, Konno T, Mishina M, Numa S (1987). Effects of substitution of putative transmembrane segments on nicotinic acetylcholine receptor function. FEBS Lett 222:56-62.
Tzartos S, Kokla A, Walgrave S, Conti-Tronconi B (1988). The main immunogenic region of human muscle acetylcholine receptor is localized within residues 63-80 of the alpha subunit. Proc Natl Acad Sci USA, in press.
Tzartos S, Rand D, Einarson B, Lindstrom J (1981). Mapping of surface structures on _Electrophorus_ acetylcholine receptor using monoclonal antibodies. J Biol Chem 256:8635-8645.
Tzartos S, Seybold M, Lindstrom J (1982). Specificity of antibodies to acetylcholine receptors in sera from myasthenia gravis patients measured by monoclonal antibodies. Proc Natl Acad Sci USA 79:188-192.
Whiting P, Cooper J, Lindstrom J (1987a). Antibodies in sera from patients with myasthenia gravis do not bind to acetylcholine receptors from human brain. J Neuroimmunol 16:205-213.
Whiting P, Esch F, Shimasaki S, Lindstrom J (1987d). Neuronal nicotinic acetylcholine receptor beta subunit is coded for by the cDNA clone alpha$_4$. FEBS Lett 219(2):459-463.
Whiting P, Lindstrom J (1986a). Purification and characterization of a nicotinic acetylcholine receptor from chick brain. Biochemistry 25:2082-2093.
Whiting P, Lindstrom J (1986b). Pharmacological properties of immunoisolated neuronal nicotinic receptors. J Neurosci 6:3061-3069.
Whiting P, Lindstrom J (1987a). Purification and characterization of a nicotinic acetylcholine receptor from rat brain. Proc Natl Acad Sci USA 84:595-599.
Whiting P, Lindstrom J (1987b). Affinity labeling of neuronal acetylcholine receptors localizes the neurotransmitter binding site to the beta subunit. FEBS Lett 213(1):55-60.
Whiting P, Lindstrom J (1988). Characterization of bovine and human neuronal nicotinic acetylcholine receptors using monoclonal antibodies. J Neurosci, in press.
Whiting P, Liu R, Morley BJ, Lindstrom J (1987b). Structurally different neuronal nicotinic acetylcholine receptor subtypes purified and characterized using monoclonal anti-

bodies. J Neurosci 7:4005-4016.
Whiting P, Schoepfer R, Swanson L, Simmons D, Lindstrom J (1987c). A monoclonal antibody to nicotinic receptors from brain identifies a functional neuronal acetylcholine receptor of PC12 cells. Nature 327:515-518.
Wilson P, Lentz T, Hawrot E (1985). Determination of the primary amino acid sequence specifying the alpha bungarotoxin binding site on the alpha subunit of the acetylcholine receptor from <u>Torpedo californica</u>. Proc Natl Acad Sci USA 82:8790-8794.
Young E, Ralston E, Blake J, Ramachandran J, Hall Z, Stroud R (1985). Topological mapping of acetylcholine receptor: Evidence for a model with five transmembrane segments and a cytoplasmic COOH-terminal peptide. Proc Natl Acad Sci USA 82:626-630.

VORONOI RECEPTOR SITE MODELS

Laurent Boulu and G.M. Crippen

College of Pharmacy, University of Michigan, Ann Arbor, Michigan 48109, U.S.A.

INTRODUCTION

The problem we address is that of understanding specific binding of small molecules to biological receptor macromolecules, such as inhibitors to enzymes. While it is of course preferable to directly determine the structure of such complexes experimentally, the required crystallographic and/or NMR studies are much more difficult to carry out than simply measuring the binding constant. We are developing methods to objectively deduce receptor site geometry and energetics, given only the binding constants for a series of compounds.

Many workers have devised methods to correlate the structure of compounds with their quantitatively measured binding to various receptors (quantitative structure-activity relationships = QSAR). Although the methods are extremely varied in detail, there are common underlying assumptions and approaches:

(1) The receptor is thought of as a relatively rigid "lock" into which the molecular "key" fits, although this is known to be not always true. (Cedergen-Zeppezauer1983a)

(2) The ligand is often assumed to take on a particular conformation when it binds (a calculated minimal energy conformation, the conformation seen in the crystal structure of the pure ligand, etc.), although this is known to be not always true. (Birdsall1983a)

(3) Structurally similar compounds are assumed to bind in analogous "modes" (internal conformation plus positioning and orientation relative to the site), although there are known dramatic exceptions to this. (Roberts1981a)

(4) The free energy of binding is formally broken down into a sum of contributions from different parts of the molecule, although significant deviations from additivity are known. (Jencks1981a)

We have dispensed with assumptions (2) and (3) in order to produce more physically realistic site models. After all, the real molecules are free to randomly explore their accessible conformation space and to enter the binding site in a variety of orientations. Since our approach does not channel the investigation into fixed *a priori* declaration of binding modes, it is better able to treat sets of ligands so chemically diverse that a common superposition or pharmacophore is not obvious. Assumption (1), however, remains for the time being in our work as well as everyone else's, with the rare exception of those using energy minimization or molecular dynamics to calculate the binding of ligands to a receptor protein of known crystal structure. Assumption (4) is still a concern for us and others who use relatively simple binding assays, where the measured affinity can be related to the free energy of binding. Much of the available data comes from such complex biological tests that the assumption is meaningless.

METHODS

Recently, we have formulated a novel approach called "Voronoi binding site models" (Crippen1987a). In order to define the geometry of the site model, one must somehow supply the x,y,z coordinates of a few so-called generating points, **g_1**, **g_2**, ... Each one of these determines a surrounding region, called a Voronoi polyhedron, defined as the set of all points closer to its generating point than to the other generating points. These regions turn out to be convex, possibly rather irregularly shaped figures filling up all space, separated from each other by planar faces. Each part of the site corresponds to a different polyhedron, where atoms of the ligand could experience energetically distinct interactions with the site. For example one Voronoi "region" might be where a carbonyl group on the ligand could form a hydrogen bond to some hydroxyl group on the real receptor, and another region might be a part of space that excludes any atom of the ligand because it represents space filled up by some of the atoms of the real receptor. Every atom of a drug molecule must lie in exactly one of these regions, and one can express the orientation and internal conformation of a particular binding mode by stating in which region each atom is found. Since the energetic contribution and the mode specification for a given atom depends only

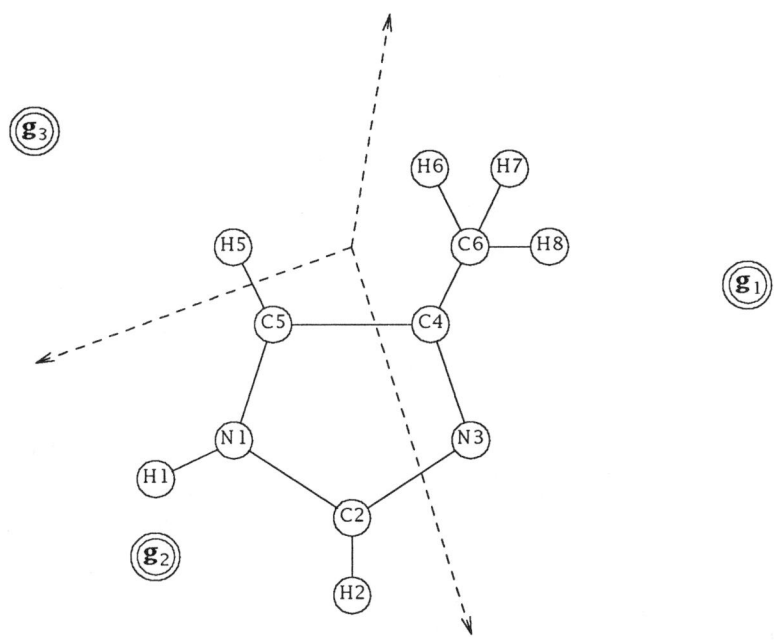

Figure 1
One possible binding mode of a small molecule in a region consisting of three (unbounded) regions. Note that the rotation of the methyl group is unspecified.

upon what region it lies in, and not where within that region it is found, a single mode may rather broadly correspond to a certain family of molecular conformations and positionings in the site. Due to the geometry of the site model and due to the structure and conformational preferences of the molecule, not every mode that can be written down is actually feasible. Some situations, illustrated in figures 2 and 3, can be decided on combinatorial knowledge of the molecule alone, avoiding any time-consuming numerical calculations. Developing an efficient database for such combinatorial information about a molecule is a

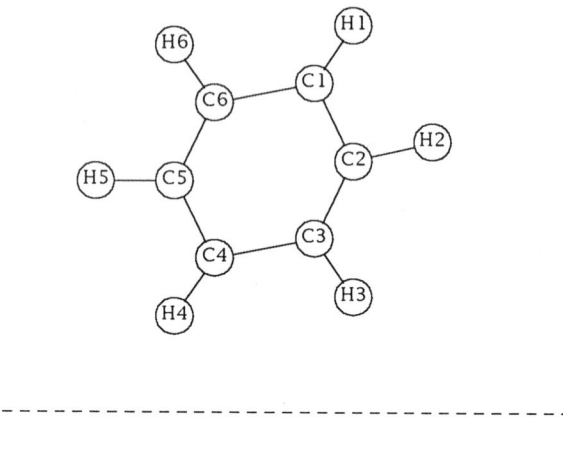

Figure 2
If atoms H1 and H4 are both in region r_1, then atoms C1 and C4 must be also, because they lie in the interior of the convex hull determined by H1 and H4, and any Voronoi polyhedron is also convex.

challenging task we intend to attempt, once we have determined whether such tests are powerful rules for eliminating infeasible modes. Other, more quantitative questions on size and shape must be solved numerically.

Secondly, we have devised a new method, compatible with this type of site model, to summarize the set of allowed conformations of a drug molecule. We call this the "linearized" representation because each atom's position is expressed as a linear combination of an overall molecular translation vector and a few unit vectors chosen in such a way as to represent the orientation of the whole molecule and the relative orientation of groups of atoms linked by rotatable bonds. Then a single exhaustive sampling of all energetically allowed conformations can be summarized in terms of the greatest and least observed scalar products between the various pairs of unit vectors.

The reason the linearized representation is so appealing is that checking the geometric feasibility of a binding mode reduces to a simple set of linear and quadratic inequalities and quadratic equalities. For example, let $\mathbf{g}_1$ be the vector coordinates of the generating point which determines the position

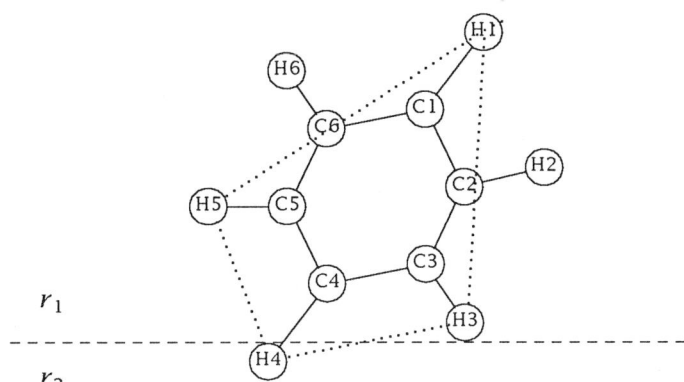

Figure 3
If atoms H3 and H5 are in region r_1, and atom H4 is in region r_2, then atom H1 must not be in r_2 as long as the figure H1-H3-H4-H5 is convex.

and shape of the Voronoi polyhedron that is region 1 in the model. Similarly, g_2 determines region 2. Let A be the linear combination of the molecular translation vector and various unit vectors which describes the position of atom A. Then the condition that A lies in region 1 rather than 2 is

$$\|A - g_1\|^2 < \|A - g_2\|^2$$

In this vector inequality, $A \cdot A$ cancels out, leaving an expression linear in the components of the unit vectors comprising A. The quadratic conditions arise from the normalization of the unit vectors and from restrictions on the dot products between pairs of unit vectors, as deduced by a thorough exploration of energetically allowed conformation space.

To test the feasibility of a mode, one can begin by looking at the proposed binding of a subset consisting of n atoms out of the total number of atoms in a ligand, on the grounds that if this much is not feasible, the whole mode can be discarded. Alternatively, if it is feasible, then we have a good guess as to the positions of these n atoms in the full binding mode. The problem consists of solving a set of nonlinear (generally underdetermined) equations $f_i(\mathbf{x}) = 0$ and inequalities $g_j(\mathbf{x}) > 0$, where $\mathbf{x}$ is the vector of unknown unit vector components. The problem is transformed into a nonlinear least squares form by

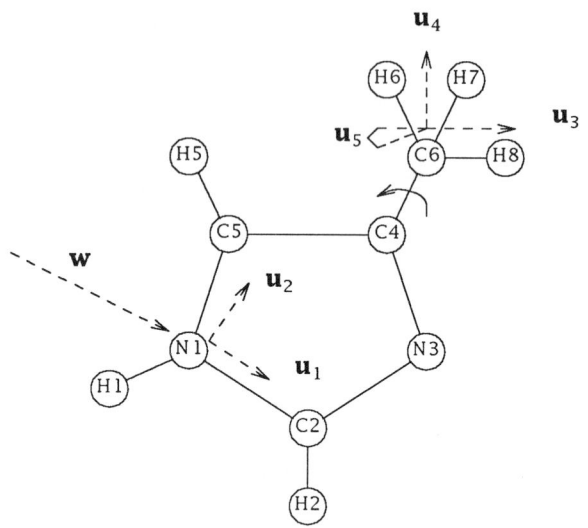

Figure 4
The overall translation of the molecule is specified by the unnormalized vector **w**, so that N1 = **w**. Then the two unit vectors $\mathbf{u}_1$ and $\mathbf{u}_2$ are all that are required to position atoms in the planar imidazole ring and their substituents. For instance N3 = $\mathbf{w}+2.5\mathbf{u}_1+0.5\mathbf{u}_2$. Locating the freely rotating hydrogens of the methyl group requires $\mathbf{u}_3$, $\mathbf{u}_4$, and $\mathbf{u}_5$. By construction, $\mathbf{u}_1\cdot\mathbf{u}_2$ is fixed, but $\mathbf{u}_1\cdot\mathbf{u}_3$ ranges from about -0.8 to +0.8.

---

replacing each inequality g_j by $f_j = min(g_j,0)$ and minimizing $\sum f_i^2$. Different Newton-like methods were tried: Newton's, Gauss-Newton's and variations on the Levenberg-Marquardt's method. Starting values of unit vectors are chosen at random over the unit sphere. The solution of the system expressing the binding of n atoms is then used as an initial guess for the one corresponding to the binding of $n+1$ atoms. Placing each subsequent atom requires an average of 1 to 10 Newton iterations (depending on the complexity of the molecule and site) if the mode so far is feasible, but we require many more iterations before we use the lack of convergence as inductive proof that the proposed mode is infeasible.

The calculated free energy of binding of a given mode is found by summing up interaction contributions for each atom that depend on the type of the atom and the region in which it finds itself. Thus we calculate energies by

$$\Delta G_{mode} = \sum_{\text{atoms } i} \sum_{\text{property } p} V_{p, atom-type_i} c_{p, region(mode, i)}$$

where there are a few atomic physicochemical properties, p, (such as hydrophobicity, molar refractivity, partial charge, etc.) with known values, V, for each atom type. Then for each property, we have a coefficient, c, assigned to each region, so that, for example, a large value of the hydrophobicity c for a particular region means it represents a hydrophobic pocket. The cs are determined empirically so as to give a good fit between observed and calculated binding. Adjusting the cs is easy because ΔG is a linear function of the cs. One advantage is that the number of adjustable parameters is kept low, since there is only one c for each property (rather than atom type) for each region. The real advantage is that even if the cs are determined by fitting the data for chloro and bromo derivatives, predictions can be made for the iodo derivative because the atomic hydrophobicity, V_p, is known in advance for iodine as well as for chlorine and bromine. We have determined all the atomic hydrophobicities in advance (Ghose1986a, Ghose1988a) by assigning detailed atom types, such as "tertiary sp^3 carbon", to all atoms of some hundreds of compounds, assuming the calculated logP must be the sum of the atomic contributions, and determining the atomic contributions by a least squares fit of observed and calculated logP. Of course, hydrophobicity is not really simply additive, but the appropriate choice of detailed atom types takes into account the constitutive effects by reflecting differences in atomic environment. The empirical fit is essentially as good as that by any other method, and the statistical significance is high, due to the very large number of compounds considered. Atomic contributions to molar refractivity turned out to be even easier to calculate by completely analogous methods because they tend to be more additive. Some of the less obvious accomplishments along the way were the development of an *unambiguous* atom classification scheme, a program to automatically classify atoms and later predict hydrophobicity and molar refractivity, interactive molecular graphics as an aid to correct input of molecules, and expansion of the Cambridge Crystallographic Data File molecular data structure to accomodate this scheme.

For all the feasible modes, the molecule is said to have a calculated binding energy corresponding to that of the energetically most favorable mode.

$$\Delta G_{calc} = \min_{modes} \{\Delta G_{mode}\}$$

In other words, the molecule is free to seek that orientation in the site and that conformation which gives the best binding energy, just like the real molecule does.

The input to the problem is the chemical structure of each drug molecule and its experimentally determined free energy of binding, including the estimated experimental error, so that each molecule m really has an allowed binding energy range from ΔG_{m-} to ΔG_{m+}. Solving the problem then consists of finding a suitable number of generating points and their coordinates, and adjusting the interaction contributions, so that the calculated binding energy falls within the experimental range for each molecule. Instead of a least-squares fit to the engergies, we require an absolute fit to the given ranges, however broad they may be.

$$\Delta G_{m-} \leq \Delta G_{m,calc} \leq \Delta G_{m+} \quad \text{for all } m$$

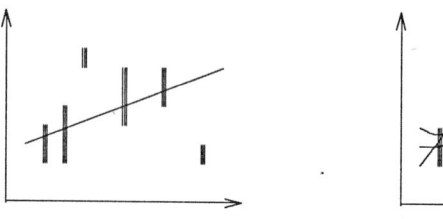

least squares fit   absolute fit

Figure 5
Qualitative illustration of the least squares fitting of data as opposed to various possible absolute fits. The object is to find some function of the independent variable (horizontal axis) that "best" agrees with the observed values of the dependent variable (vertical axis). Vertical lines represent the experimentally determined range for an observation, i.e. the error bars.

The more customary "least squares" philosophy concentrates on adjusting a small number of parameters so as to achieve a good average fit to the observations, paying little attention to the given experimental error estimates. Any deviations in the fit are attributed to unknown random causes, even if the calculated value lies far from the experimentally determined range.

In contrast, we advocate building site models which agree *completely* with the experimentally determined binding constants (up to the given experimental error limits, of course), but in some sense express a whole range of possible solutions. This is in the spirit of Jaynes' maximum entropy approach, which has been proven to be the only consistent way to use information inductively. (Shore1980a) That has the advantage of compactly simulating simple site features, leaving undetermined regions appropriately vague, yet allowing each atom to contribute to binding, as in the atomic physicochemical parameter work.

Since we use the linear free energy assumption, some linear combination of the interaction contributions must lie in the observed binding free energy range for each molecule. For a given molecule, the linear combination holds only for a limited region of energy parameter space where a particular mode is most favorable, but it changes to a different expression outside that region where another mode become optimal. Getting the calculated binding energy for each molecule to agree with its observed range then amounts to optimizing a *piecewise* linear function of the interaction parameters. Such a function can have local minima corresponding to failure of agreement between observed and calculate binding for some of the molecules. Our approach so far (Crippen1987a) has been to locally optimize the function from a number of randomly chosen starting sets of interaction parameters. This may miss a solution, or it is unable to prove there is no possible solution, or at the very least, it is rather inefficient when there are many regions.

Although there are situations where one can prove that more site regions are required than molecules (!), (Ghose1983a) economy of description is desirable. Connected with that is the principle of finding the lowest-resolution or most noncommittal model of the site that will account for the data. The Voronoi site models are very attractive in this respect. Sometimes the models are very simple, and consequently the binding modes are described appropriately vaguely. It often happens that the binding data do not uniquely determine all interaction energy parameters, but rather there are families of solutions with certain still adjustable degrees of freedom that specify a particular member of the family. Suppose we select the member of the solution set that maximizes the information theory entropy of the set of molecules:

$$H = - \sum_{molecules} \sum_{modes} p \log p$$

where p is the probability that a particular molecule will bind in a particular mode. Assuming a Boltzmann equilibrium distribution of energy states, p is proportional to $exp[-\Delta G_{bind}/RT]$, where ΔG_{bind} is the calculated binding energy of that mode relative to that of the optimal mode. This would tend to determine all interaction energy parameters such that many modes of each molecule would have calculated binding energies nearly as good as that of the optimal one, which implies that the model would tend to be noncommittal about which mode is the optimal and hence predicted one.

As it now stands, the investigator proposes the relative positions of the generating points (thus determining the sizes and shapes of the corresponding regions), and if a satisfactory site model can be constructed on that basis, the hypothesis is substantiated. One systematic approach to this process that we have already successfully used, is to determine whether or not a single featureless region suffices. If so, this amounts to a simplified Free-Wilson model where only molecular composition is important, independent of the attachment position for the various substituents. If not, try a two region model, and so on until the data have been accounted for. The difficulty is that it is hard to prove there is no model involving n regions that fits the data, as soon as $n > 3$. There are useful rules for what sorts of regions are necessary, some of which we have already published, (Crippen1984a) but a more systematic rule book needs to be developed. For instance, in order to account for two enantiomers in three dimensions having non-overlapping binding energy ranges, we need at least four regions meeting at a common vertex. Another approach would be to choose how many regions there are to be, but then to vary the coordinates of their generating points so as to produce agreement with the experimentally observed binding energy ranges.

RESULTS

So far, we have successfully treated a data set consisting of twelve biphenyl derivatives binding to human prealbumin. (Crippen1987a) We were able to develop a simple site model from examining five of these compounds and their observed binding constants, and then we successfully predicted the binding of six others, but one other prediction failed.

REFERENCES

Cedergen-Zeppezauer, E., (1983). Crystal Structure Determination of Reduced Nicotinamide Adenine Dinucleotide

Complex with Horse Liver Alcohol Dehydrogenase Maintained in its Apo Conformation by Zinc-Bound Imidazole. Bioch. 22:5761-5772.

Birdsall, B., Roberts, G.C., Feeney, J., Dann, J.G., and Burgen, A.S.V., (1983). Trimethoprim Binding to Bacterial and Mammalian Dihydrofolate Reductase: A Comparison by Proton and Carbon-13 Nuclear Magnetic Resonance. Bioch. 22:5597-5604.

Roberts, G.C.K., Feeney, J., Burgen, A.S.V., and Daluge, S., (1981). The Charge State of Trimethoprim Bound to *Lactobacillus casei* Dihydrofolate Reductase. FEBS Lett. 131:85-88.

Jencks, W.P., (1981). Proc. Natl. Acad. Sci. USA 78:4046.

Crippen, G.M., (1987). Voronoi Binding Site Models. J. Comput. Chem. 8:943-955.

Ghose, A.K. and Crippen, G.M., (1986). Atomic Physicochemical Parameters for Three-Dimensional Structure- Directed Quantitative Structure-Activity Relationships I. Partition Coefficients as a Measure of Hydrophobicity. J. Comp. Chem. 7:565-577.

Ghose, A.K., Pritchett, A., and Crippen, G.M., (1988). Atomic Physicochemical Parameters for Three Dimensional Structure Directed Quantitative Structure-Activity Relationships III: Modeling Hydrophobic Interactions. J. Comp. Chem. 9:80-90.

Shore, J.E. and Johnson, R.W., (1980). IEEE Trans. Inf. Theory IT-26:26.

Ghose, A.K. and Crippen, G.M., (1983). Combined Distance Geometry Analysis of Dihydrofolate Reductase Inhibition by Quinazolines and Triazines. J. Med. Chem. 26:996-1010.

Crippen, G.M., (1984). Deduction of Binding Site Structure from Ligand Binding Data. Ann. New York Acad. Sci. 439:1-11.

A Structural and Dynamic Model for the Nicotinic Acetylcholine Receptor

Edward M. Kosower

Biophysical Organic Chemistry Unit, School of Chemistry, Sackler Faculty of Exact Sciences, Tel-Aviv University, Ramat-Aviv, Tel-Aviv 69978 ISRAEL and Department of Chemistry, State University of New York, Stony Brook, New York 11794-3400 USA

ABSTRACT

Folding of the five polypeptide subunits ($\alpha_2\beta\gamma\delta$) of the nicotinic acetylcholine receptor (AChR) into a functional structural model is described. The principles used to arrange the sequences into a structure include: (1) Hydrophobicity → membrane crossing segments (2) amphipathic character → ion-carrying segments (ion channel with single group rotations) (3) molecular shape (elongated, pentagonal cylinder) → folding dimensions of exobilayer portion (4) choice of acetylcholine binding sites → specific folding of exobilayer segments (5) location of reducible disulfides (near agonist binding site) → additional specification of exobilayer arrangement (6) genetic homology → consistency of functional group choices (7) noncompetitive antagonist labeling → arrangement of bilayer helices. The AChR model is divided into three parts (a) exobilayer: 11 antiparallel β-strands from each subunit (b) bilayer: 4 hydrophobic and 1 amphiphilic α-helices from each subunit and (c) cytoplasmic: one (folded) loop from each subunit.

The exobilayer strands can form a closed "flower" (the "resting state") which is opened ("activated") by agonists bound perpendicular to the strands. Rearrangement of the agonists to a strand-parallel position and partial closing of the "flower" leads to a desensitized receptor. The actions of acetylcholine and succinoyl and suberoyl bischolines are clarified by the model. The opening and closing of the exobilayer "flower" controls access to the ion channel which is composed of the 5 amphiphilic bilayer helices. A molecular mechanism for ion flow in the channel is given. The unusual photolabeling of intrabilayer serines in α, β and δ, but not in γ-subunits near the binding site for non-competitive antagonists (NCAs) is explained. The dynamic behavior of the AChR channel and many experimental results can be interpreted in terms of the model.

INTRODUCTION

Full sequence data for the nicotinic acetylcholine receptor (AChR) of Torpedo californica (2,333 amino acids) and Mus musculus (mouse)(2,314 amino acids) and partial data for Torpedo marmorata(α-subunit), Gallus domesticus (chicken) (α-, γ- and δ-subunits), Bos taurus (calf) (α-, β-, γ- and ε-(fetalγ) subunits) and Homo sapiens (human)(α- and γ-subunits) have appeared. Neural α-subunit from Rattus rattus (rat) AChR is different from the muscle subunit. (refs. in Kosower, 1987). We derive a structural AChR model from the sequences of the subunits and a mechanism for the action of the AChR.(Maelicke,1984) The AChR(Kosower,1987) and sodium channel models (Kosower,1985) illustrate the relationship of molecular structure to activity.

Construction of models is important in analyzing any complex process and in guiding the experimenter. I have tried to consider all possible information in building an AChR model,a "holistic" approach.(Kosower,1987) The present abbreviated discussion considers (a) single group rotation (SGR) theory (b) AChR models: bilayer portion (c) models of exobilayer portion of AChR α-subunits (d) dynamics of the AChR (e) binding site for non-competitive antagonists(NCA) (f) tests of model (g)agonists and antagonists (h) desensitization/resensitization (i) conclusions.

SINGLE GROUP ROTATION THEORY

I developed a general approach to modeling receptor structure in the absence of complete sequence and/or structural information.(Kosower,1982) In essence, one considers (a) the nature of the amino acid groups which might interact with the molecule being bound and (b) the motions which bound molecules or amino acid side chains might undergo. Positional change results from rotation of one conformation into another; such rotations are termed single group rotations (SGR). Various binding classes of amino acids and the corresponding SGR are illustrated in Fig.1.

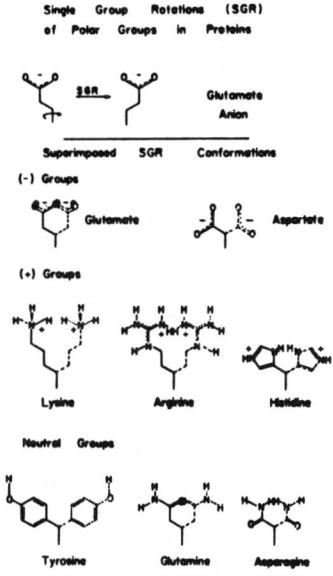

Fig.1. Single group rotations (SGRs) for polar amino acid side chains of proteins. Superimposed SGRs (drawing the group with a dashed line after SGR) over the original conformation are shown.(Reprinted, with permission, from Kosower, 1987)

A model binding site for acetylcholine is based on agonist-receptor interactions of three amino acid side chains: (1) (-)-charged amino acid (electrostatic interactions with the (+)-trimethylammonio group) (2) a (+)-ε-ammonium ion of a lysine (hydrogen bonding to the acetyl carbonyl) and (3) a second (-)-charged amino acid (electrostatic interaction and hydrogen bonding with the lysine (+)-ε-ammonium ion). (Fig.2) A biological effect could arise from the combination via conformational change such as single group rotation (SGR) in either/or both the ligand and side chains to a rearranged ligand-receptor complex. SGRs may be involved in desensitization of the active ACh.AChR complex.

A cation might traverse a membrane via a protein in which the bilayer portions are α-helices. The cation is attracted by negative charges(glu/asp). For the cation to

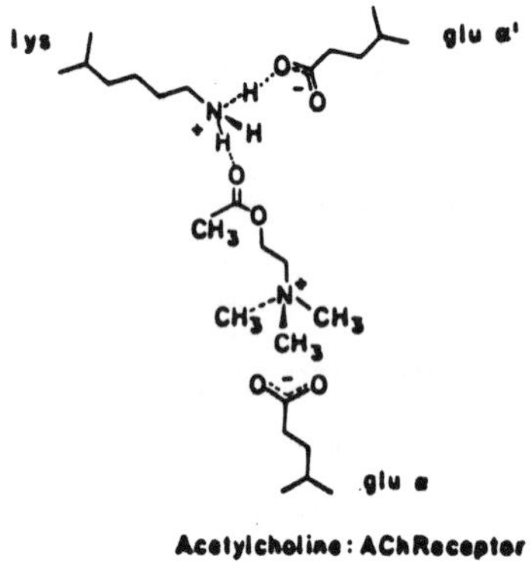

Acetylcholine: AChReceptor

Fig.2. A model combining site for acetylcholine constructed from three amino acid side chains. (Reprinted, with permission, from Kosower, 1987)

progress, we weaken the attraction of negative charges for the cation with another positive charge. A (+)-ε-ammonium ion of a lysine located one helix turn away undergoes SGR, the positive charge is closer to the negative charge, and release of the cation facilitated. Another SGR moves the charges apart. Extending this idea to a series of negatively and positively charged amino acids along an α-helix leads to a functional arrangement capable of serving as an ion channel. The AChR ion channel was predicted to have the sequence, lys(1), glu(5), lys(8), glu(12)..(Kosower,1982)

ACETYLCHOLINE RECEPTOR MODELS : BILAYER PORTIONS

The AChR has five subunits in the order,αβαγε(Hamilton, 1985). Image processing of electron micrographs of organized layers of AChR show five regions of increased density arranged in pentagonal symmetry.(Kistler et al,1982; Brisson, 1986; Giersig et al,1986) Three structural models have been suggested for the nAChR but only the Kosower model explains the whole AChR.(Kosower,1987)

Various analyses based on the hydrophobicity (refs. in Kosower,1987) of amino acids are convenient, but not sufficiently accurate, for folding from primary sequences.Amphipathic elements are missed. Folding schemes based on thermodynamic parameters empirically derived for soluble proteins are inappropriate for membrane proteins. (Wallace et al,1986) Other criteria must therefore be used to construct specific structures.

Our approach is illustrated in a summary scheme.

SCHEME

Physical and chemical data* >>> Exobilayer portion ---> Exobilayer model

```
Amino      | Functional ---> Ion     |
acid ------| model**      Channel|
sequence   |              Elements|   AChR
           |Hydro-                 |-->Bilayer ----------> Bilayer Model
           |phobicity  Hydrophobic|   Helices
           |Measure --> Helices    |
           |                       |
```

Physical and chemical data*** >>> Cytoplasmic portion ---> Cytoplasmic model

* Electron microscopy, x-ray scattering, functional model and single group rotation (SGR) theory, acetylcholine binding sites, thiol labeling, Raman data, genetic homology, deletion and mutation data
** Ion channel mechanism and structure based on SGR theory, genetic homology, deletion and mutation data, noncompetitive antagonist labeling data
*** Electron microscopy, immunological data, phosphorylation sites

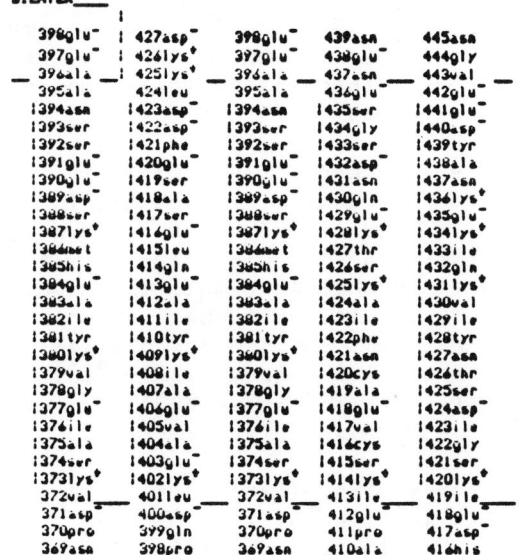

Fig.3. The ion channel elements, one from each subunit, of the acetylcholine receptor.(Reprinted, with permission, from Kosower, 1987)

A key channel element in the α-subunit (373-391) of the AChR (Kosower, 1983a) is an amphiphilic segment. One additional ion channel element was identified in each of the β (401-424), γ (413-395) and δ (419-442) subunits. (Kosower, 1983b, 1984) The ion channel elements were assumed to be α-helices, with appropriately located charged side chains interacting via SGR. The aligned ion channel elements are shown together in Fig.3. From the four hydrophobic segments in each subunit, a 24-amino acid α-helical portion is placed within the bilayer, accounting for 600/2333 (25.7%) of the AChR protein.

MODEL OF EXOBILAYER PORTION OF AChR α-SUBUNIT

The criteria used include (1) Two binding sites must be present, since AChR ion channel activation depends upon two molecules of ACh. (2) The ACh binding sites (certain +,- pairs, e.g. a lysine ε-ammonium ion and a negatively charged carboxylate group (glu/asp), chosen for electrostatic attraction and hydrogen bonding to ACh) must be found (3) A reducible disulfide bond is near an ACh binding site as shown by affinity alkylations with bromoacetylcholine (one thiol per α-subunit) (4) The exobilayer region must extend 55-70Å from bilayer as implied by electron microscopy (5) There must be a high antiparallel β-sheet content as suggested by Raman spectroscopy.

We have therefore arranged strands of 17-18 amino acids (56-60Å) as folded antiparallel β-sheets (X1-X11, Fig.4), including apposed lys^+ and glu^- (or asp^-) and with one pair of half binding sites near the disulfide. Twists in the strands have not been incorporated into the model. Surprisingly, there were not too many alternative arrangements for the strand structure shown. Four binding sites were found, two near the bilayer (1) $172glu^-$, $179lys^+$ (2) $107lys^+$, $97asp^-$ and two near the 192-193 disulfide (3) $166asp^-$, $185lys^+$ and (4) $115lys^+$, $89asp^-$.

The spatial organization of the exobilayer strands is given by (1) the location of the binding site groups (strands X5, X6, X9, X10) and the 192-193 disulfide (X11) are on the side of the ion channel (2) the X6 strand is close to the X7 strand (3) strand X8 is closely linked to X7 by a disulfide (4) strands X1-X4 and X7-X8 are away from the side of the ion channel (5) X4 is close to X5. To satisfy criterion (2) about the proximity of X6 and X7, a

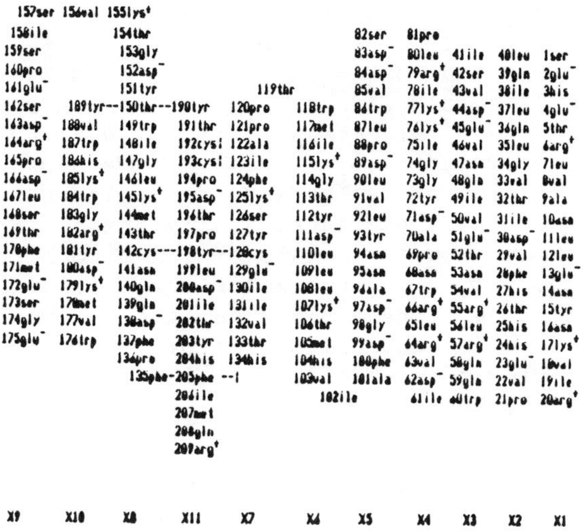

Fig.4. The exobilayer sequence of the α-subunit according to the criteria explained in the text. (Reprinted, with permission, from Kosower, 1987)

third rank of strands with X1 and X2 must exist. Remarkably, the arrangement resembles that found in the most refined EM images.

The excellent genetic homology between subunits in the exobilayer sequence(Changeux and Popot,1984) suggests that exobilayer portions of the β-, γ- and δ-subunits are similar in structure to the α-subunit, except for the α-cys-cys disulfide. (β, γ: 128cys-142cys, δ-: 130cys-144cys)

No well-defined model of the cytoplasmic portion of the AChR can be formulated at present. Certain details of the composition have been described elsewhere.(Kosower,1987)

DYNAMICS OF THE AChR

The model should exhibit: (1) a resting state which requires two molecules of agonist ligand for opening (2) competition of antagonists and agonists for binding sites (3) binding sites for non-competitive blocking agents(anesthetics) (4) partial agonists (5) ion channel function.

The exobilayer portion consists of a pentagonal arrangement of strands carrying a substantial number of charged groups within a "cup" or "flower". The oppositely charged sides("petals") should attract one another. A bending angle of approximately $7°$ with respect to the vertical would be sufficient for exclusion of an agonist. We propose that the resting state of AChR is "poised" as a closed ("resting") form, into which cations cannot easily enter. The left side of Fig.5 shows how the petals of the flower can close. Agonists (e.g., two ACh) can enter and interact with <u>both sides</u> of the cup, opening the cup and allowing ions to diffuse to the ion channel at the bilayer, as shown in Fig.5. The many charged groups in the exobilayer cup are illustrated elsewhere. (Kosower,1987)

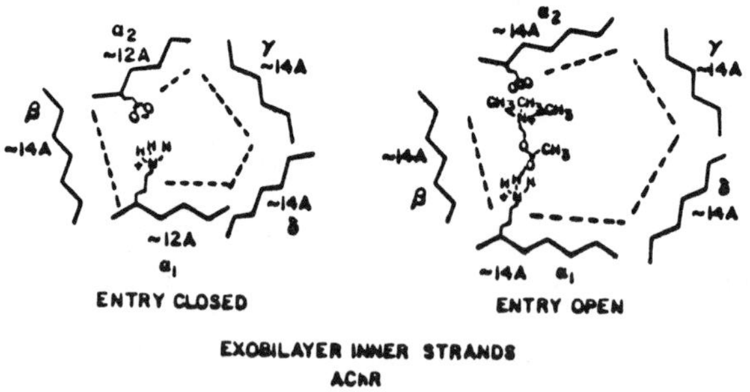

Fig.5. Schematic drawings of the resting(closed) and active (open) forms of the acetylcholine receptor.
(Reprinted, with permission, from Kosower, 1987)

The channel elements (Fig.3) are schematized in Fig.6. The charged groups are indicated by positive or negative signs, and are divided into seven levels. The entry of sodium ion is shown. The model of the ion channel leads to a smooth molecular mechanism for ion flow.

BINDING SITE FOR NON-COMPETITIVE ANTAGONISTS

An important group of antagonist molecules, including histrionicotoxin, phencyclidine and chlorpromazine, block the ion channel without competing with ACh. Remarkably, irradiation of ^{3}H-chlorpromazine (Giraudat et al, 1985, 1987) or ^{3}H-methyltriphenylphosphonium ion (Hucho, 1986) and <u>Torpedo marmorata</u> AChR labels &262 ser. The photolabel was found in homologous positions on the α and β, but not on the γ-subunit. The agents used for labeling as well as

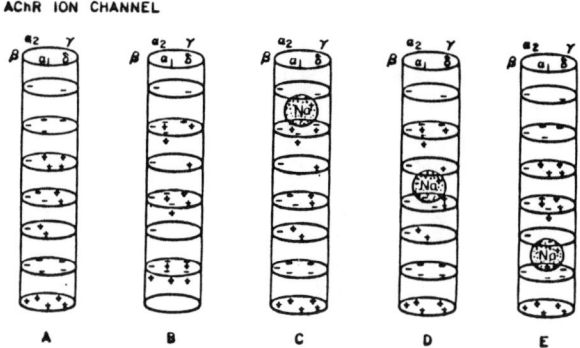

Fig.6. A schematic model for the acetylcholine receptor ion channel, illustrating acquisition and passage of a sodium ion. (Reprinted, with permission, from Kosower, 1987)

the NCA's are positively charged. For each helix that is labeled, a negatively charged residue (glu⁻) is found two turns closer to the cytoplasm. The labeled positions and the associated glu⁻ for each subunit are α [254 ser, 241 glu], β [254ser, 247 glu] and δ [262 ser, 255 glu]. In the γ-subunit, which is not labeled, the residue at the position two turns nearer the cytoplasm is the uncharged glutamine (γ250 gln). The implications of the NCA labeling for structure are described elsewhere.(Kosower, 1987)

In order to fix an image of the acetylcholine receptor, a schematic drawing of the overall structure according to the model is shown in Fig.7. The relationship of the exobilayer strands with binding site groups and 192-193 disulfide link to the α-subunit bilayer helices and the associated cytoplasmic sequence is shown.

TESTS OF MODEL

The model may be probed through (1) measures of evolutionary persistence of exobilayer and ion channel "active" groups (2) effects of genetically engineered deletions and mutations (3) response to agonists and antagonists (4) behavior of reduced receptors (5) possibility of producing rapid current fluctuations. An extended discussion of the evolutionary persistence of "active" groups is found elsewhere.(Kosower,1987)

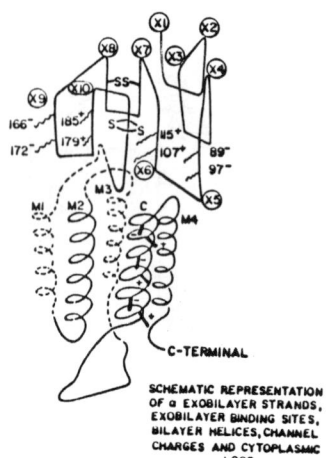

Fig.7. A schematic illustrating the relationship of exobilayer strands (X1-X10, X11 with cys-cys-SS bond not marked) to the bilayer helices (M1-M3, C = channel, M4) and the cytoplasmic extensions in an α-subunit. The binding site groups are named, and the charges on the ion channel element are indicated. (Reprinted, with permission, from Kosower, 1987)

In all, 298 exobilayer "active" amino acids were considered (the α-subunit was counted twice). About 79% perfect homology (completely identical) for the active amino acids together with 7% functionally plausible substitutions sums to a persistence of 86%. The persistence differs from subunit to subunit the highest, α (94%), the lowest β (67%). Histidine is a functionally plausible substitution for lysine at a putative upper binding site. At pH 7, with carboxylate groups in the vicinity, the his would be protonated and could be hydrogen bonded to a ligand.

It should be noted that rat "neural" AChR is included in the comparison. As impressive is the homology noted above, one should also consider the persistence of putative binding sites in those cases for which amino acids have changed. A closer look at the rat neural α sequence suggests that the four potential ACh binding sites are present but in a somewhat different arrangement than proposed for the Torpedo α-subunit. That functional homology does not equal sequence homology has been established by electrophysiological measurements on hybrid AChR receptors derived from subunits of more than one species (White et al, 1986; Sakmann et al,1985) or the fetal and adult forms of AChR of one species (Mishina et al, 1986).

The homologies for channel active amino acids (total 182) of all species with the channel amino acids of Torpedo are extensive and almost complete for the 7 α-subunits (92% identity,96% including functionally plausible substitutions (f.p.s.)) and 3 δ-subunits (76% identity,100% with f.p.s.), but less so for the β- (71% including f.p.s.) and γ- (81% including f.p.s.) subunits.

The evolutionary persistence of the 480 exobilayer and channel "active" amino acids is 88% (including f.p.s.), a figure which seems sufficiently higher than expected from the overall homology.

Agonists and Antagonists: Activation of the receptor requires two molecules of ACh. Our model has four binding sites but only two can be occupied simultaneously.

Perhaps the most striking success of the model is the ready explanation for the activity of long bis-cationic agonists. Succinoyl-bis-choline (SucCh) and suberoyl-bis-choline(SubCh) fit almost perfectly into the receptor.SucCh fits across the space between asp97 (α_1) and asp97(α_2) and SubCh can associate with glu172(α_1) and glu172 (α_2). The short bis-cation, hexamethonium (Hex), is not long enough to open the cup but binds well enough to interfere (as an antagonist) with the binding of ACh. Longer bis-cations, decamethonium (Dec) and trans-bis-Q, fit into the same site as SucCh.

DESENSITIZATION AND RESENSITIZATION

One of the striking properties of the AChR is conversion to an inactive ("desensitized") form after activation by agonist (Maelicke,1984) even for pure receptor in vesicles. In our model, desensitization occurs when agonist stabilizes the closed form of the cup. A three-dimensional representation of the three states (resting, active and desensitized) is shown in Fig.8. With respect to the exobilayer strand direction, parallel rather than perpendicular binding should result in desensitization. Parallel binding might be somewhat stronger than perpendicular binding since the agonist would be bound to multiple groups from both α_1 and α_2 subunits), a molecular explanation for the "tight-binding" conformer. Computer graphics indicates that 111asp might be a suitable group for parallel binding to a positive charge in the agonist, even for anatoxin-a.(Aronstam, Witkop,1981;Koskinen,Rapoport,1985;Petsko and Kosower, unpublished) Resensitization could occur without activation by partial opening of the cup and dissociation of the agonist. A powerful result is a straightforward explanation for the desensitization which succeeds activation by a variety of agonists. The model accounts for the lower rate of reduction of native AChR by dithiothreitol in presence of desensitizing quantities of agonists.

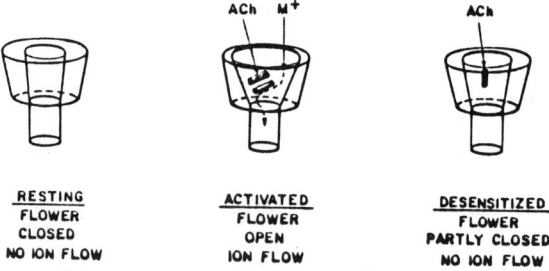

Fig.8. Three-dimensional model ("flower") representing in a simple way the operation of the acetylcholine receptor.
(Reprinted, with permission, from Kosower, 1987)

CONCLUSIONS

The most important sources for data and ideas used in constructing the model were (1) the sequence (2) the overall structure (electron microscopy) (3) thiol labeling (4) hydrophobicity of segments (5) binding site model and ion channel selection (single group rotation theory) (6) genetic homology and evolutionary stability of "active" amino acids (7) ligand structures (8) plausible physical ideas about the operation of the receptor (9) labeling by noncompetitive antagonists.

Experimental approaches to testing the proposed model include (1) cross-linking putative proximate groups (2) preparation of antibodies to peptides (3) molecular graphics studies of the exobilayer region (G.A.Petsko and E.M. Kosower, work in progress) (4) structural determination by x-ray crystallography (5) identification of labeled sites especially those which have been cross-linked. The crosslinking of proximate carboxylic acid and amino groups followed by enzymatic degradation to small peptides could be carried out with chromatographic separation (HPLC) and mass spectrometric determination of peptide composition could provide enough information to evaluate the three-dimensional structure of the exobilayer portion of the receptor. The present model should provide a useful basis for design of experiments, for theoretical analysis of biological channels and for the design of agents designed to affect the operation of the acetylcholine receptor.

REFERENCES

Aronstam RS, Witkop B (1981) Proc Nat'l Acad Sci USA 78: 4639-4643.
Brisson A (1986) in "Nicotinic Acetylcholine Receptor: Structure & Function" (ed) Maelicke A, NATO-ASI Series Vol H-3 Berlin: Springer pp 1-6
Finer-Moore J, Stroud RM (1984) Proc Nat'l Acad Sci USA 81: 155-159
Giersig M, Kunath W, Sack-Kongehl H, Hucho F (1986) in "Nicotinic Acetylcholine Receptor:Structure and Function" (ed) Maelicke A, NATO-ASI Series Vol H-3, Berlin: Springer pp 7-17
Giraudat J, Dennis M, Heidmann T, Chang J-Y, Changeux J-P (1986) Proc Nat'l Acad Sci USA 83:2719-2723
Giraudat J, Dennis M, Heidamnn T, Haumont P-Y, Lederer F, Changeux J-P (1987) Biochemistry 26:2410-2418.
Hamilton SL, Pratt DR, Eaton DC (1985) Biochemistry 24:2210-2219.
Hucho F (1986) Eur J Biochem 158:211-226.
Kistler J, Stroud RM, Klymkowsky MW, LaLancette R, Fairclough RH (1982) Biophys J 37:371-383
Koskinen AMP, Rapoport,H (1985) J Med Chem 28:1301-1309.
Kosower EM (1982) Abstr Int'l Sympos "Structure and Dynamics of Nucleic Acids and Proteins" La Jolla, Calif. 5-9 September pp.52-53
Kosower EM (1983a) Biochem Biophys Res Comm 111:1022-1029.
Kosower EM (1983b) FEBS Lett 155:245-247.
Kosower EM (1984) FEBS Lett 172:1-5
Kosower EM (1985) FEBS Lett 182:234-242.
Kosower EM (1987) Eur J Biochem 168:431-449
Maelicke A (1984) Angew Chem Int Ed 23:195-221.
Mishina M, Takai T, Imoto K, Noda M, Takahashi T, Numa S, Methfessel C, Sakmann B (1986) Nature 321:406-411.

Popot J-L, Changeux J-P (1984) Physiolog Revs 64:1162-1239.
Sakmann B, Methfessel C, Mishina M, Takahashi T, Takai T, Kurasaki M, Fukuda K, Numa S (1985) Nature 318:538-543.
Wallace B, Cascio M, Mielke DL (1986) Proc Nat'l Acad Sci USA 83:9423-27
White MM, Mayne KM, Lester HA, Davidson N (1986) Proc Nat'l Acad Sci USA 82:4852-4856.

CONFORMATIONAL DIFFERENCES BETWEEN AGED AND NON-AGED ORGANOPHOSPHORYL CONJUGATES OF CHYMOTRYPSIN

N. Steinberg, J. Grunwald., E. Roth, R. August, E. Haas, Y. Ashani and I. Silman

Depts. of Neurobiology (N.S., E.R., I.S.) and Chemical Physics (R.A., E.H.), Weizmann Institute of Science, Rehovot 76100; Israel Institute for Biological Research, Ness-Ziona 70450 (J.G., Y.A.), and Dept. of Life Sciences, Bar-Ilan University, Ramat Gan 52100 (E.H.).

INTRODUCTION

Many serine hydrolases, such as acetylcholinesterase (EC 3.1.1.7; AChE) and chymotrypsin (EC 3.4.21.1; Cht) are inhibited by organophosphorus (OP) esters which form a stoichiometric (1:1) covalent conjugate with the active-site serine (Aldridge and Reiner, 1972). Inhibition can often be effectively reversed by treatment of the conjugate with suitable nucleophilic reagents which detach the OP group from the serine hydroxyl (Froede and Wilson, 1971). Alternatively, such conjugates may undergo a dealkylation reaction, commonly termed "aging", which converts the inhibited enzyme into a non-reactivatable form (Hobbiger, 1955; Berends et al., 1959). Aging occurs especially rapidly in the case of OP-AChE conjugates in which the OP moiety contains a secondary alkyl group such as isopropyl (e.g. sarin, DFP) or pinacolyl (e.g. soman); it thus renders therapy of intoxication by such OP's notoriously difficult, since the quaternary oximes which serve as active-site-directed reactivators of OP-inhibited AChE (Froede and Wilson, 1971) become ineffective after aging has occurred (Loomis and Salafsky, 1963).

Irrespective of the mechanism of aging, it involves, as already mentioned, the net loss of an alkyl group from the bound OP moiety, resulting in the introduction of an

O⁻ negative charge into the active site, as depicted in Scheme I:

$$\text{EH} + \text{XP(O)(OR')R} \underset{\text{reactivation}}{\overset{-\text{HX}}{\rightleftarrows}} \text{EP(O)(OR')R} \overset{-R'}{\underset{\text{aging}}{\dashrightarrow}} \text{EP(O)(O}^-\text{)R}$$

SCHEME I

where R = alkyl, aryl, alkyloxy or aryloxy; R' = alkyl or aryl, and X = F, Cl, p-nitrophenoxy or dialkylaminoethanethiol. This formation of a P-O⁻ bond at the active site has only recently been demonstrated directly, by ^{31}P-nmr spectroscopy, for a number of serine hydrolases, including Cht (van der Drift et al., 1985; Grunwald et al., 1985, 1988).

Kinetic studies with neutral and negatively charged OP esters, analogous to the OP-enzyme conjugates depicted schematically in Scheme I, indicate that the negative charge can retard nucleophilic displacement at the phosphorus atom by no more than 50-100-fold (Kirby and Younas, 1970). The electrostatic barrier imposed by the negatively charged oxygen cannot, therefore, fully explain the unusual resistance to reactivation of the aged conjugates of the various enzymes studied. A possible contribution to this resistance might originate in a conformational change occurring concomitantly with the aging process; indeed, Amitai et al. (1982) utilized suitable pyrene-containing OP's to prepare fluorescent OP conjugates of AChE and to demonstrate conformational differences between aged and non-aged conjugates by techniques of fluorescence spectroscopy.

It was earlier suggested by Masson and Goasdue (1986) that the increased resistance of the aged OP conjugates to reactivation, relative to the homologous non-aged conjugates, might be reflected in an increased resistance to conformational change in general, and to denaturation in particular. These authors offered experimental evidence, based upon the transverse urea-gradient electrophoresis technique of Creighton (1979), to support this contention in the case of OP conjugates of human serum cholinesterase.

Cht is a particularly suitable enzyme for spectroscopic studies on the physicochemical basis of the aging reaction since it is available in large quantities in highly purified form, and its sequence and three-dimensional structure have been fully worked out (Hartley, 1964; Blow et al., 1969). In the following we compare aged and non-aged pyrene-containing OP conjugates of Cht by optical spectroscopic techniques, and show that a marked change in the environment of the fluorescent probe occurs upon aging. We further demonstrate that aging is accompanied by an enhanced resistance to unfolding of the protein in solutions of guanidine hydrochloride (Gu.HCl).

EXPERIMENTAL

Preparation and Characterization of the OP-Cht Conjugates: The pyrene-containing organophosphates, PBEPF and PBPDC, were synthesized as described previously (Amitai et al., 1982). The OP-Cht conjugates, PBEP-Cht and PBP-Cht, were obtained by dropwise addition of an acetonitrile solution (0.2-1 mM) of the appropriate OP to a stirred solution of Cht (0.5-1 mg/ml) in double distilled water at room temperature. A pH of 7.4 was maintained during the reaction by addition of 0.02 N NaOH, and the fraction of organic solvent in the reaction mixture did not exceed 10%. The course of inhibition was followed by monitoring Cht activity according to Cunningham and Brown (1956). Small amounts of residual enzymic activity were abolished by addition of excess DFP. The OP-Cht conjugates obtained were lyophilized, redissolved in double distilled water and separated from excess fluorophore by chromatography on Sephadex G-10. The stoichiometry for the aged (PBP-Cht) and non-aged (PBEP-Cht) conjugates obtained using PBPDC and PBEPF was found to be 0.97±0.05 and 0.94±0.03, respectively, of bound fluorophore per active site.

Spectroscopy: Absorption spectra were measured on a Zeiss PMQII spectrophotometer. Fluorescence spectra were obtained with a Hitachi-Perkin Elmer MPF 44A spectrofluorometer.

Circular dichroism spectra were recorded with a Jasco J-500C spectropolarimeter. The absorption anisotropy factor, g_{ab}, is defined by $g_{ab} = (\varepsilon_1 - \varepsilon_r)/\varepsilon$, where ε_1 and ε_r

are the extinction coefficients for left- and right-handed circularly polarized light, respectively, and ε is their average.

Circularly polarized luminescence was measured with an instrument constructed at the Weizmann Institute (Steinberg and Gafni, 1972). The emission anisotropy factor, g_{em}, relates to the conformation of the fluorophore in the excited electronic state from which emission occurs in the same way that g_{ab} relates to the conformation in the ground state (Steinberg, 1978). Thus it is defined by $g_{em} = (I_l - I_r)/(I/2)$, where I_l and I_r are the intensities of the left- and right-handed circularly polarized luminescence, and I is the intensity of the total luminescence emitted.

Fluorescence decay measurements were performed on an instrument constructed at the Weizmann Institute (Hazan et al., 1974). The fluorescence decay data were analysed according to Grinvald and Steinberg (1974).

All spectroscopic measurements described below were performed in 0.1 M NaCl-0.01 M phosphate, pH 7.0, at room temperature.

RESULTS

Aged and non-aged pyrene-containing OP conjugates of Cht were prepared by adopting the same strategy utilized earlier for preparation of the corresponding conjugates of AChE (Scheme II).

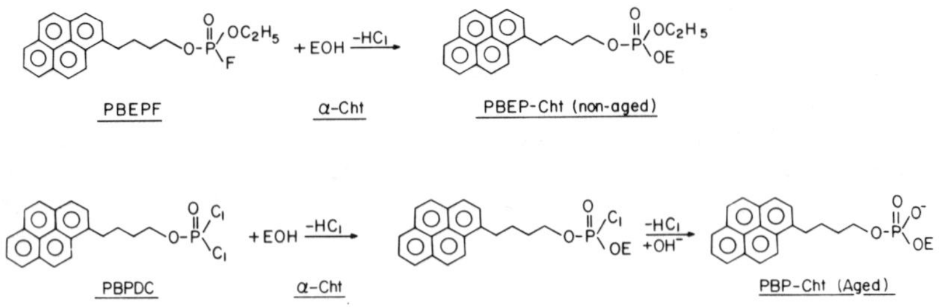

SCHEME II

Briefly, 1-pyrenebutyl ethyl phosphorochloridate (PBEPF) was used to obtain the non-aged conjugate, 1-pyrenebutyl ethylphosphoryl-Cht (PBEP-Cht), and 1-pyrenebutyl phosphorodichloridate (PBPDC) was used to obtain the aged conjugate, 1-pyrenebutyl hydroxyphosphoryl-Cht (PBP-Cht), as a result of rapid nucleophilic substitution by OH⁻ of the Cl atom not directly involved in reaction of the OP with the enzyme (Wins and Wilson, 1974).

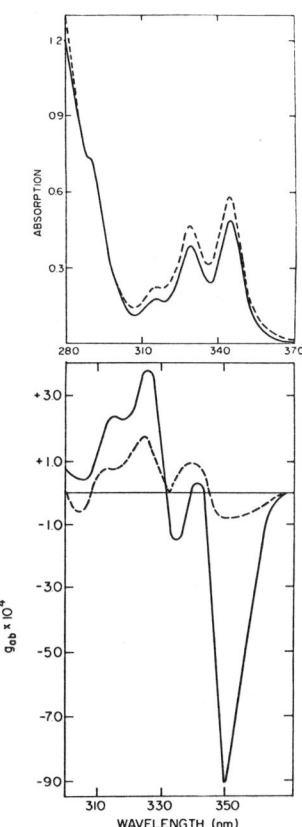

Figure 1. Absorption and circular dichroism spectra of PBEP-Cht (————) and PBP-Cht (— — —). The concentration of the conjugates was 4×10^{-5} M and the optical path in the absorption measurements was 0.5 cm.

The absorption spectra of the non-aged conjugate, PBEP-Cht, and of the aged conjugate, PBP-Cht, are shown in the upper panel of Fig. 1. For both conjugates the pyrene absorption band is structured, with maxima at 315, 329 and 345 nm, and at 315, 329 and 346 nm for the non-aged and aged conjugates respectively. The extinction coefficient of PBP-Cht at 346 nm is ca. 15% higher than that of PBEP-Cht at 345 nm.

The circular dichroism (CD) spectra of the two conjugates are presented in the lower panel of Fig. 1. The CD spectra of PBEP-Cht and of PBP-Cht differ markedly from each other, especially in the vicinity of the pyrene absorption maximum (ca. 350 nm), where the absorption anisotropy factor is much more negative for the non-aged form than for its aged counterpart. Thus the ground-state conformation of the pyrene probe within the active site of the enzyme must differ markedly in the two conjugates.

Steady-state fluorescence measurements for the two conjugates are displayed in the upper panel of Fig. 2. At a concentration of 1×10^{-6} M both conjugates yield a similar emission spectrum, but the quantum yield for PBP-Cht is ca. 20% lower than for PBEP-Cht.

Fluorescence decay measurements revealed two different decay rates for both PBEP-Cht and PBP-Cht. At a concentration of 1×10^{-6} M (λ_{ex} = 315 nm, λ_{em} > 370 nm), the non-aged conjugate, PBEP-Cht exhibits the following values of lifetimes (τ) and amplitudes (α): τ_1 = 98.6±0.4 nsec, τ_2 = 6.2±0.6 nsec; α_1 = 0.105±0.001, α_2 = 0.018±0.001; the corresponding decay parameters for PBP-Cht were τ_1 = 87.7±0.3 nsec, τ_2 = 6.1±0.5 nsec; α_1 = 0.116±0.001, α_2 = 0.023±0.001. The relative quantum yields calculated from these data for the aged and non-aged conjugates are in agreement with those calculated from the steady-state data and prove that the observed quenching is dynamic.

Circularly polarized luminescence (CPL) spectra for PBEP-Cht and for PBP-Cht are displayed in the lower panel of Fig. 2. The CPL spectra of the two conjugates, which reflect the asymmetry of the pyrene probe in its excited state (Steinberg, 1978), can be seen to be very different, inasmuch as the emission anisotropy factor, g_{em}, differs not only in magnitude, but also in sign. The difference in magnitude is indicative of a change in the excited state

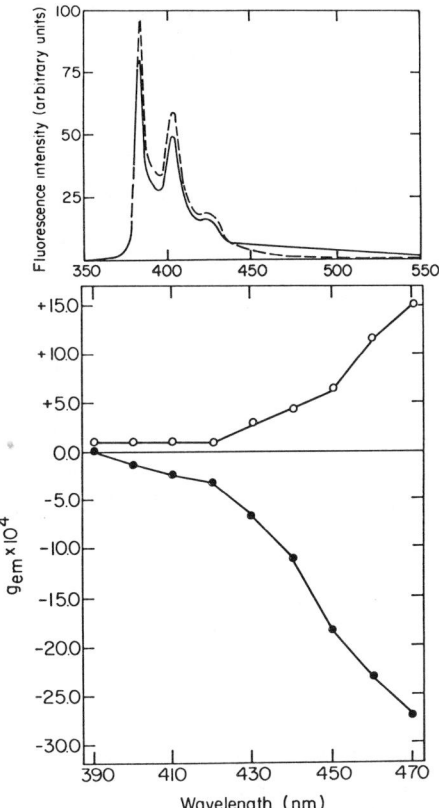

Figure 2. Fluorescence and circularly polarized luminescence (CPL) spectra of aged and non-aged OP conjugates of Cht.

Upper panel: Emission spectra of PBEP-Cht (— — —) and of PBP-Cht (———). Excitation was at 345 nm, and the concentration of the conjugates was 1×10^{-6} M.

Lower panel: CPL spectra of PBEP-Cht (o—o—o) and of PBP-Cht (●—●—●). Excitation was at 335 nm, and the concentration of the conjugates was 4×10^{-5} M. The experimental error in g_{em} was $\pm 0.5 \times 10^{-4}$.

conformation of the OP-Cht conjugate upon aging, while the opposite sign implies that the induced chirality of the fluorophore within the active sites of the aged and non-aged conjugates is of opposite handedness.

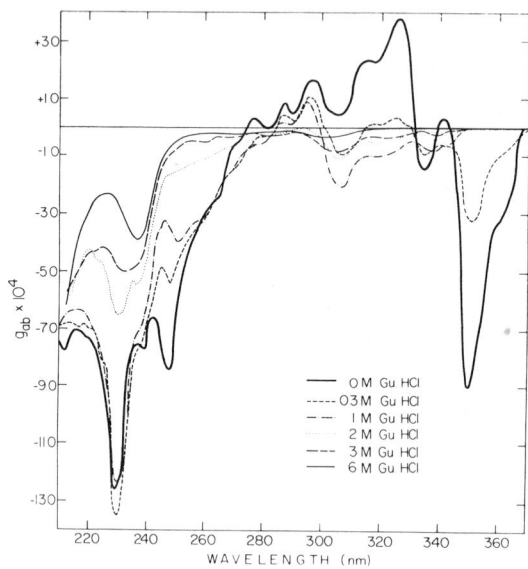

Figure 3. Circular dichroism spectrum of PBEP-Cht as a function of Gu.HCl concentration. The concentration of the conjugate was 4×10^{-5} M.

Figs. 3 and 4 show the CD spectra of PBEP-Cht and PBP-Cht as a function of Gu.HCl concentration. It can be seen that the effect of Gu.HCl differs markedly for the two conjugates. The CD spectrum of PBEP-Cht exhibits four conformational transitions: The first, at 0.3-0.4 M Gu.HCl, at 310-340 nm; the second, at about 0.7 M Gu.HCl, near 350 nm; the third, at 1-2 M Gu.HCl, in the range of 290-310 nm; and the fourth, at 2-3 M Gu.HCl, in the range of 215-290 nm. In contrast, PBP-Cht exhibits only two conformational transitions: The first, at ca. 0.7 M Gu.HCl, is observed at about 350 nm; the second, at 2-3 M Gu.HCl, is observed in the whole spectral range of 215-370 nm. The optical activity above 310 nm originates in the pyrene chromophore; it thus reflects the conformation of the active site, whereas the transition at ca. 230 nm may be ascribed primarily to the peptide linkages of the whole protein. In both conjugates conformational changes in the region of pyrene absorption occur at Gu.HCl concentrations much lower than required for total inactivation of the enzyme (2-3 M). However, reactivation studies carried out with 3-pyridinealdoxime methiodide (3-PAM) (Cohen and Erlanger, 1960) demonstrated that the ability of PBEP-Cht to undergo reactivation is maintained as long as total denaturation has not been achieved. All conformational transitions observed above 310 nm, and at Gu.HCl concentrations below 2 M, thus appear to result from different orientations of the pyrene probe within the active site, rather than from disruption of the site itself. There is, however, a marked difference between the tendencies of PBEP-Cht and PBP-Cht to undergo such orientational changes. The non-aged conjugate undergoes more such reorientations, the first occurring at a Gu.HCl concentration

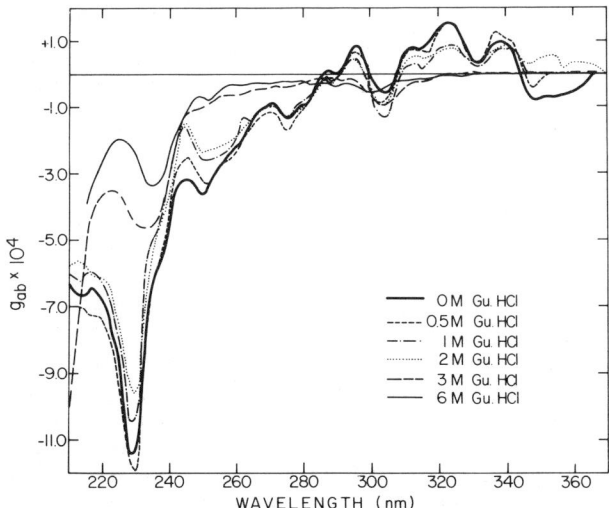

Figure 4. Circular dichroism spectrum of PBP-Cht as a function of Gu.HCl concentration. The concentration of the conjugate was 4×10^{-5} M.

as low as 0.3 M; in contrast, the first such reorientation observed for PBP-Cht occurs only at 0.7 M Gu.HCl. These results clearly support the contention that the aged form has undergone a conformational stabilisation in its active site relative to the non-aged conjugate.

DISCUSSION

The CD and CPL measurements presented above clearly indicate significant differences between PBEP-Cht and PBP-Cht with respect to the orientation of the pyrene probe within the active site in both the ground state and in the excited state. The steady-state fluorescence data, showing a lower quantum yield for the aged than for the non-aged conjugate, taken together with the fluorescence decay data (performed in the presence and absence of oxygen), which show that quenching is dynamic, suggest that the interaction of the pyrene moiety with the polypeptide chain is significantly stronger in PBP-Cht than in PBEP-Cht. This conclusion is supported by collisional quenching data (not presented) which show that the accessibility of the active site to external quenchers in the aged and non-aged conjugates is approximately the same. It is also consistent with the experiments in which the CD spectra of the two conjugates were monitored as a function of Gu.HCl concentration. Whereas the CD spectrum of PBEP-Cht displays multiple transitions in the active-site region, starting at Gu.HCl concentrations as low as 0.3-0.4 M, only one such transition is observed for PBP-Cht, at 0.7 M Gu.HCl. Thus, the local flexibility of the pyrenebutyl moiety within the active site is considerably larger in the non-aged form than in its aged counterpart, allowing for more orientations of the probe within the active site. It is also worth noting that the aging reaction leads to a more general stabilization of the OP-Cht conjugate. Thus the band at ca. 230 nm, which may be ascribed primarily to the polypeptide chain, is largely abolished at 2 M Gu.HCl in the case of PBEP-Cht, whereas it takes 3 M Gu.HCl to achieve a similar effect in the case of PBP-Cht.

Aging is a process common to the conjugates of several serine hydrolases (Aldridge and Reiner, 1972). It appears, furthermore, to be a catalytic process (Aldridge and Reiner, 1972; Bender and Wedler, 1972) which should, therefore, involve the active-site histidine of such enzymes. Indeed, evidence in support of this possibility has already been offered both for Cht and AChE (Toia and Casida, 1979; Beauregard et al., 1981; see also Sterri, 1977). A plausible explanation for the known resistance to oxime reactivation, together with the concomitant conformational stabilization which we have demonstrated, might thus lie in an ionic interaction between the negative charge of the P-O⁻ group produced by the aging reaction (Van der Drift et al., 1985; Grunwald et al., 1985, 1988) and the imidazole group of the active-site histidine - His 57 in the case of Cht (Blow et al., 1969). We are currently attempting to obtain direct experimental evidence to support this proposal by X-ray crystallographic comparison of suitable aged and non-aged OP-Cht conjugates, using for this purpose crystals obtained by soaking crystals of native γ-Cht in solutions of the appropriate OP's (Su et al., 1988).

ACKNOWLEDGEMENTS

This work was supported by Grants DAMD17-83-G-9548 and DAMD17-87-GC-7037 from the US Army Medical Research and Development Command.

REFERENCES

Amitai G, Ashani Y, Gafni A, Silman I (1982). Biochemistry 21: 2060-2069.
Aldridge WN, Reiner E (1972). Enzyme Inhibitors as Substrates. North-Holland, Amsterdam.
Beauregard G, Lum J, Roufogalis BD (1981). Biochem. Pharm. 30: 2915-2920.
Bender ML, Wedler FC (1972). J Am Chem Soc 94: 2101-2109.
Berends F, Posthumus CH, Van der Sluys I, Deierkauf FA (1959). Biochim Biophys Acta 34: 576-578.
Blow DM, Birktoft JJ, Hartley BJ (1969). Nature 221: 337-340.
Cohen W, Erlanger, BF (1960). J Am Chem Soc 82:3928-3934.
Cunningham LW, Brown CS (1956). J Biol Chem 221: 287-299.
Creighton TE (1979). J Mol Biol 129: 235-264.

Froede HC, Wilson IB (1971). Enzymes, 3rd ed., 5: 87-114.
Grinvald A, Steinberg IZ (1974). Anal Biochem 59: 583-598.
Grunwald J, Ashani Y, Segall Y, Waysbort D, Steinberg N, Silman I (1985). Soc Neurosci Abst 15: 132.
Grunwald J, Segall Y, Shirin E, Weysbort D, Steinberg N, Silman I, Ashani Y (1988). Submitted for publication.
Hartley BS (1964). Nature 201: 1284-1287.
Hazan G, Grinvald A, Maytal M, Steinberg IZ (1974).
Rev Sci Instrum 45: 1602-1604.
Hobbiger F (1955). Brit J Pharmacol 10: 365-362.
Kirby AJ, Younas M (1970). J Chem Soc B: 1165-1172.
Loomis AT, Salafsky BJ (1963). Toxicol Appl Pharmacol 5: 685-701.
Masson P, Goasdoue JL (1986). Biochim Biophys Acta 869: 304-313.
Steinberg, IZ (1978). Methods Enzymol 49: 179-198.
Steinberg IZ, Gafni A (1972). Rev Sci Instrum 43: 409-413.
Steinberg N, Van der Drift ACM, Grunwald J, Segall Y, Shirin E, Haas E, Ashani Y, Silman I (1988). Submitted for publication.
Sterri SH (1977). Biochem. Pharm. 26: 656-658.
Su CT, Harel M, Grunwald J, Ashani Y, Sussman JL, Silman I (1988). In preparation.
Toia RF, Casida JE (1979). Biochem Pharm 28: 211-216.
Van der Drift ACM, Beck, HC, Dekker WH,
Hulst AG, Wils ERJ (1985). Biochemistry 24: 6894-6903.
Wins, P, Wilson IB (1974). Biochim Biophys Acta 334: 137-145.

STRUCTURAL AND IMMUNOCHEMICAL PROPERTIES OF FETAL BOVINE SERUM ACETYLCHOLINESTERASE

B.P. Doctor, K.K. Smyth, M.K. Gentry, Y. Ashani, C.E. Christner, D.M. De La Hoz, R.A. Ogert, and S.W. Smith, Division of Biochemistry, Walter Reed Army Institute of Research, Washington, DC 20307-5100 USA

Acetylcholinesterase (EC 3.1.1.7, AChE) is known to be associated with cells involved in cholinergic synaptic transmission. It exists in multiple molecular forms, e.g., globular (G1,G2,G4) and asymmetric forms (A4,A8,A12) (Massoulie and Bon, 1982; Brimijoin, 1983; Massoulie et al., 1984). The complex polymorphism of AChE may be related to the different tissues and environments where this enzyme originates. Pharmacological studies and enzyme kinetic analysis, on the other hand, have demonstrated that different forms of AChE possess similar catalytic properties (Brimijoin, 1983). Different AChE forms may thus share regions of common structure or domains, e.g., the esteratic site and the peripheral anionic site(s). Recent reports (MacPhee-Quigley et al., 1985; Hall and Spierer, 1986; Schumacher et al., 1986; Lockridge et al., 1987; Prody et al., 1987; Smyth et al.,1988) indicate that the esteratic regions of several cholinesterases share extremely high degrees of amino acid sequence homology. Moreover, homologies detected by monoclonal antibodies (mAbs) between different AChEs isolated from various sources further substiantiate the existence of such common domains (Doctor et al., 1983; Rakonczay and Brimijoin, 1985).

In this report we describe the results of the elucidation of the structure of the epitope for mAb AE-2 (Fambrough et al., 1982) and compare its anticholinesterase properties with a second mAb, 25B1 (Gentry et al., 1988; Ashani et al., 1988). Both of these mAbs inhibit catalytic activity of fetal bovine serum acetylcholinesterase (FBS-AChE) and may be used as suitable molecular probes for mapping the topography of the AChE region which is associated with enzyme activity and may allow the potential location of aspartic acid, serine, and histidine residues which could be involved in the "charge triad". (For a recent review see Steitz and Shulman, 1982).

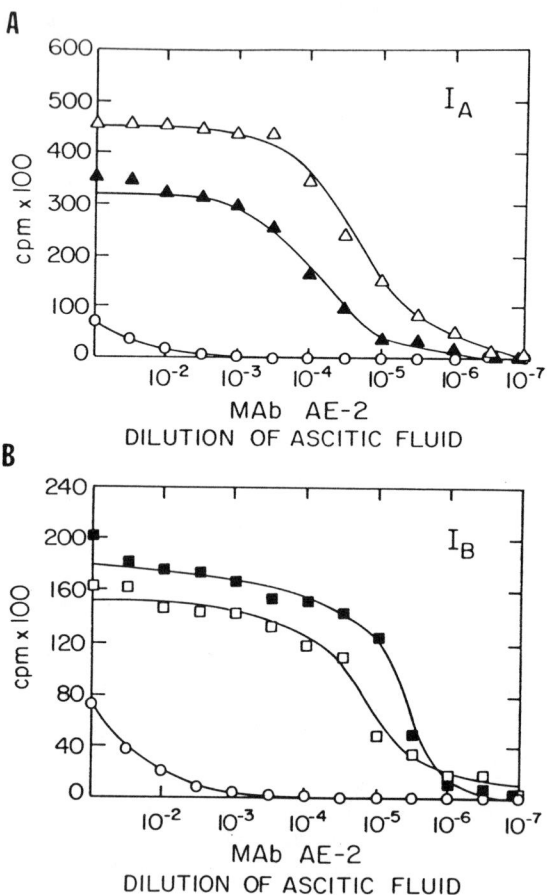

Figure 1: Reactivity of mAb AE-2 with native and DFP-inhibited FBS-AChE (A) and Torpedo 11S AChE (B). Cross reactivity of mAb AE-2 was measured using solid phase RIA. The source of mAb AE-2 used was mouse ascitic fluid.
Figure 1A: ∆----∆, native FBS-AChE; ▲----▲, DFP-inhibited FBS-AChE; o----o, binding of AE-2 to non-specific protein (soybean trypsin inhibitor). Figure 1B: ---- , Torpedo native 11S AChE; ---- , DFP-inhibited Torpedo 11S AChE; o----o, non-specific protein.

PRIMARY STRUCTURE OF CHOLINESTERASES

```
                                                                               (-------------------*)
TORPEDO    1  DDHSELLVNTKSGKVMGTRVPVLSSHISAELGIPFAEPPVGMRFRREPKKPWSGVWNASTYPNNCQQYVDEQFPGFSGSEMHNPNREWSEDCLYLNIWVPSRPKSTTVWVIYGG
FBS           EGPEDPELLVMVRGGELRGLRLMAPRGPVSAFLGIPFAEPPVGPRFLPPEPKRPWPGVLNATAFGSVCYQVDTLYPFGETEMWNPNRELSEDCLYLNWTPYPRPSSPTVLGIYGG
DROS          VCGVIDRLVQTSSGPVRGRSVTVQQREVHVTGIPYAKPPVEDLRFRKPVPAEPWHGVLDATGLSATCVQERYEYFPGFSGEEIWNPNTNVSEDCLYINVWAPAKARLRHGILIWIYCC
HUM BUChE     EDDIIATKNGKVRGMNLTVFGGTVTAFLGIPYAQPPLGRLRFKKPQSLTKWSDIWNATKYANSCCQNIDQSFPGFHGSEMWNPNTDLSEDCLYLNVWIPAPKPKNATVLIWIYGG
         121
              GFYSGSSTLDVYNGKYLAYTEEVLVSLSYRVGAFGFLALHGSQEAPGNVGLLDQRMALQWVHDNIQFFGGDPKTVTIFGESAGGASVGMHILSPGSRDLFRRAILQSGSPNCPWASVSV
              GFYSGASSLDVYDGRFLVQAEGTVLVSMNYRVGAFGFLALPGSREAPGNVGLLDQRLALQSVQENVAAFGGDPTSVTLFGESAGAASVGMHFXSPGSRXXXXAVLQSGAPNGPNATVGV
              GFMTGSATLDIYNADIMAAVGNVIVASFQYRVGAFGFLHLPEMPEAPCNVGLWDQALAIRWLKDNAHAFGGNPEWMTLFGESAGSSSVNAQLMSPVTRGLVKRGMMQSCTMNAPWSHMTS
              GFQTGTSSLHYYDGKFLARVERVIVVSMNYRVGALGFLALPGNPEAPGNMGLFDQQLALQWVQKNIAAFGGNPKSVTLFGESAGAASVSLHLLSPGSHSLFTRAILQSGSFNAPWAVTSL
         241- (--------)
              AEGRRAVELGRNLNCLNSDEELIHCLREKPQELIDVEWNVLPFDSIRFRSFVPVIDGEFFPTSLESMLNSGNFKKTQILLGVNKDECGSFFLLYGAPCFSKDSESSKISREDFMSGVKL
              CEAQYVYVELLNPXXXXXXXMVHCLKARPAQDLVDHEWRVLPQEHVFRFSFVPVIDGEFFPXXXEXXXXGQFKKTQPALVPAKDECSYFLVYGAPGFSKXXXXKIARGNFXXXXXV
              EKAVEIGKALINDCNCNASMLKTNPACMRSVDAKTISVQQW--NSYSGILSFPSAPTIDGAFLPADPMTLMKTADLKDYDILMGNVRDEGTYFLLYDFIDFDKDDATALPRDKYLEIMNN
              YEARNRTLNLAKLTGCSRENETEIIKCLRNKDPQEILLNEAFVYPYQTPLSVNFGPYDGDFLTDMPDILLELGQFKKTQILVGVNKDECTAFLVYGAPGFSKDNNSIITRKEFQEGLKI
         361
                                                                      (-----------------------#)
              SVPHANDLGLDAVTLQYTDWMDDNNGIKNRDGLDDIVGDHNVICPLMHFVNKYTKFGNGTYLYFFNHRASNLVWPEWMGVIHCYEIEFVFGLPLVKELNYTAEEEALSRRIMHYWATFAK
              CVPQASDLAAEAVVLHYTDWLHPEDPARXXEALSDVVGDHNVVCPVAQLAGRXXXXVAYIFEHRASTLSWPEWMGVPHGYEIEFIFGCLPLEPSLNYTIEERTFAQRLMRYWANFAR
              IFGKATAQEREAIIFQYTSWEG NPGYQNQQQIGRAVGDHFFTCPTNEYAQALAERGASVHYYFTHRTSTSLWGEWMGVLHGDEIEYFPGQPLNNSLQYRPVERELGCKRMLSAVIEFAK
              FFPGVSEFGKESILFHYTDWVDDQRPENYREALCDVVGDVNFICPLAEFFTKKFSEWGNNAFFYYFEHRSSKLPWPEWMGVMHCYEIEFVFGCLPLERRDNYTKAEEILSRSIVKRWANFAK
         481
                                (---------------------)
              TCNPNEPHSQESKWPLFTTKEQKFIDLNTEPMKVHQRLRVQMCVFWNQFLPKLLNATETIDEAERQWKTEFHRWSSYMMHWKNQFDHY-SRHESCAEL
              TCDPNDPRKAKQPWPLYTTKXXXXXXVPQASLAQACAFWNRFLPKLRNATDTLDEAERQWKAEFHRWSSYMVHWKNQFDHY-SKQDRCSDL
              TCNPAQ---DGEECPNFSKEDPVYYIFSTDDKEKLARGPAARCSFWNDYLPKVRSWACTCDCDSGSASISPRLQLLGIAALYICAALRTKRVF
              YCNPNETQNNSTSWPVFKSTEQKYLTLNTESTRIMTKLRAQQCRFWTSFFPKVLEMTGNIDEAEWEWKAGFHRWNNYMMDWKNQFNDYTSKKESCVGL
```

Fig. 2. Amino acid sequences (single letter code) of Torpedo californica, fetal bovine serum (partial sequence), and drosphila AChEs and human serum BuChE. The bracketed sequence on top shows the disulfide loops. The underlined amino acids show homology among all ChEs. * = positions of aspartic acid residues common to all sequences. # = positions of common histidine residues. @ = active site serine residue. X = amino acid residues not determined. The sequences are aligned to maximize the homology. The cyanogen bromide fragment (aa 24-85) of FBS-AChE which contains the epitope for mAb AE-2 is underlined. The amino acid sequence homology between Torpedo and FBS-AChE in this region is shown by underlining under the Torpedo sequence.

Fambrough et al. (1982) demonstrated that monoclonal antibodies generated against human erythrocyte membrane AChE showed varying degrees of cross reactivity when tested against neuromuscular junction AChE in several species. Of these mAbs, AE-2 appeared to show the widest range of cross reactivity. Ralston et al. (1985) showed that mAb AE-2 also cross reacted with FBS-AChE. Therefore, we selected this mAb to further examine the extent of cross reactivity with AChE isolated from other species. Figures 1A and 1B show the comparison of mAb AE-2 binding to native FBS-AChE, Torpedo 11s AChE, and DFP-inhibited enzymes. These results demonstrate that mAb AE-2, raised against human erythrocyte AChE, shows cross reactivity with Torpedo AChE to the same extent as with the FBS-AChE. Recently Sorensen et al. (1987) showed that AE-2 cross reacted equally with human erythrocyte and adult or fetal brain AChE. We have observed similar cross reactivity of AE-2 with bovine erythrocyte and bovine brain AChE.

TABLE 1

CROSS REACTIVITY OF mAb AE-2 WITH FBS-AChE AND TORPEDO 11S ACHE AS DETERMINED BY DOT ELISA

ANTIGEN	nmol Ag Required for Positive Assay	
	FBS-AChE	TORPEDO 11S AChE
Native	2.0×10^{-4}	6.1×10^{-4}
Native-DFP	3.0×10^{-4}	6.1×10^{-4}
DRA*	4.5×10^{-3}	7.6×10^{-5}
Deglycosylated-DRA	7.9×10^{-5}	N.D.
Trypsin Treated	9.0×10^{-3}	N.D.
CNBr Treated	5.0×10^{-3}	N.D.

*Denatured, reduced and alkylated; N.D., Not determined

Further characterization of the cross reactivity of mAb AE-2 was carried out using the Dot-ELISA technique (Pappas et al., 1983). The results are shown in Table 1. The results show that binding to antibody is maintained after denaturation and deglycosylation of AChE and also after fragmentation by trypsin and cyanogen bromide. Although the affinity was low with fragmented enzyme, it demonstrates that AE-2 recognizes the structural peptide region of the molecule. This permitted us to isolate, characterize, and determine the amino acid sequence of the epitopic region of mAb AE-2. This was done as follows:

Approximately 200 nmoles of purified FBS-AChE was reduced, denatured, and alkylated after labeling the active site serine with ^{3}H-diisopropyl phosphorofluoridate. The enzyme was subjected to cyanogen bromide cleavage, and fragments were separated by gel-filtration on a Sephadex G-50 column (1.5 x 100 cm). The fractions giving a positive Dot-ELISA with mAb AE-2 were pooled and subjected to further separation by HPLC on µbondapack C_4 and C_{18} columns using phosphate buffer, acetonitrile, and TFA-acetonitrile solvent systems. Only one of all the possible CNBr fragments of FBS-AChE gave a positive Dot-ELISA with mAb AE-2. This purified CNBr-cleaved fragment was subjected to gas-phase amino acid sequence analysis. The sequence is shown in Figure 2. Also shown is approximately 90% of the sequence of the FBS-AChE molecule. (Smyth et al., 1988) The sequences of Torpedo californica 11S AChE (Schumacher et al., 1986), Drosphila AChE (Hall and Spierer, 1986) and human serum butyrylcholinesterase (BuChE, Lockridge et al., 1987; Prody et al., 1987) are shown for comparison. The epitopic region for mAb AE-2 is located between aa 24 and 85 from the amino terminus of the molecule and is underlined in the figure.

To further characterize this 62-amino acid peptide, four synthetic peptides, mimicking the amino acid sequences 24-41 (1), 40-48 (2), 53-69 (3), and 69-84 (4) were tested for their ability to compete with FBS-AChE for binding to mAb AE-2 using a competition assay. (Formal et al., 1984) The results are shown in Figure 3. Peptides 3 and 4 competed with FBS-AChE, and thus it can be concluded that the mAb AE-2 epitope is located, at least in part, between aa 53 and 84. It should be noted that cysteine in position 69 forms a disulfide bond with cysteine 96 (MacPhee-Quigley et al., 1987; Lockridge et al., 1987). The significance of The disulfide loop at this location will be discussed later.

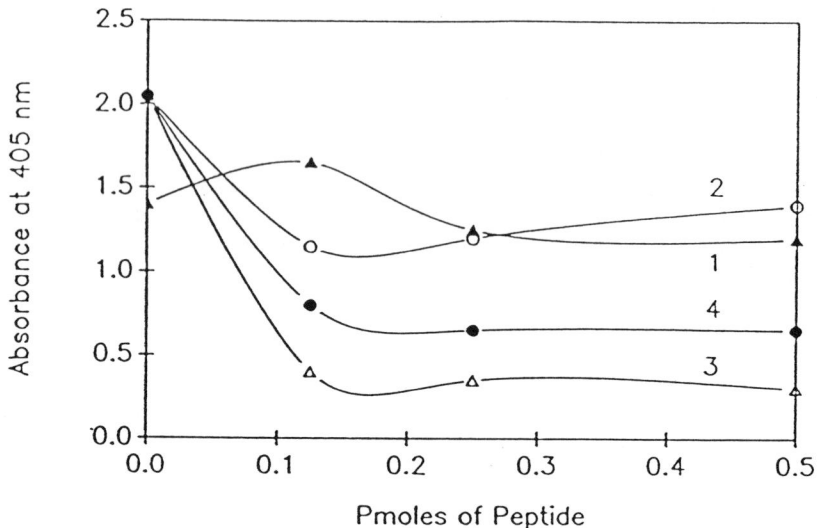

Figure 3. Competition by synthetic peptides to binding of mAb AE-2 to FBS-AChE. The synthetic peptides 1-4 (see text and Fig. 2 for their sequences) were used in an ELISA assay to determine their ability to compete with FBS-AChE for binding to mAb AE-2. Peptides 3 and 4 competed with the enzyme for binding to mAb AE-2.

As previously shown (Sorenson et al., 1987), mAb AE-2 inhibits the catalytic activity of several mammalian AChEs. Figure 4 shows the effect of the binding of mAb AE-2 to Torpedo 11s AChE and FBS-AChE on the ability of these enzymes to hydrolyze acetylthiocholine. It appears that binding of mAb AE-2 to Torpedo AChE has essentially no effect on its catalytic activity; however, it inhibited up to 75% of the catalytic activity of FBS-AChE. Antibody AE-2 showed no affinity for eel 11s AChE or human serum BuChE and did not effect their catalytic activity. On the other hand, mAb 25B1, raised against DFP-inhibited FBS-AChE, completely inhibited the catalytic activity of FBS-AChE. These results indicate a different mechanism of interaction as compared to that of AE-2. In addition 25B1 inhibited the catalytic activity of bovine brain and bovine erythrocyte AChE. However, it failed to inhibit the catalytic activity of Torpedo, eel, and human erythrocyte AChEs, or human serum

BuChE. It failed to bind to denatured FBS-AChE, indicating that the epitope is conformational rather than a structural region of the enzyme. The antigen-antibody complex between FBS-AChE and mAb AE-2 is much weaker than with mAb 25B1, as judged by ease of dissociation of the former upon dilution.

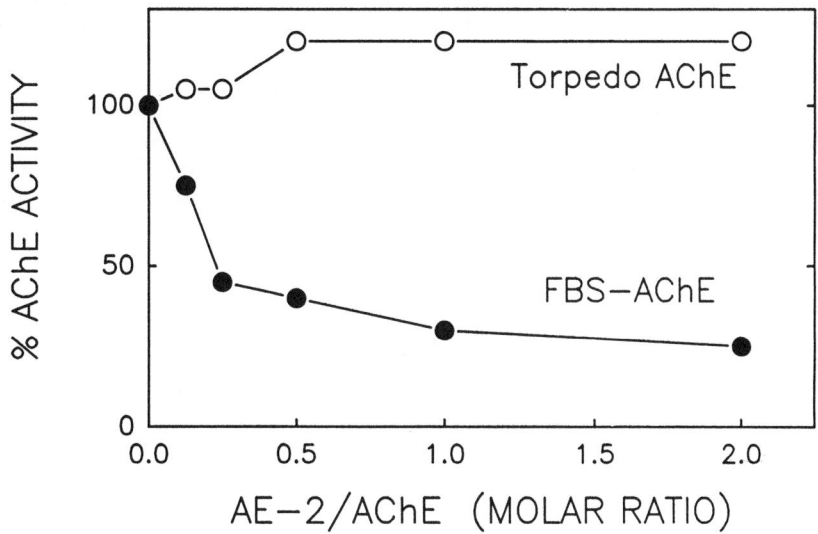

Figure 4. The effect of mAb AE-2 on catalytic activity of FBS-AChE and Torpedo 11S AChE; 0.0625 nmoles of enzyme were used for each data points. The amount of antibody was varied from Ag:Ab molar ration of 0.1 to 5.0. The total incubation volume was 0.25 ml in phosphate buffer pH 8.0. The antigen-antibody mixture was incubated for 1 hr at room temperature. The enzyme activity was determined by the method of Ellman et al. (1961). The Ag:Ab complex was not precipitated since the enzyme assay before and after high speed centrifugation yielded the same results. The mAb AE-2 used in this experiment was purified to homogeneity by HPLC and affinity chromatography ●——● FBS--AChE; o——o Torpedo 11S AChE.

Another difference between these two monoclonal antibodies is the almost complete lack of binding of ^{3}H-DFP to FBS-AChE in presence of a stoichiometric amount of mAb 25B1, whereas there was no effect in the presence of mAb AE-2. The results shown in Table 2 demonstrate that when FBS-AChE was preincubated with amounts of mAb 25B1 sufficient to inhibit approximately 50% of catalytic activity, and then incubated with ^{3}H-DFP, the extent of radiolabelling of enzyme was 50% of the control (without antibody). When the amount of mAb 25B1 was increased to cause approximately 95+% inhibition of catalytic activity, the extent of radiolabelling was proportionately decreased to approximately 2-3% of the control. Similar experiments using mAb AE-2 instead of 25B1 showed no difference in radiolabelling of FBS-AChE with ^{3}H-DFP in the presence or absence of that antibody.

TABLE 2

COMPETITION BETWEEN mAbs AND DFP FOR BINDING TO FBS-AChE

REACTANTS	^{3}H-DFP BOUND TO FBS-AChE CPM	INHIBITION OF AChE BY mAb %
DFP + AChE	21,000	--
DFP + AChE + AE-2	20,800	10
DFP + AChE + AE-2	22,500	50
DFP + AChE + 25B1	11,500	50
DFP + AChE + 25B1	900	>95

Enzyme and mAb were incubated for 16 hr. at 25°. ^{3}H-DFP (approximately 50% molar excess) was added and incubated for additional 24 hr. The amount of DFP bound to FBS-AChE was determined by gel filtration on Biorad P-6 and measurement of radioactivity in void volume (front) peak.

These results indicate that although both these mAbs inhibit the catalytic activity of FBS-AChE, the mechanism of inhibition appears to be different, possibly either due to the fact that they recognize different epitopes or that mAb AE-2 forms an easily reversible complex with FBS-AChE, while mAb 25B1 forms an irreversible complex. Thus, both mAbs may be utilized as molecular probes for studying the structure of the enzyme regions involved in substrate hydrolysis, as illustrated by the following approach.

The epitope for mAb AE-2 appears to be located near the amino terminus and is composed of the structural peptide moiety of the FBS-AChE molecule. The observation by Sorensen et al(1987) and by us that binding of propidium, a reversible peripheral anionic site ligand, is affected by mAb AE-2, implies that it is probable that the epitopic region also contains the peripheral anionic site. In addition, the amino-terminus region of the molecule which contains the epitope for mAb AE-2 may be folded in such a fashion that a portion of it is reaching the active site serine-200 area. Closer examination of amino acid sequences and the homologies between them (see Fig. 2) shows that there are seven aspartic acid residues located at similar locations in all the cholinesterase molecules. The active site serine residue is also located at the same position, and there are two histidine residues located at similar positions. If the mechanism of catalytic hydrolysis of acetylcholine by cholinesterases involves the charge triad, as with α-chymotrypsin or subtilisin, then one may assume that the serine, aspartic acid, and histidine involved can be located at similar positions in all cholinesterase molecules. Serine-200, histidine-442 and aspartic acid-95 appear to be the best candidates for the following reasons: these three amino acids are located at similar positions in all cholinesterases; there is a very high degree of amino acid sequence homology surrounding them; both aspartic acid and histidine are located within a disulfide loop and aspartic acid 95 is located next to cysteine.

The epitopic region for mAb AE-2, at least in part, appears to be located between aa 53 and 84. Within this region, cysteine-69 is disulfide bonded to cysteine-96; therefore, this disulfide loop region must be involved in binding to mAb AE-2. As discussed above, if aspartic acid-95 is involved in the charge triad, then it would have to be located very close to the active site serine-200. This in part may explain why the catalytic activity of FBS-AChE is partially inhibited when mAb AE-2 is bound to it. This observation is further substantiated by preliminary observations that mAb 25B1 blocks the binding of mAb AE-2 to FBS-AChE.

In addition to classical approaches such as x-ray diffraction and 2D-NMR studies of AChE to elucidate the three dimentional structure of the catalytic site(s), the use of molecular probes such as mAbs provides additional information for mapping the topography of various domains of protein molecules. Such information will facilitate the use of computer simulated modelling and design of drugs for prophylaxis and treatment of toxicity caused by organophosphates.

ACKNOWLEDGMENT

This work was done while one of the authors, Y. Ashani, held a National Research Council-WRAIR associateship.

REFERENCES

Ashani Y, Raveh L, Gentry MK, Doctor BP (1988). Anticholinesterase properties of a monoclonal antibody raised against phosphorylated fetal bovine serum AChE. FASEB J 2(5)A1748.

Brimijoin S (1983). Molecular forms of acetylcholinesterase in brain, nerve and muscle: nature, localization and dynamics. Prog Neurobiol 21:291-322.

Doctor BP, Camp S, Gentry MK, Taylor SS, Taylor P (1983). Antigenic and structural differences in the catalytic subunits of the molecular forms of acetylcholinesterase. Proc Natl Acad Sci USA 80:5767-5771.

Ellman GL, Courtney KD, Andres V Jr, Featherstone RM (1961). A new and rapid colorimetric determination of acetylcholinesterase activity. Biochem Pharmacol 7:88-95.

Fambrough DM, Engel AG, Rosenberry TL (1982). Acetylcholinesterase of human erythrocytes and neuromuscular junctions: Homologies revealed by monoclonal antibodies. Proc Natl Acad Sci USA 79:1078-1082.

Formal SB, Hale TL, Kapfer C, Cogan JP, Snoy PJ, Chung R, Wingfield ME, Elisberg BL, Baron LS (1984). Oral vaccination of Monkeys with an invasive Escherichia coli K-12 hybrid expressing Shigella flexneri 2a somatic antigen. Infect Immun 46:465-469.

Gentry MK, De La Hoz DM, Ogert RA, Ashani Y, Doctor BP (1988). Inhibition of catalytic activity of acetylcholinesterases by monoclonal antibodies. FASEB J 2:(#5)A1357.

Hall LMC, Spierer P (1986). The Ace locus of Drosophila melanogaster: Structural gene for acetylcholinesterase with an unusual 5' leader. EMBO J 5:2949-2954.

Lockridge O, Adkins S, La Du BN (1987). Location of disulfide bonds within the sequence of human serum cholinesterase. J Biol Chem 262:12945-12952.

Lockridge O, Bartels CF, Vaughan TA, Wong CK, Norton SE, Johnson LL (1987).Complete amino acid sequence of human serum cholinesterase. J Biol Chem 262:549-557.

MacPhee-Quigley K, Taylor P, Taylor SS (1985). Primary structures of the catalytic subunits from two molecular forms of acetylcholinesterase. J Biol Chem 260:12185-12189.

McTiernan C, Adkins S, Chatonnet A, Vaughan TA, Bartels CF, Kott M, Rosenberry TL, La Du BN (1987). Brain cDNA clone for human cholinesterase. Proc Natl Acad Sci USA 84:6682-6686.

Massoulie J, Bon S (1982). The molecular forms of cholinesterase and acetylcholinesterase in vertebrates. In Cowan WM, Hall ZW, Kandel ER (eds): "Annual Review of Neuroscience", (Vol 5), Palo Alto: Annual Reviews Inc., pp 57-106.

Massoulie J, Bon S, Lazar M, Grassi J, Marsh D, Meflash K, Toutant JP, Vallette F, Vigny M (1984).

Pappas MG, Hajkowski R, Hockneyer WT (1983). Dot enzyme-linked immunosorbent assay (Dot-ELISA): A micro technique for rapid diagnosis. J Immuno Meth 64:205-214.

Prody CA, Zevin-Sonkin D, Genatt A, Goldberg O, Soreq H (1987). Isolation and characterization of full-length cDNA clones coding for cholinesterase from fetal human tissues. Proc Natl Acad Sci USA 84:3555-3559

Ralston JS, Rush RS, Doctor BP, Wolfe AD (1985). Acetylcholinesterase from fetal bovine serum. J Biol Chem 260:4312-4318.

Rakonczay Z, Brimijoin S (1985). Immunochemical differences among molecular forms of acetylcholinesterase in brain and blood. Biochim Biophys Acta 832:127-134.

Schumacher M, Camp S, Maulet Y, Newton M, MacPhee-Quigley K, Taylor SS, Friedmann T, Taylor P (1986). Primary structure of Torpedo californica acetylcholinesterase deduced from its cDNA sequence. Nature 319:407-409.

Smyth KK, De La Hoz DM, Christner CE, Rush RS, De La Hoz F, Doctor BP (1988). FASEB J 2:(#6)A1745.

Sorenson K, Brodbeck U, Rasmussen AG, Norgaard-Pedersen B (1987). An inhibitory monoclonal antibody to human acetylcholinesterases. Biochim Biophys Acta 912:56-62.

Steitz TA and Shulman RG (1982). Crystallographic and NMR studies of the serine proteases. Ann Rev Biophys Bioeng 11:419-441.

RECEPTOR-EFFECTOR COUPLING PROCESSES PROBED BY MONOCLONAL ANTIBODIES.

Enrique Ortega, R. Schweitzer-Stenner and Israel Pecht

Department of Chemical Immunology
The Weizmann Institute of Science
Rehovot, 76100, Israel

INTRODUCTION

The plasma membrane provides cells with a boundary which separates them from the environment and confers them with their individuality. This fundamental partition and screening function of the membrane is inherently also setting the requirements for communication and the controlled transport across this barrier. Identification of membrane components involved in transmembrane signalling and resolution of their mode of action is a major biochemical and biophysical research challenge. One obvious type of reagent to be used for investigating these membrane components are their respective functional ligands, for example, a hormone as a probe for its specific receptor, or a toxin such as bungarotoxin for the acetylcholine receptor. An alternative type of reagent, which is widely employed for investigating biological structure and function are specific, mono- or polyclonal antibodies raised against receptors and their associated membrane components. The advantages of antibodies as reagents for probing, analysis, and modulation of biological activities have been amply documented: Monoclonal antibodies (mAbs) with exquisite specificity can be raised to well defined epitopes of the examined component (Kohler and Milstein, 1975; Brodsky et al., 1979; Lindstrom, 1979; Goding, 1986; Greaves, 1984). Moreover, such mAbs can be selected for different applications, e.g., for rather high affinities, or for binding to epitopes different from the ligand binding sites. Alternatively, they can be screened for their capacity to modulate in several ways the functions of an examined cell. Different mAbs may be selected and used to analyse structurally distinct subunits or domains of a given membrane component. Last, but definitely not the least useful application, mAbs have and are being employed for the isolation and purification of membrane components.

The capacity of antibodies to modulate (i.e., enhance or suppress) the activity of its antigen or antigen carrying cell is obviously one of their more interesting and important features. A range of monoclonal antibodies have been raised in our laboratory which are employed in an effort to identify and study membrane components involved in the coupling of the IgE-mediated stimulus to the secretion from mast cells (Gomperts and Fewtrell, 1986). These cells provide a useful and widely employed prototype system for investigating the nature and sequence of biophysical and biochemical processes which are initiated by triggering via a membrane-residing immunoglobulin-E which serves as a specific detector for its antigen. The mast cell system is a particularly interesting and convenient one for a number of reasons; The cellular response is fast and amenable to several monitoring procedures. There is a rodent mast cell line (RBL-2H3) which can be grown in tissue culture and provide a homogenous source of cells, also in amounts required for biochemical investigation of cellular components. Indeed, the receptor for the Fcε (FcεR) domains of IgE has been isolated from this cell line and characterized (Metzger et al., 1986). On top of the intrinsic fundamental interest of this system, furtherance of the understanding of the processes involved in mast-cell stimulation may also carry significant practical implications for the treatment of a wide-spread human-health problem of the immediate type hypersensitivity.

The initial step triggering the cascade of events, which culminates in secretion, is aggregation of the high affinity, monovalent Fcε receptors (FcεR) (Ishizaka and Ishizaka, 1968; Segal et al., 1977). Already this first event lacks a satisfactory quantitative definition in terms of the minimal number of aggregated receptors required to constitute a signal, the resolution of the physical constraints (e.g. distance among FcεRs in an aggregate) or the minimal life-time needed for the aggregate to exist for exertion of the trigger (Menon et al., 1984). Following FcεR aggregation, several biochemical events have been reported to be initiated in mast cells: These include;

activation of proteases, hydrolysis of polyphosphoinositides and transient increases in cellular concentrations of cAMP and of free Ca^{+2} ions (Ishizaka and Ishizaka, 1984; Metzger et al., 1986). The role of the Fcε receptor in initiating these events is still a mute point. The best characterized FcεR is that of the RBL-2H3 cell line (Froese, 1984, Metzger et al., 1986). It was shown to consist of three polypeptide subunits; The α chain, which is a glycoprotein of ca.50 kD molecular mass, is the IgE binding subunit and its topology involves two immunoglobulin-like domains exposed on the extracellular surface, as well as a transmembrane anchoring polypeptide stretch and a cytoplasmic tail (Kinet et al., 1987; Shimizu et al., 1988). The two other polypeptides co-isolated with the α subunit are the single 33 kDa β chain and two 7-9 kDa disulfide linked γ chains. These three subunits are non-covalently associated in the plasma membrane. This FcεR complex is practically the only membrane component clearly identified so far in the triggering cascade and this has been achieved due to its affinity for IgE. Significantly, none of the subunits of the FcεR have been found so far to have any intrinsic activity, such as enzymatic or otherwise, which could be related to the observed initial events triggered upon aggregation (Metzger et al., 1986).

RESULTS AND DISCUSSION

Monocolonal antibodies specific to epitopes of the α chain of FcεR were frequently encountered when the screening was based on the mAbs capacity to induce RBL-cells to mediator secretion. Three such anti FcεR-mAbs were selected, produced in large amounts and purified. Analysis of their affinities and stoichiometries of binding to the FcεR were carried out by monitoring the interaction of radiolabelled intact mAbs, their respective Fab fragments and of IgE with the cells. As illustrated in Figure 1, the linear Scatchard plots of the results of binding experiments done with the Fabs and IgE clearly show that homogeneous reaction partners are involved; i.e., the monoclonal Fab bind each to its defined epitopes on the cells as does the IgE bound monovalently via its Fcε domain. Furthermore, we employed the IgE binding data to determine the number of FcεR per cell in each batch for which we had also determined the number of Fab fragments of each of the three mAbs bound per cell.

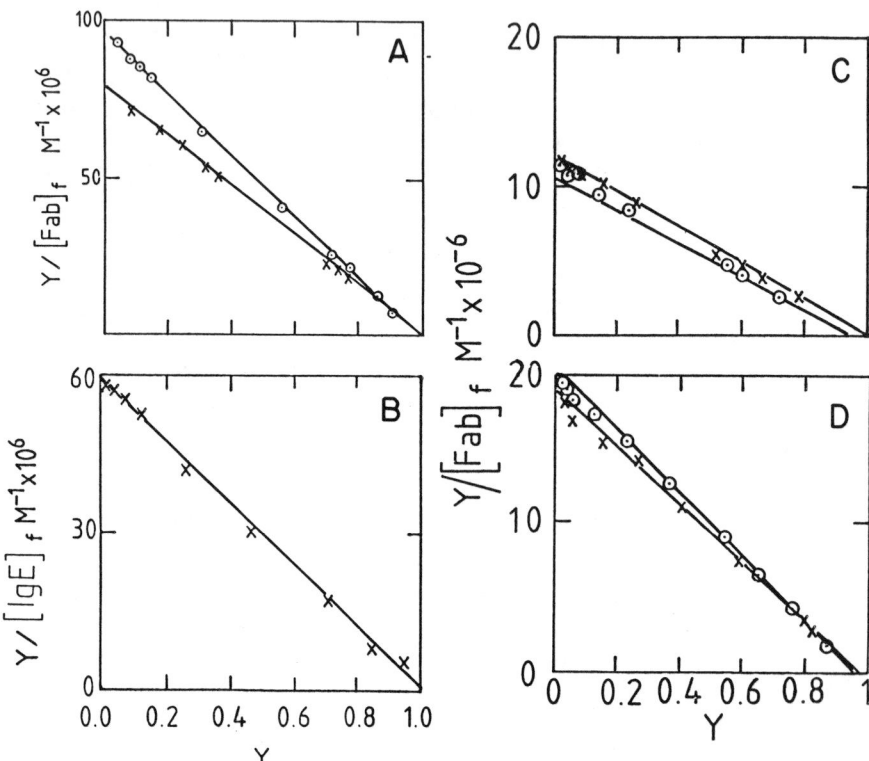

Figure 1: Scatchard plots of the binding of Fab fragments prepared from mAbs specific for the FcεR and of IgE to RBL-2H3 cells. RBL-2H3 cells were incubated with increasing concentrations of ^{125}I-labelled IgE or anti-FcεR mAbs Fab fragments for 1 hour at 25°C (O) or 37°C (X). The cell-bound radioactivity was determined after separating the cells from its supernatant containing the unbound antibody by sedimentation through layer of fetal calf serum. A non-linear fitting procedure was employed to calculate the affinity (Kr) and the total conc. of receptors. The lines were drawn according to the parameters obtained from the fitting procedure. A: H10-Fab; B: IgE; C: F4-Fab; D: J17-Fab.

The maximal anti-FcεR Fabs and IgE concentrations bound at saturation were found to be equal. Since the high affinity FcεR binds a single IgE, all these Fabs have also a single binding epitope per FcεR. This 1 Fab : 1 FcεR stoichiometry of binding implies that the intact mAbs, all three being of the IgG class, would bind two such receptor entities, hence being capable of dimerizing them.

In Figure 2, the cellular secretory dose response to the three mAbs F4, H10 and J17 is presented together with results of binding measurements of these mAbs and of IgE to the cells.

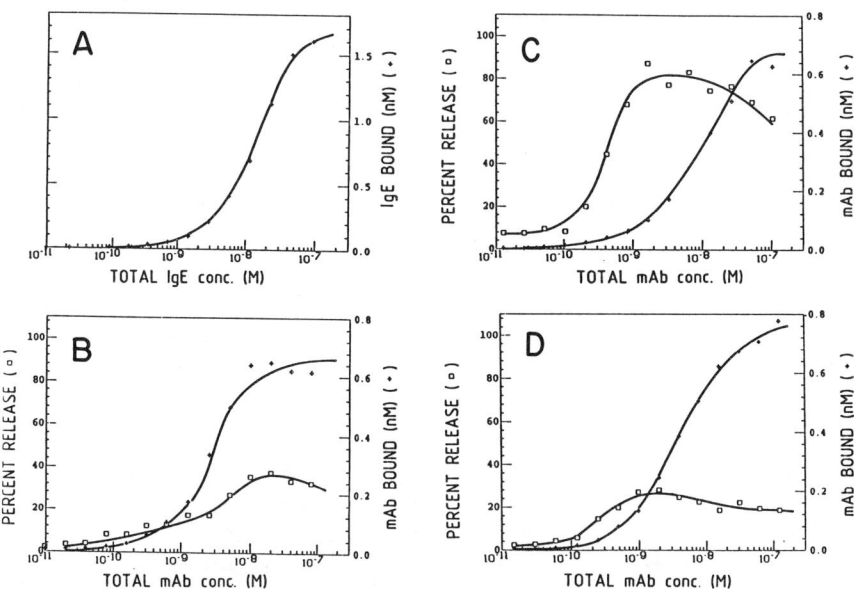

Figure 2. - Secretory dose-responses and binding of anti-FcεR mAbs to RBL-2H3 cells: Identical monolayers of RBL-2H3 cells were treated with increasing concentrations of ^{125}I-labelled mAbs or IgE, for 30 min at 37°C. Secretion was then determined in aliquots of the supernatants by the release of β-hexosaminidase. The monolayers were washed, and the amount of mAb or IgE bound to the cells was determined. A: IgE; B: mAb H10; C: mAb F4; D: mAB J17.

Already a preliminary examination of the data presented in Figure 2, reveals that rather marked differences exist in the extent of mediators secreted by the cells in response to each of the three mAbs: F4 causes rather high secretion, whereas H10 and J17 induce only intermediate to low ones. One possible reason for these marked differences in secretory response could simply be that distinct binding properties of the mAbs would lead to disparate degrees of dimer formation on the cells' surfaces and hence to different extents of stimulation. To examine this possibility, a quantitative analysis of the mAbs binding data using a relatively simple model which describes the divalent binding equilibrium of the mAbs to the monovalent FcεRs freely diffusing in the cells plasma-membrane, has been carried out. The input into this analysis included the intrinsic binding constants for each of the mAbs, as determined from their Fabs fragments binding; the total number of binding sites per cell, as determined from IgE binding; and the binding data obtained for each intact mAb (all performed on identical cell batches). Thus, the equilibrium constant for FcεR - dimer formation on the cells membrane (Kd in equation 1) by each mAb was the free parameter calculated in the fitting procedure.

(1) $\text{Fc}\varepsilon\text{R} + \text{mAB} \xrightleftharpoons{K_r} [\text{Fc}\varepsilon\text{R} - \text{mAb}] + \text{Fc}\varepsilon\text{R} \xrightleftharpoons{K_d} [\text{Fc}\varepsilon\text{R}-\text{mAb}-\text{Fc}\varepsilon\text{R}]$

The results of these computations provided us with values for the fraction of the total FcεRs per cell that underwent dimerization upon binding of each mAb.

Comparison of the secretory dose-response curves with the above described computed fraction of FcεRs dimerized by each of the three mAbs as a function of their concentrations makes it evident that the different secretory response to each of the mAbs cannot be rationalized only on the basis of the formation of different amounts of FcεR dimers. Rather, one observes that the discrepancy is made even more extreme: F4 causes its maximal secretory response (ca 75%) already when 10% or less of the receptors are dimerized, while H10 which can yield practically 100% dimers, causes only some 35% net release as its maximal response. Hence, the rational that we have come to consider as the most plausible one is that the relative orientational constraints; which each mAb exerts on the two FcεR brought together upon forming a dimer, are different and characteristic for the examined mAb. Distinct relative orientations of FcεRs in the produced dimers provide stimuli which differ in their amplitudes or efficiency of coupling to the succeeding steps in the cascade.

This rational is further supported by separate experiments where Fab fragments of each of the three mAbs were first bound to the cells and, obviously, did not cause any response. Then, upon their cross-linking by anti-mouse IgG antibodies, practically the same maximal response was obtained for all three Fabs. These results were thus informative in two different ways: First, they clearly showed that Fab fragments of all three mAbs were ineffective by themselves and could exert triggering only upon being cross-linked by anti-mouse antibodies. Secondly, the observed similar maximal mediators secretion suggested that the cause for different response to the intact mAbs lies in their IgG structure, namely in spatial and/or configurational constraints which their binding to the FcεR impose upon dimer formation. We are pursuing further, independent experimental approaches in order to examine our proposed rational and to provide better scrutiny for its validity. As already stated above, the triggering signal to quite a diverse range of cells is an aggregative process of receptors. Hence, our finding may be of general significance for other cells where the initial activating signal involves cross-linking of membrane-residing components.

Amongst other mAbs raised to probe the functional and possibly some structural correlates of mast cells, we shall dwell here on one further mAb found to be an effective inhibitor of the IgE (and hence FcεR) mediated secretion. This mAb was, however, found to bind and cause its inhibitory effect via an hitherto unknown plasma membrane component. We named this mAb G63 and in Figure 3, its binding isotherm to the RBL-2H3 cells is shown. From such binding experiments we determined that the number of epitopes to which G63 can bind per cell is 1.6×10^4, i.e., markedly less than that of FcεR which is present at several times 10^5 per cell.

There was no detectable effect of G63 binding on that of IgE interaction with the FcεR. Moreover, full saturation of the cells with IgE did not affect binding of G63. These results excluded the possibility that G63 exerts its inhibition by perturbing the initial signalling step at the FcεR site.

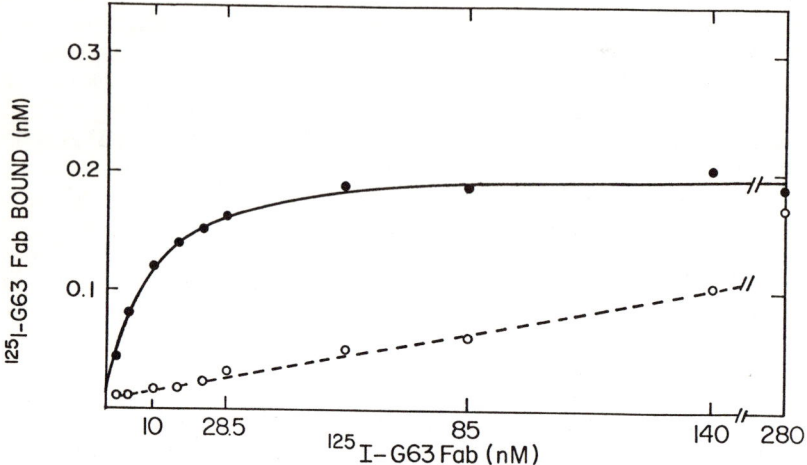

Figure 3: - Binding of ^{125}I labeled Fab fragments of mAb G63 to RBL-2H3 cells. RBL cells were incubated with increasing conc. of ^{125}I-G63-Fab, for one hour at 4°C. Following this time, the cells were separated from the supernatants containing unbound antibody by sedimentation through fetal calf serum and the cell-bound radioactivity was determined. Non-specific binding (O) was estimated by a titration in the presence of saturating conc. of unlabeled mAB G63. The solid symbols (●) represent the specific (total minus non-specific) binding.

In Figure 4 the inhibitory capacity of G63 on antigen -induced secretion from RBL -2H3 cells is shown. Evidently, inhibition is most pronounced at the lower concentration of stimulant, where it can reach some 60% suppression. It is however clearly inhibitory at saturating levels of secretagouges. We examined several of the early biochemical processes activated by FcεR cross-linking and which couple it to secretion (e.g. transient increase in cytoplasmic free Ca^{+2} ions) we found that they are also inhibited to similar degrees by G63. Thus, the antigen recognized by G63 should be having a role in the very early events of coupling the initial signal of FcεR aggregation to cell secretion.

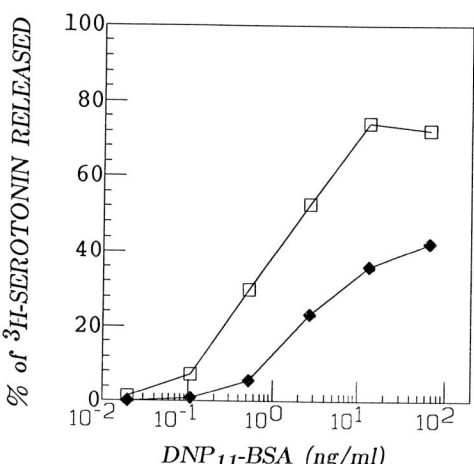

Figure 4: - Inhibition of ^{3}H-serotonin release by mAb G63. Monolayers of RBL-2H3 cells were loaded with ^{3}H-serotonin and primed with a DNP-specific IgE. Following washing, the cells were preincubated with 3×10^{-8}M mAb G63 (■) or in buffer alone (□), for 30 minutes at room temperature. Secretion was induced by DNP$_{11}$-BSA. After 30 min at 37°C, secretion was assessed in the supernatants by the release of ^{3}H-serotonin.

ACKNOWLEDGMENTS

The generous support of grants-in-aid from The Council for Tobacco Research - USA., Inc. (# 1818) and The Hermann and Lilly Schilling Stiftung for Medical Research, FRG are gratefully acknowledged.

REFERENCES

Brodsky FM, Parham P, Barnstable CJ, Crampton MJ, Bodmar WF (1979). Monoclonal antibodies for analysis of the HLA system. Immunol Rev 47:3-61.
Froese A (1984). Receptors for IgE on mast cells and basophils. Prog Allergy 34:142-187.

Goding J (1986). Monoclonal antibodies, principles and practice. 2nd edition, London: Academic Press.

Gomperts BD, Fewtrell CMS (1985). The mast cell: A paradigm for receptor and exocytotic mechanisms. In Cohen P, Houslay M (eds): "Molecular Mechanisms of Transmembrane Signalling" Amsterdam: Elsevier, pp. 377-409.

Greaves MF (1984). Ed. Monoclonal antibodies to receptors. In "Receptors and Recognition". Series B, Vol. 17, London: Chapman and Hill.

Ishizaka K, Ishizaka T (1968). Induction of erythema wheal reactions by soluble antigen -IgE antibody complexes in man. J Immunol 101:68-78.

Ishizaka K, Ishizaka T (1984). Activation of mast cells for mediator release through IgE receptors. Prog Allergy 34:188-235.

Kinet JP, Metzger H, Hakimi J, Kochan JA (1987). cDNA presumtively coding for the α subunit of the receptor with high affinity for IgE. Biochemistry 26: 6922-6927.

Kohler G, Milstein C (1975). Continuous cultures of fused cells secreting antibody of predefined specificity. Nature 256: 495-497.

Lindstrom J (1979). Autoimmune response to acetylcholine receptors in Myasthenia Gravis and its animal model. Adv Immunol 27: 1-32.

Menon, AK, Holowka D, Baird B (1984). Small oligomers of immunoglobulin E (IgE) cause large-scale clustering of IgE receptors on the surface of rat basophilic leukemia cells. J Cell Biol 98:577-583.

Metzger H, Alcaraz G, Hohman R, Kinet JP, Pribluda V, Quarto R (1986). The receptor with high affinity for immunoglobulin E. Ann Rev Immunol 4: 419-470.

Segal DM, Taurog JD, Metzger H (1977) Dimeric immunoglobulin E serves as a unit signal for mast cell degranulation. Proc Natl Acad Sci USA 74: 2993-2997.

Shimizu A, Tepler I, Benfey PH, Bernstein EH, Siraganian RP, and Leder P (1988). Human and rat mast cell high affinity IgE receptors: characterization of putative α chain gene products. Proc Natl Acad Sci USA 85:1907-1911.

STRUCTURAL AND FUNCTIONAL DIVERSITY OF MUSCARINIC ACETYL-CHOLINE RECEPTOR SUBTYPES

J. Ramachandran*, E.G. Peralta, A. Ashkenazi, J.W. Winslow, D.H. Smith and D.J. Capon

Department of Developmental Biology and Molecular Biology, Genentech, Inc., South San Francisco, CA 94080 (*Present Address: Neurex Corporation, 3760 Haven Avenue, Menlo Park, CA 94025)

INTRODUCTION

Muscarinic acetylcholine receptors (mAChR) are present in neurons of the central and peripheral nervous systems as well as in cardiac and smooth muscle, and in exocrine glands. These receptors are involved in regulating a diverse set of physiological functions through activation of guanine nucleotide binding (G) proteins. The biochemical responses induced by the interaction of acetylcholine and other muscarinic agonists with mAChR in target cells include inhibition of adenylyl cyclase activity, stimulation of phosphoinositide (PI) turnover and activation of cardiac K^+ channels. The distinct biochemical and pharmacological properties of mAChR in different tissues and the discovery of selective antagonists such as pirenzepine (Hammer et al., 1980) led to the concept of mAChR subtypes. The molecular basis of this subtype diversity remained obscure until the recent elucidation of the primary structures of mAChR from porcine brain (Kubo et al., 1986a) and porcine atrial tissue (Kubo et al., 1986b; Peralta et al., 1987a) by molecular cloning. These studies provided the first unambiguous evidence that mAChR isolated from the brain and the heart are coded by related but distinct genes. Comparison of the amino acid sequences deduced from the cDNA sequences coding for the M_1 subtype mAChR isolated from porcine brain (PM_1) and the M_2 subtype purified from porcine atrial tissue (PM_2) revealed only 38% amino acid identity between the two molecules. The finding that the entire coding sequence of the porcine M_2 receptor is present in a single exon in the gene (Peralta et al., 1987a) enabled Peralta et al. (1987b) to isolate the genes for human mAChR subtypes by screening a human genomic library. This investigation revealed the presence of genes coding for two novel subtypes designated HM_3 and HM_4 in addition to the genes coding for the human subtypes (HM_1 and HM_2) corresponding to PM_1 and PM_2 mAChR. The structural features of the four different human mAChR subtypes as well as some biochemical properties of the recombinant receptors expressed in mammalian cells are discussed in this article.

STRUCTURAL FEATURES OF mAChR SUBTYPES

The mAChRs belong to the class of integral membrane glycoproteins that includes visual rhodopsins and several hormone and neurotransmitter receptors which utilize a G protein for transducing the light or agonist-binding stimulus. Hydropathicity analysis of the amino acid sequences of mammalian visual rhodopsins (Nathans et al., 1986; Zuker et al., 1986) and β-adrenergic receptors (Dixon et al., 1986; Yarden et al., 1986) by the method of Kyte and Doolittle (1982) indicates that these proteins adopt a membrane topology similar to that of bacteriorhodopsin for which a structure consisting of seven hydrophobic membrane-spanning domains was revealed by electron diffraction studies (Henderson and Unwin, 1975) and limited proteolysis (Ovchinnikov, 1982). In Fig. 1 the hydropathicity profiles of porcine M_1 and M_2 mAChR are compared with those of the profiles for β-adrenergic receptors from turkey erythrocyte (Yarden et al., 1986) and hamster lung (Dixon et al., 1986). It is readily apparent that the mAChR also adopt a structure consisting of seven transmembrane domains with the amino terminus containing the concensus sites for N-glycosylation in the extracellular space and the carboxyl terminus on the cytoplasmic side of the plasma membrane.

Similar analysis of the deduced amino acid sequences of the four human mAChRs revealed that all four adopt a structure with seven hydrophobic, potential transmembrane domains (Fig. 2). The high degree of structural identity (65%) among the four mAChRs found in the membrane spanning regions and the short connecting loops on the cytoplasmic side suggests that the cleft created upon insertion of these hydrophobic domains in the membrane may serve as the acetylcholine binding pocket. The conservation of two aspartic acid residues among all four subtypes in the second

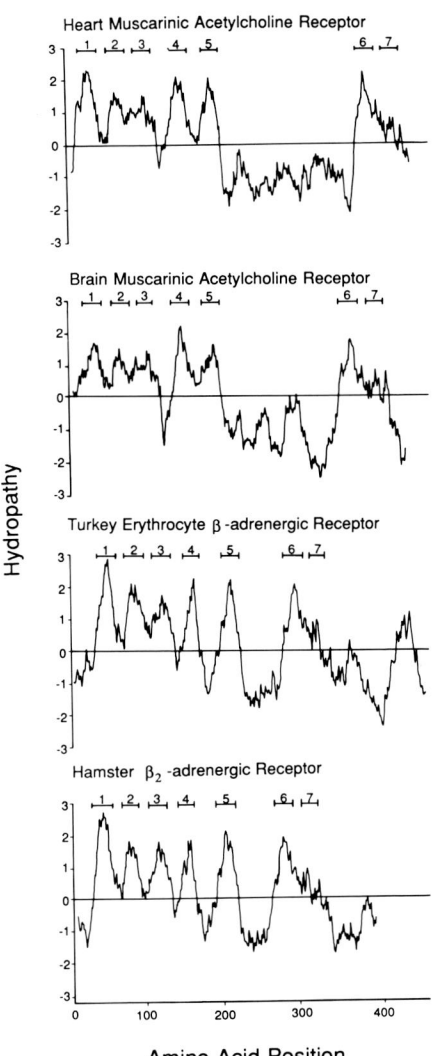

Figure 1. Hydropathicity analysis of the predicted amino acid sequences of mAChR and β-adrenergic receptors.

and third transmembrane domains is compatible with a direct role for these negatively charged residues in binding acetylcholine, a quaternary amine. The aspartic acid residue in the second transmembrane region is also conserved in the β-adrenergic receptors and substitution of this aspartate with asparagine by site-directed mutagenesis resulted in a mutant receptor with vastly reduced affinity for agonists, strongly suggesting an important role for this residue in binding to the catecholamines (Chung et al., 1988).

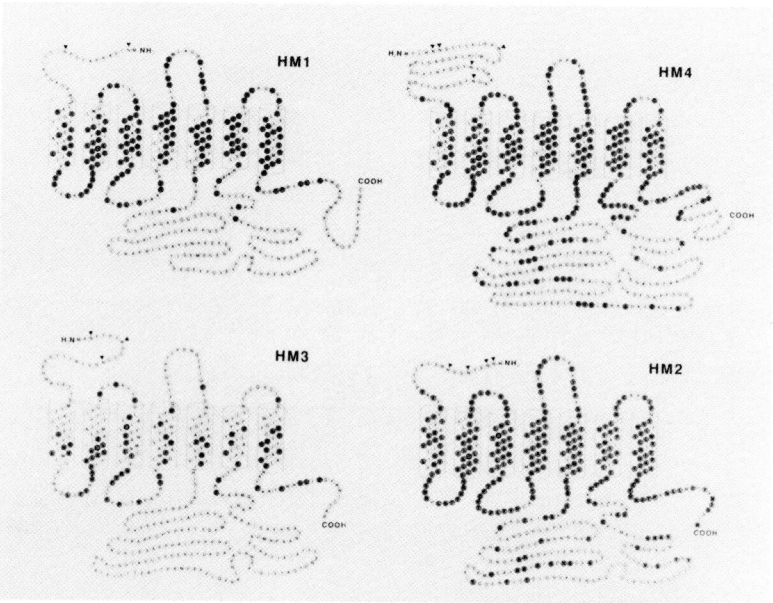

Figure 2: Predicted transmembrane domain structure and amino acid sequence comparison of the four human mAChRs. The shaded residues superimposed on the structural models indicate the following comparisons: HM1, amino acid identities shared between all four human mAChR subtypes; HM2, identities between HM2 and HM3; HM3, identities shared between all four human mAChRs as well as the human, hamster and avian β-adrenergic receptors; HM4, identities shared between HM1 and HM4. The solid triangles indicate potential N-linked glycosylation sites found in the predicted extracellular regions of the receptors. (Taken from Peralta et al., 1988.)

The largest differences among the mAChR subtypes are found in the third cytoplasmic loop bridging the fifth and sixth transmembrane segments (Fig. 2). The length of this loop differs significantly among the four subtypes, consisting of 156, 181, 184 and 241 residues in HM_1, HM_2, HM_3, and HM_4, respectively. There are only six residues in the 5-6 loop of HM_1 which are also present in all three of the other subtypes. Although the sizes of the 5-6 loops show the greatest disparity between HM_1 and HM_4, there is 22% amino acid identity between these subtypes in this loop region. The overall sequence identities among the four subtypes are: HM_1-HM_2, 39%; HM_2-HM_3, 55%; HM_3-HM_4, 37%; HM_4-HM_1, 43%; HM_4-HM_2, 35%; and HM_3-HM_1, 40%. In the 5-6 loop region HM_2 and HM_3 exhibit 21% identity. On the basis of these sequence identities, HM_1 and HM_4 appear to be more closely related to each other than to HM_2 or HM_3. Similarly, HM_2 and HM_3 are structurally more closely related to each other than to HM_1 or HM_4. The exceptional degree of divergence between mAChR subtypes in the 5-6 loop region may be related to their capacity for differential coupling to distinct biochemical effector systems as discussed below.

Although there are considerable differences between the different mAChR subtypes of the same species, there is hardly any difference between the same subtypes of mAChR from different species. Thus HM_1 mAChR is 98.9% identical to PM_1 receptor and HM_2 mAChR is 97.4% identical in amino acid sequence to PM_2.

FUNCTIONAL DIVERSITY OF mAChR SUBTYPES

Although the unique pharmacological properties of mAChRs in different tissues is well documented, the functional characteristics of each subtype could not be discerned owing to the presence of more than one subtype in several tissues. With the availability of molecular clones encoding each of the subtypes, functional analysis of the pure subtypes became feasible through expression of each subtype in mammalian cells (Peralta et al., 1987a; 1987b; Ashkenazi et al., 1987). Chinese hamster ovary (CHO) cells were stably transfected with a vector capable of directing the expression of the porcine atrial receptor (PM_2) cDNA with the mouse dihydrofolate reductase (DHFR) gene as a selectable marker (Peralta et al., 1987a). Cell populations expressing various levels of PM_2 were selected by their resistance to different concentrations of the DHFR inhibitor methotrexate (Ashkenazi et al., 1987). Equilibrium binding of the muscarinic antagonist [^{3}H] quinuclidinyl benzilate (QNB) to transfected cells selected for their resistance to 500 nM methotrexate revealed a single class of high affinity receptors with an apparent K_D of 75 pM and a capacity of 1.5×10^6 sites/cell (Fig. 3A). Kinetic analysis of the association (Fig. 3B) and dissociation (Fig. 3C) reactions in the same population of cells yielded a similar value (apparent K_D = 78 pM). All four human subtypes bound [^{3}H]QNB with high affinity as shown by saturation binding experiments using human embryonic kidney cells transfected with mAChR genes in an expression plasmid (Peralta et al., 1987b). The apparent K_D ranged from 22.8 pM for HM_1 to 112 pM for HM_4 with intermediate values for HM_3 (39.1 pM) and HM_2 (83.3 pM). Competition binding studies using [^{3}H]QNB as the radioligand and various muscarinic antagonists showed that the classic antagonist atropine bound with high affinity to all four subtypes (Fig. 4). There were significant differences in the binding of the selective antagonists pirenzepine which distinguishes between the cerebral and atrial subtypes (Hammer et al., 1980) and AF-DX116 which distinguishes the atrial and glandular subtypes (Giraldo et al., 1987). Recombinant HM_1 binds pirenzepine with 25-fold greater affinity than the recombinant HM_2, a difference comparable to that observed between M_1 receptors of rat brain cortex and hippocampus and M_2 receptors of rat atria (Birdsall and Hulme, 1983) supporting the assignment of the human M_1 and M_2 subtypes. The recombinant human M_2 mAChR displayed the highest affinity for AF-DX116 and the M_1 subtype showed the lowest affinity. HM_3 and HM_4 mAChRs exhibited affinities intermediate between those of HM_1 and HM_2 for both antagonists. Similar binding studies using the muscarinic agonists carbachol and oxotremorine as competing ligands showed that heterogeneity of agonist binding is an intrinsic property of each mAChR subtype. That this multiplicity of agonist binding reflects conformational states of the receptor arising from interactions with G proteins was demonstrated in the case of recombinant porcine M_2 mAChR by conversion of the high affinity binding sites for carbachol to a low affinity state in the presence of the GTP analog, GTP gamma-S (Ashkenazi et al., 1987). The proportion of high affinity sites ranged from 30.1% for M_1 and 28.4% for M_2 to 6.5% for M_4. In the human embryonic kidney cells in which these human mAChRs were expressed, HM_3 appeared to have no high affinity sites

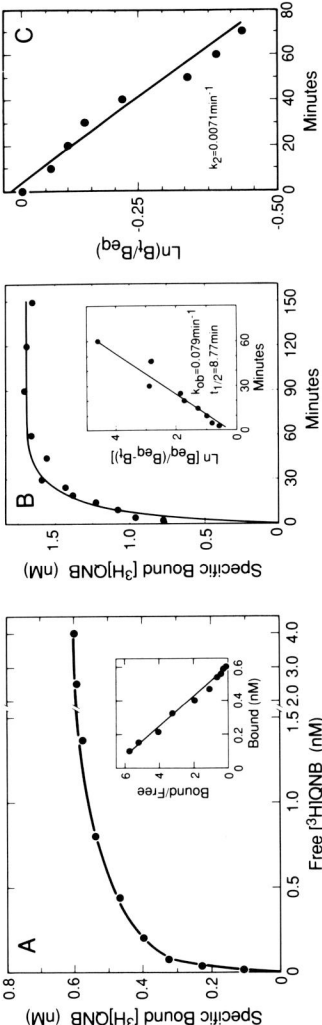

Figure 3. Binding of [^{3}H]QNB to CHO cells transfected with the porcine atrial M$_2$ mAChR cDNA. (A) Saturation analysis of binding. Inset: Scatchard analysis of the saturation data. (B) Time course and association kinetics of [^{3}H]QNB binding. Inset: Kinetic analysis of the binding data. (C) Dissociation of specific bound [^{3}H]QNB. Experimental details are described in Ashkenazi et al. (1987).

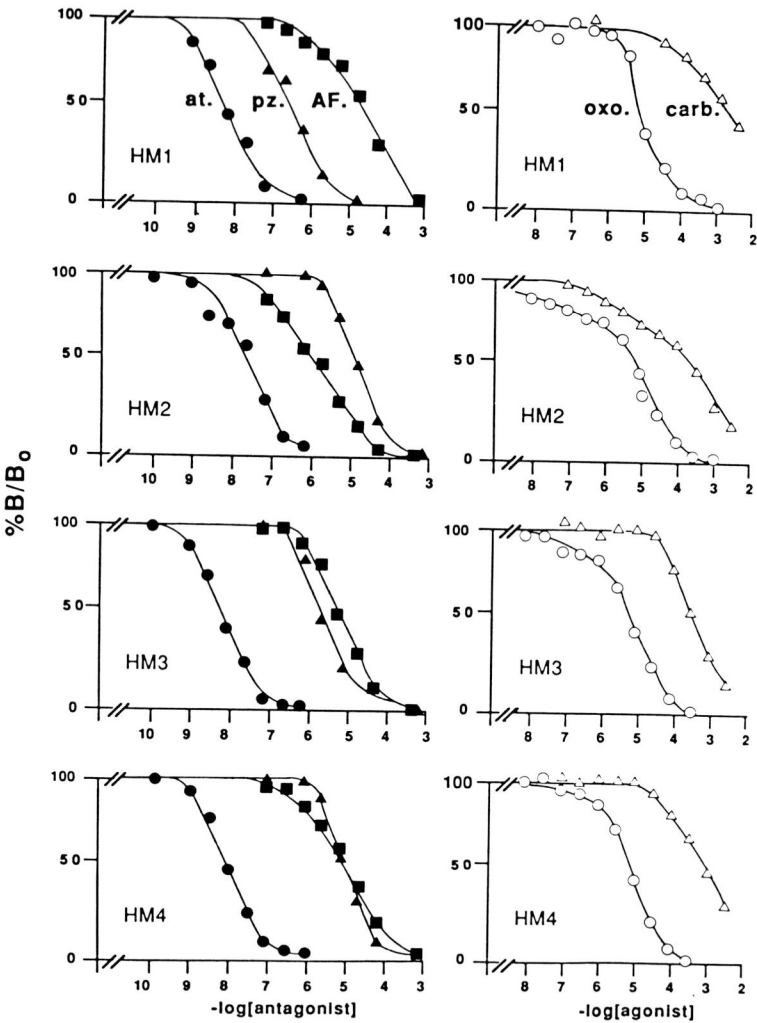

Figure 4. Competition displacement of [^{3}H]QNB binding by muscarinic antagonists atropine (●), pirenzepine (▲) and AF-DX116 (■), (left panel) and agonists carbachol (Δ) and oxotremorine (O), (right panel). (Taken from Peralta et al., 1987b.)

for carbachol and 5.1% high affinity sites for oxotremorine. The variability in the proportion of high affinity agonist binding may be a reflection of the ability of the mAChR subtypes to interact with the endogenous G proteins in this cell line.

Tissue-specific expression of mAChR subtypes was demonstrated by Northern blot analysis of polyadenylylated mRNA isolated from rat tissues and probed with subtype specific nucleotide sequences coding for amino acid residues contained in the 5-6 loop region of each of the human mAChR (Peralta et al., 1987b). All four subtypes were detected in whole rat brain providing definitive evidence for the transcription of the novel M_3 and M_4 subtype genes and supporting a role for each subtype in central nervous system function. In contrast M_2 mAChR mRNA was found only in the heart and hybridization with the M_4 probe was observed only with pancreatic mRNA among the non-neural tissues analyzed. The latter finding suggests that M_4 corresponds to the 'glandular' mAChR subtype described by Giraldo et al. (1987) and Korc et al. (1987) and is supported by the significantly larger size reported for the pancreatic mAChR (Hootman et al., 1985), as expected for M_4 (590 residues vs 460-479 residues for HM_1, HM_2 and HM_3). In the established neuronal cell line NG108-15, expression of mAChR mRNA of the M_3 subtype alone was detectable.

Analysis of the coupling of PM_2 receptors expressed in CHO cells to biochemical responses revealed that carbachol inhibited adenylyl cyclase activity by activation of endogenous Gi since pertussin toxin (PTX) which is known to ADP-ribosylate Gi and abolish coupling with mAChRs, was very effective in eliminating the carbachol-induced adenylyl cyclase inhibition (Fig. 5) (Ashkenazi et al., 1987). Carbachol induced half maximal inhibition of adenylyl cyclase activity at a concentration of $7.1 \times 10^{-8}M$ and at higher concentrations also stimulated phosphoinoside (PI) hydrolysis (Fig. 5). Half maximal stimulation of PI hydrolysis occurred at $6 \times 10^{-6}M$ carbachol. The muscarinic antagonists atropine and pirenzepine which have high and low affinity for the M_2 subtype of mAChR inhibited both biochemical actions of carbachol with the same affinity characteristic of each antagonist (Fig. 5). The concentration of pertussin toxin needed to abolish the stimulation of PI hydrolysis in PM_2 transfected CHO cells was nearly 100-fold higher than that required for abolishing

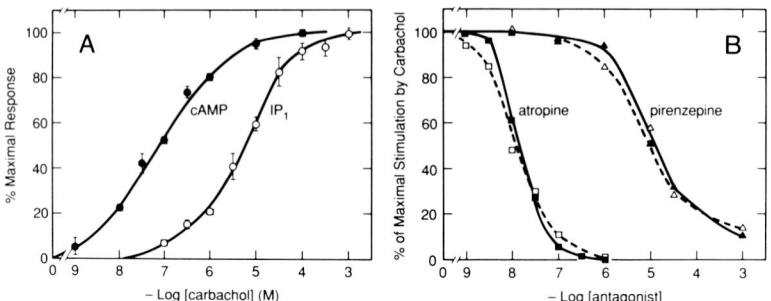

Figure 5. Effect of carbachol on cAMP formation and PI hydrolysis in transfected CHO cells (A) and inhibition of these actions by atropine and pirenzepine (B). (Taken from Ashkenazi et al., 1987.)

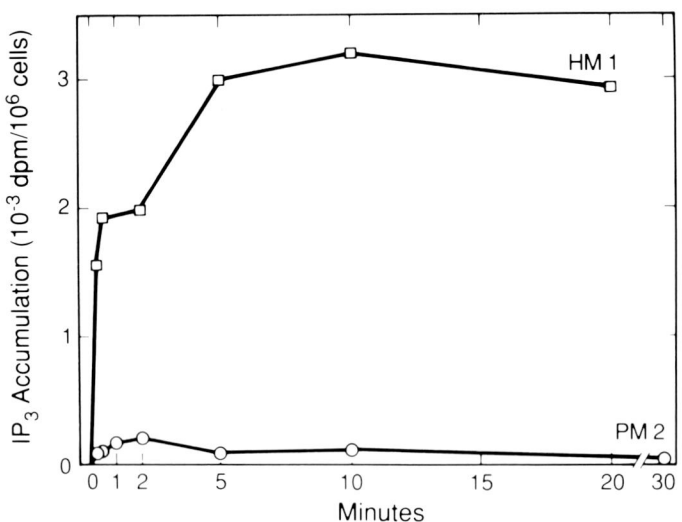

Figure 6. PI hydrolysis induced by carbachol in CHO cells transfected with HM_1 cDNA or PM_2 cDNA. IP_3 accumulation was measured as described previously (Ashkenazi et al., 1987).

the adenylyl cyclase inhibition, strongly suggesting that the PM_2 receptors interact with two different G proteins in the CHO cells to elicit the two biochemical responses. When CHO cells were transfected with an expression plasmid carrying the HM_1 cDNA, the cells showed a strong PI response to carbachol (Fig. 6).

The preferential coupling of PM_2 to inhibition of adenylyl cyclase compared to PI hydrolysis and the efficient coupling of HM_1 to PI hydrolysis in the same cellular context suggests that each mAChR subtype may have evolved the structural features to activate particular cell responses in a selective manner (Ashkenazi et al., 1987; Fukuda et al., 1987). On the basis of the structural similarities discussed in the previous section, it appears that the M_1 and M_4 subtypes are involved primarily in activating PI hydrolysis whereas the principal coupling of the M_2 and M_3 subtypes appears to be with the adenylyl cyclase system. However, the ability of a specific receptor subtype to recognize various effector systems is differential, rather than exclusive. Thus, each receptor subtype may mediate more than one biochemical effect with different potencies, and the physiological significance of the sum of these effects may be determined by the cellular context in which they are evoked, i.e., the specific receptor subtypes and G proteins present, their relative concentration, as well as numerous other factors related to the differentiation of the specific tissue or cell type. The cloning and stable expression of single receptor subtypes in various cell systems is a powerful approach to the study of such cellular events and will contribute to a better understanding of the significance of multiple receptor subtypes.

REFERENCES

Ashkenazi A, Winslow JW, Peralta EG, Peterson GL, Schimerlik MI, Capon DJ, Ramachandran J (1987). An M_2 muscarinic receptor subtype coupled to both adenylyl cyclase and phosphoinositide turnover. Science 238:672-675.

Birdsall N, Hulme E (1983). Muscarinic receptor subclasses. Trends Pharmacol Sci 4:459-463.

Chung F-Z, Wang C-D, Potter PC, Venter JC, Fraser CM (1988). Site-directed mutagenesis and continuous expression of human β-adrenergic receptors. J Biol Chem 263:4052-4055.

Dixon RAF, Kobilka BK, Strader DI, Benovic JL, Dohlman HG, Frielle T, Bolanowski MA, Bennett CD, Rands E, Diehl RE, Mumford RA, Slater EE, Segal IS, Caron MG, Lefkowitz RJ, Strader CD (1986). Cloning the gene and cDNA for mammalian β-adrenergic receptor and homology with rhodopsin. Nature 321:75-79.

Fukuda K, Kubo T, Akiba I, Maeda A, Mishina M, Numa S (1987). Molecular distinction between muscarinic acetylcholine receptor subtypes. Nature 327:623-625.

Giraldo E, Hammer R, Ladinsky H (1987). Distribution of muscarinic receptor subtypes in rat brain ad determined in binding studies with AF-DX116 and pirenzepine. Life Sci 40:833-840.

Hammer R, Berrie C, Birdsall N, Burger ASV, Hulme E (1980). Pinzerpine distinguishes between different subclasses of muscarinic receptors. Nature 283:90-92.

Henderson R, Unwin P (1975). Three-dimensional model of purple membrane obtained by electron microscopy. Nature 257:28-33.

Hootman S, Picardo-Leonard T, Burnham D (1985). Muscarinic acetylcholine receptor structure in acinar cells of mammalian exocrine glands. J Biol Chem 260:4186-4194.

Korc M, Ackerman MS, Roeske WR (1987). A cholinergic antagonist identifies a subclass of muscarinic receptors in isolated rat pancreatic acini. J Pharm Exp Ther 240:118-122.

Kubo T, Fukuda K, Mikami A, Maeda A, Takahashi H, Mishina M, Haga K, Ichiyama A, Kangawa K, Kojima M, Matsuo H, Hiron T, Numa S (1986a). Cloning, sequencing and expression of complementary DNA encoding the muscarinic acetylcholine receptor. Nature 323:411-416.

Kubo T, Maeda A, Sugimoto K, Akida I, Mikami A, Takahashi H, Haga K, Ichiyama A, Kangawa K, Matsuo H, Hirose T, Numa S (1986b). Primary structure of porcine cardiac muscarinicacetylcholine receptor deduced from the cDNA sequence. FEBS Lett 209:367-373.

Kyte J, Doolittle R (1982). A simple method for displaying the hydrophatic character of a protein. J Mol Biol 57:105-132.

Nathans J, Thomas D, Hogness, D (1986). Molecular genetics of human color vision. Science 232:193-200.

Ovchinnikov Y (1982). Rhodopsin and bacteriorhodopsin: Structure-function relationships. FEBS Lett 148:179-190.

Peralta EG, Winslow JW, Peterson GL, Smith DH, Ashkenazi A, Ramachandran J, Schimerlik MI, Capon DJ (1987a). Primary structure and biochemical properties of an M_2 muscarinic receptor. Science 236:600-605.

Peralta EG, Ashkenazi A, Winslow JW, Smith DH, Ramachandran J, Capon DJ (1987b). Distinct primary structures, ligand-binding properties and tissue-specific expression of four human muscarinic acetylcholine receptors. EMBO J 6:3923-3929.

Peralta EG, Winslow JW, Ashkenazi A, Smith DH, Ramachandran J, Capon DJ (1988). Structural basis of muscarinic acetylcholine receptor subtype diversity. In Levine RR, Birdsall NJM, North A, Holman M, Watanake A, Iversen LL (eds) Trends Pharmacol Sci Supplement, February, "Subtypes of Muscarinic Receptors III", Elsevier Publishers, Cambridge, UK, pp 6-11.

Yarden F, Rodriquez H, Wong SK-F, Brandt DR, May DC, Burnier J, Hawkins RN, Chen EY, Ramachandran J, Ullrich A, Ross EM (1986). The avian β-adrenergic receptor: Primary structure and membrane topology. Proc Natl Acad Sci USA 83:6795-6799.

Zuker C, Cowman A, Rubin G (1985). Isolation and structure of a rhodopsin gene from D. melanogasten. Cell 40:851.

MODELLING THE CHOLINERGIC BINDING SITE: CONSIDERATIONS

Jonathan M. Gershoni

Department of Biophysics
The Weizmann Institute of Science
Rehovot, Israel

"...it seems improbable that...is unlikely,...is most probably...thus it is tempting to assume that the acetylcholine binding site is... An alternative possibility is..."

(Noda et al. Nature 299:793, 1982)

Clearly, where biochemical data are missing, speculation must be cautiously made. However, often the caution is forgotten in favor of accepting the "tempting" over the "unlikely" alternative. The efforts in trying to model a cholinergic binding-site have been many and the story that has evolved is both interesting and instructive. The purpose of this discussion is not to provide new data beyond those already published, but rather to review the sequence of events which has led to such confusion in this field.

In 1966 Karlin and Bartels demonstrated for the first time that the cholinergic receptor is sensitive to reducing agents such as 1,4-dithiothreitol and concluded that not only is the receptor a protein but also that the cholinergic binding-site is somehow associated with a cystine disulfide. This led to the construction of affinity alkylating reagents such as 4-(N-maleimido) benzyltrimethyl ammonium iodide (MBTA). Application of radioactive MBTA indicated that the relevant alkylated cysteine is on the alpha-subunit of the reduced nicotinic acetylcholine receptor (AChR) and is situated 1 NM away from the cholinergic binding-site (reviewed in Karlin,

1980).

This finding has become a dogma and underlies most of the subsequent models to be proposed for the AChR binding-site. It is therefore worthwhile to consider the significance of these observations. First, whereas the relevant cysteine is proven to be within the alpha-subunit, it does not necessarily follow that the cholinergic negative subsite must actually reside on this same subunit. True, the stoichiometry of ligand binding is two ligands per one receptor thus possibly reflecting the stoichiometry of the receptor subunits: alpha$_2$beta gamma delta. This does strengthen the argument for the alpha-subunit. However, the binding of MBTA to the AChR is normally only 1:1, arguing an assymetry between the two alpha-subunits. Moreover, the proximity of the relevant cysteine to the binding-site can only be demonstrated in the DTT-denatured receptor and its orientation in the native conformation can only be assumed.

None-the-less, the proximity of a cysteine residue to the functional binding-site has become the first constraint for model building.

The next turning point was the elucidation of the sequence of the alpha-subunit. (Noda et al., 1982). Here, Noda and his co-workers report that of the 437 amino-acid residues, seven are cysteines. Considering hydrophobicity and a unique glycosylation site at Asn-141 it was concluded that cysteines 128, 142, 192, and 193 are extracellular and thus likely candidates for MBTA alkylation. Whereas the theme of tandem cysteines is extremely common in proteins, providing the linking of two loops, Noda et al. concluded that the region for ligand binding should be associated with Cys-128 and Cys-142.

This conclusion, which seemed "tempting to assume", immediately gained support. Evidence was found that the MBTA labeled cysteine was Cys-142 (Cahill and Schmidt, 1984). A synthetic peptide, alpha 125-147, was found to have affinity for alpha-neurotoxins (McCormick and Atassi, 1984) and a number of models were generated showing that the area defined by these two cysteines could be compatible with ligand binding (e.g. Smart et al., 1984; Luyten, 1986). Thus a second dogma evolved: the cholinergic binding site is within the sequence alpha 125-147.

Here too some re-consideration is warranted. Most surprising is that the postulated area is one of the most highly conserved sequences in the entire gene family of ligand regulated channels. Thus, very high homology exists between alpha, beta, gamma and delta subunits for this region and there is significant similarities also with the corresponding region in both the glycine and GABA receptors (Grenninglon et al., 1987: Schofield et al., 1987). One would expect that in view of the fact that only the alpha-subunit is MBTA labeled, a unique sequence would be more appropriate for the binding site. Indeed, the tandem cysteines 192 and 193 have been recognized as the "hallmark" of alpha-subunits throughout nature.

Most compelling though is the ever increasing body of experimental evidence supporting that the region alpha 180-200 directly binds alpha-neurotoxins with affinities similar to those of the complete alpha-subunit.

In 1981 Haggerty and Forehner demonstrated that SDS - denatured alpha-subunits are unique in their ability to bind alpha-bungarotoxin (BTX). This finding was in fact the first demonstration that true binding activity can be associated with the alpha-subunit. However, in order to map the binding site within the alpha-subunit it was anticipated that a flexible and sensitive methodology should be developed. Ligand-overlay of protein blots has proven to be such a technique (Gershoni, 1987a).

First, the alpha-subunit was reconfirmed as being the only subunit that binds cholinergic ligands after receptor dissociation (Gershoni et al., 1983). Proteolysis of the alpha-subunit was then found to generate fragments that retain their capacity to bind BTX (Wilson et al., 1984). By systematically analysing these fragments, the binding site could be mapped to the area of alpha 180-200 (Neumann et al., 1985, 1986). Since then, a number of conclusions could be made that were subsequently confirmed: 1) It appeared that Cys 192/193 were most likely to be MBTA alkylated (Neumann et al., 1985) and indeed was proven to be so (Kao et al., 1984). 2) The previous assumption that alpha 125-147 was the binding-site inspired the hypothesis that all four extracellular cysteines were clustered together, via disulfide bridging (Kao et al., 1984; Cirado et al., 1985; Boulter et al., 1985; Luyten, 1986). However, this was found not to be the case (Neumann et al.,

1985, 1986; Kao and Karlin 1986; Criado et al., 1986; Mosckovitz and Gershoni, 1988).

Thus we are in a new situation. We have convincingly shown that the speculated binding-site is in fact most probably not a major element required for toxin recognition. Furthermore, by direct biochemical analyses the area alpha 180-200 can be proven to be independently capable of alpha-neurotoxin binding. The structure of BTX has been solved (Love and Stroud, 1986), and so the naive biochemist might expect that a docking experiment between the 20 amino acid sequence binding-site and the toxin should be feasible. Clearly this would be an enormous undertaking and so more constraints are still necessary.

In an effort to produce such constraints, we have devised a system which allows the production of large amounts of binding sites and their manipulation. Bacterial expression of the binding domain makes this area amenable to the flexibility provided by recombinant DNA technology (Gershoni, 1987b; Aronheim et al., 1988).

Thus, three practical questions have been formulated:

1) Which amino-acid residues are actually involved in contact formation?
2) How are these functional residues oriented in space?
3) What aspects of the ligands participate in the recognition process?

By generating a variety of sites, both randomly and site mutated, novel binding sites have been produced. These are being studied and the specific contribution of individual residues is being evaluated. Furthermore, NMR analyses of ligand binding to the site are being conducted in collaboration with Dr. Gil Navon of Tel Aviv University. By combining biochemical with biophysical data, a clearer understanding of the mechanism of toxin binding should emerge.

In the meantime, the mere identification of the toxin binding site has led to the development of a new type of drug. The sequence alpha 184-200 is a mimic binding site capable of binding alpha-neurotoxins with considerable affinity. It was therefore conceived that such mimics could act in vivo as molecular decoys providing protection against alpha-neurotoxins. This hypothesis of molecular "decoyance" has recently been tested and found experimental support (Gershoni and Aronheim, 1988).

And so, as apposed to a tempting speculation, a well supported alternative has emerged. Through systematic experimentation constraints will be provided so to enable relevant modelling of the cholinergic binding site. From such models, a more effective class of molecular decoys should be realized.

ACKNOWLEDGMENT

I thank Rachel Samuel for preparing this manuscript and Rachel, Yoav, Ami, Bella, Ileana and Jean-Michel for doing the work.

REFERENCES

Aronheim A, Eshel Y, Mosckovitz R, Gershoni JM. J Biol Chem (in press).
Boulter J, Luyten W, Evans K, Mason P, Ballivet M, Goldman D, Strengelin S, Martin G, Heinemann S, Patrick J (1985). J. Neurosci 5:2545-2552.
Cahill S, Schmidt J (1984). Biochem Biophys Res Comm 122:602-608.
Criado M, Sarin V, Fox JL, Lindstrom J (1985). Biochem Biophys Res Comm 128:864-871.
Criado M, Sarin V, Fox JL, Lindstrom J (1986). Biochem 25:2839-2846.
Gershoni JM (1987a). Electrophoresis 8:428-431.
Gershoni JM (1987b) Proc Natl Acad Sci USA 84:4318-4321.
Gershoni JM, Aronheim A (1988). Proc Natl Acad Sci USA 85:4087-4089.
Gershoni JM, Hawrot E, Lentz TL (1983). Proc Natl Acad Sci USA 80:4973-4977.
Grenningloh G, Rienitz A, Schmitt B, Methfessel C, Zensen M, Beyreuther K, Gundelfinger ED, Betz H. Nature 328:215-220.
Haggerty JG, Forehner SC (1981). J Biol Chem 256:8294-8297.
Kao PN, Dwork AJ, Kaldany RJ, Silver ML, Wideman J, Stein S, Karlin A (1984). J Biol Chem 259:11662-11665.
Kao PN, Karlin A (1986). J Biol Chem 261:8085-8088.
Karlin A (1980). in: "The Cell Surface and Neuronal Function" (CW Cotman, G Poste, GL Nicolson, eds) Elseveir/North Holland Biomedical Press pp 192-244.

Karlin A, Bartles E (1966). Biochem Biophys Acta 126:525-535.
Love RA, Stroud RM (1986). Protein Engineering 1:37-46.
Luyten WHML (1986). J Neurosci Res 16:51-73.
McCormick DJ, Atassi MZ (1984). Biochem J 224:995-1000.
Mosckovitz R, Gershoni JM (1988) J Biol Chem 263:1017-1022.
Neumann D, Barchan D, Safran A, Gershoni JM, Fuchs S (1986) Proc Natl Acad Sci USA 83:3008-3011.
Neumann D, Gershoni JM, Fridkin M, Fuchs S (1985). in: "Molecular Basis of Nerve Activity" (Changeux JP, Hucho F, Maelicke A, Neumann E, eds) de Gruyter, Berlin pp 273-282.
Neumann D, Gershoni JM, Fridkin M, Fuchs S (1985). Proc Natl Acad Sci USA 82:3490-3493.
Noda M, Takahashi H, Tanabe T, Toyosato M, Furutani Y, Hirose T, Asai M, Inayama S, Miyata T, Numa S (1982). Nature 299:793-797.
Schofield PR, Darlison MG, Fujita N, Burt DR, Stephenson FA, Rodriguez H, Rhee LM, Ramachandran J, Reale V, Glencorse TA, Seeburg PH, Barnard EA. Nature 328:221-27.
Smart L, Meyers HW, Hilgenfeld R, Saenger W, Maelicke A (1984). FEBS Lett 178:64-68.
Wilson PT, Gershoni JM, Hawrot E, Lentz TL (1981) Proc Natl Acad Sci USA. 81:2553-2557.

SEQUENCE SIMILARITIES BETWEEN HUMAN ACETYLCHOLINESTERASE AND RELATED PROTEINS: PUTATIVE IMPLICATIONS FOR THERAPY OF ANTICHOLINESTERASE INTOXICATION

Hermona Soreq and Catherine A. Prody

Department of Biological Chemistry, The Life Sciences Institute, The Hebrew University, Jerusalem 91904, Israel

INTRODUCTION

Cholinesterases (ChEs) are highly polymorphic carboxylesterases of broad substrate specificity, involved in the termination of neurotransmission in cholinergic synapses and neuromuscular junctions. ChEs belong to the B type carboxylesterases on the basis of their sensitivity to inhibition by organophosphorous (OP) poisons (Heymann, 1980) and are primarily classified according to their substrate specificity and sensitivity to selective inhibitors into acetylcholinesterase (AChE, acetylcholine acetylhydrolase, EC 3.1.1.7) and butyrylcholinesterase (BuChE, acylcholine acylhydrolase, EC 3.1.1.8) (Massoulie and Bon, 1982). Further classifications of ChEs are based on their charge, hydrophobicity, interaction with membrane or extracellular structures and multisubunit association of catalytic and non-catalytic "tail" subunits (Silman and Futerman, 1987; Soreq and Gnatt, 1987).

The severe clinical symptoms resulting from OP intoxication (Koelle, 1972) are generally attributed to their inhibitory interaction with ChEs (Aldridge and Reiner, 1972). OPs are substrate analogues to ChEs; the labeled OP diisopropylfluorophosphate (DFP) was shown to bind covalently to the serine residue at the active esteratic site region of ChEs, that is common to all of the carboxylesterases (Dayhoff et al., 1983; Lockridge and LaDu, 1986). However, the binding and inactivation capacity of OPs on ChEs is considerably higher than their effect on other serine hydrolases. Furthermore, even within species the inhibition of ChEs by different OPs tends to be highly specific to particular ChE types (Austin and Berry, 1953). In order to improve the designing of therapeutic and/or prophylactic drugs to OP intoxication, it is therefore desirable to reveal and compare the primary amino acid sequence and three dimensional structure of Human AChE to that of human BuChE, as well as to the homologous domains in other serine hydrolases. Elucidation of these sequences can deepen the understanding of the mode of functioning of ChEs and the amino acid residues involved in this functioning. In the following, we report the analysis of sequence similarities between human AChE and related proteins, based on molecular cloning, DNA sequencing and computer analysis of the derived sequences.

RESULTS

To search for cDNA clones encoding human AChE, oligodeoxynucleotide probes were synthesized according to the amino acid sequences in evolutionarily conserved and divergent peptides from electric fish AChE (Schumacher et al., 1986) as compared with human serum BuChE (Prody et al., 1986, 1987; Lockridge et al., 1987). These synthetic oligodeoxynucleotide probes were used for a comparative screening of cDNA libraries from several human tissue origins.

Previous biochemical analyzes relvealed that in the fetal human brain, the ratio AChE:BuChE is close to 20:1 (Zakut et al., 1985). In contrast, we found the cDNA library from fetal human liver to be relatively rich in BuChEcDNA clones (Prody et al., 1987). We therefore searched for cDNA clones that would interact with selective oligodeoxynucleotide probes, designed according to AChE-specific peptide sequences in cDNA libraries from fetal brain origin, and particularly from brain basal ganglia that are highly enriched with cholinoceptive cell bodies. Positive clones were then examined for their relative abundance in brain-originated cDNA libraries, as compared with liver. Brain-enriched cDNAs were further tested for their capacity to hybridize with the OPSYN oligodeoxynucleotide probes, previously designed according to the concensus amino acid sequence at the active esteratic site of ChEs (Prody et al., 1986). Finally, the confirmed clones were hybridized with BuChEcDNA and found to be highly homologous to it.

DNA sequence analysis followed by computerized alignment of the encoded primary amino acid sequences of human AChE and BuChE demonstrated, as expected, that the functional similarity among ChEs reflects genetic relatedness. The active site peptide of human AChE, as deduced from the AChEcDNA clones, revealed 17 out of 21 amino acid residues identical to those of either human BuChE or Torpedo AChE (Figure 1). Lower level of similarity (12 out of 21 amino acid residues) was observed in comparison with Drosophila AChE (Hall and Spierer, 1986). Esterase 6 from Drosophila (Myers et al., 1988) displayed 10 identical residues out of these 21, and several serine proteases - 3 or 4 identical residues only (Figure 1). This comparison draws a distinct line between serine proteases and the family of carboxylesterases, and more particularly - the highly conserved ChEs.

Hu. BuChE	NPKSV--TLFGESAGAASVSLHL
Hu. AChE	DPTSV--TLFGESAGAASVGMHL
Tor. AChE	DPKTV--TIFGESAGGASVGMHI
Dros. AChE	NPEWM--TLFGESAGSSSVNAQL
Dros. Est 6	EPENV--LLVGHSAGGASVHLEM
Pig Elastase	-GNGVRSGCQGDSGGPLV--CQK
Bov. Chymotrypsin	-ASGV-SSCQGDSGGPLV--CQK
Bov. Prothrombin	EGK-RGDACEGDSGGPFVMKSPY
Bov. Factor X	DTQPE-DACQGDSGGPHV--TRF
Hu. Plasminogen	G--T--DSCQGDSGGPLV--CFE
α lytic Protease	IQTNV-CAEPGDSGGSL

Active Site Homology   Overall Homology of
of HuAChE to: HuAChE to:

Hu. BuChE - 85%   Hu. BuChE: 51%
Tor. AChE   78%   Tor. AChE 56%
Dros. AChE 52%   Dros. AChE 31%

Figure 1. Comparison of ChE active site region sequences with other serine hydrolases. The star indicates [^{3}H]-DFP-labeled or active site serine. Amino acid sequence data were based on DNA sequencing of human AChEcDNA clones and follow reports by Prody et al. (1986), Schumacher et al. (1986), Hall and Spierer (1986) and Myers et al. (1988) regarding the other active site sequences. Note the considerable difference between the levels of sequence similarities within the ChE family (upper part) and other serine hydrolases (lower part).

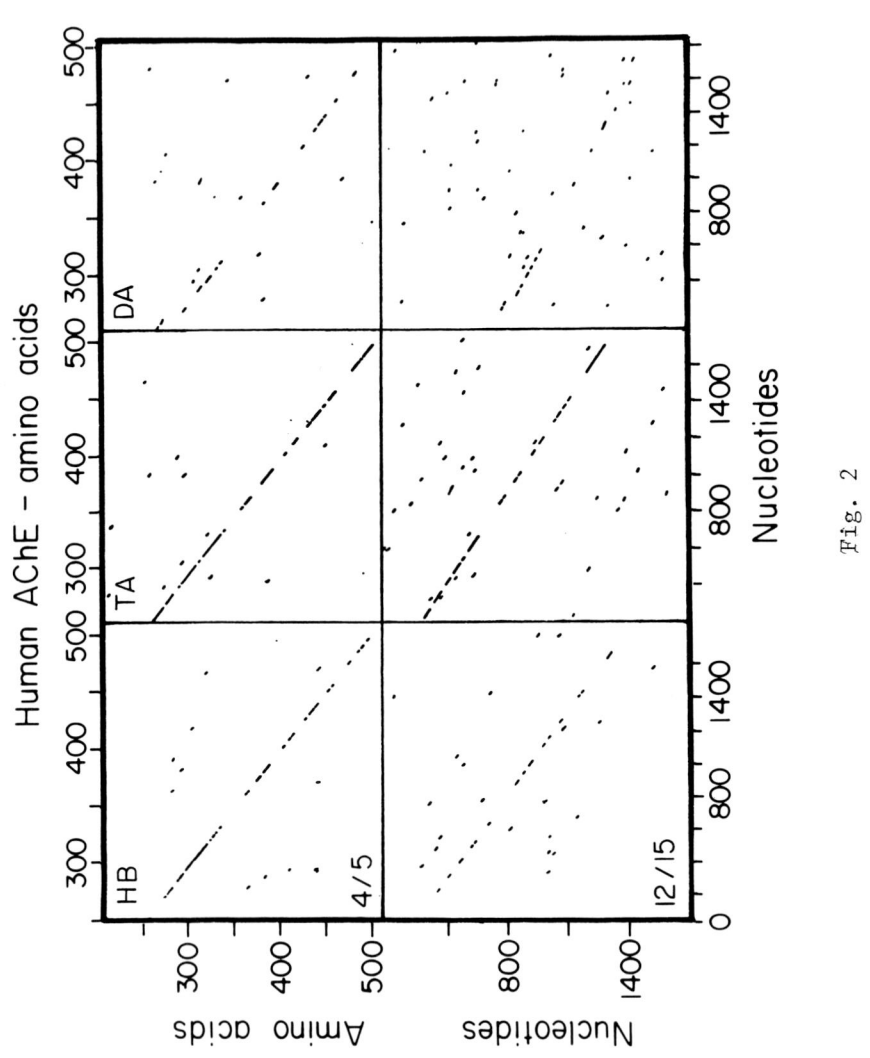

Fig. 2

The coding region in human AChEcDNA and the inferred amino acid sequence of the human AChE protein were compared with the parallel sequences of human BuChEcDNA (Prody et al., 1986,1987; McTiernan et al., 1987), of AChEcDNA from Torpedo (Schumacher et al., 1986) and of the more evolutionarily remote AChEcDNA from Drosophila (Hall and Spierer, 1986) (Figure 2). This analysis revealed several peptide regions and DNA sequence domains that are highly conserved in all of the ChEs and displayed clearly the higher level of divergence between human and Drosophila AChEs, as opposed to the extensive similarities between human AChE and BuChE and Torpedo AChE. A higher level of conservation was found at the amino acid level (Figure 2, up) than at the nucleotide level (Figure 2, down), in complete agreement with previous observations (Prody et al., 1987; Soreq and Gnatt, 1987). Significant homology was also observed with the DNA and the amino acid sequence of bovine thyroglobulin, in corroboration of previous findings (Schumacher et al., 1986; Soreq and Gnatt, 1987).

To further examine the molecular properties of the human AChE protein encoded by the newly isolated cDNA clones, we subjected it to hydrophobicity analysis according to Hopp and Woods (1981). The results of this analysis are presented in Figure 3, together with parallel analyzes of the homologous sequences of human BuChE, Torpedo AChE and Drosophila AChE. The hydrophobicity patterns predicted by this analysis reveal, in all four cases, putatively globular proteins with very short regions of limited hydrophobicity that appear in the same highly conserved positions in the entire family.

Figure 2. Amino acid (Up) and nucleotide (Down) similarities between the coding regions in most of the human AChEcDNA sequence (Note that a part coding for the N-terminal 74 amino acids is still missing), and the parallel regions in the cDNAs coding for human BuChE (Prody et al., 1987), Torpedo AChE (Schumacher et al., 1986) and Drosophila AChE (Hall and Spierer, 1986). Regions of homology were searched for by the dot matrix approach (Maizel and Lenk, 1981). Match values that yielded clear homology regions and minimal background noise are presented: 12 out of 15 conservative matches for nucleotide sequence and 4 out of 5 conservative matches for amino acid residues. Nucleotides are numbered in the 5'-to-3' direction and amino acids in the N`-to-C' direction for all of the sequences.

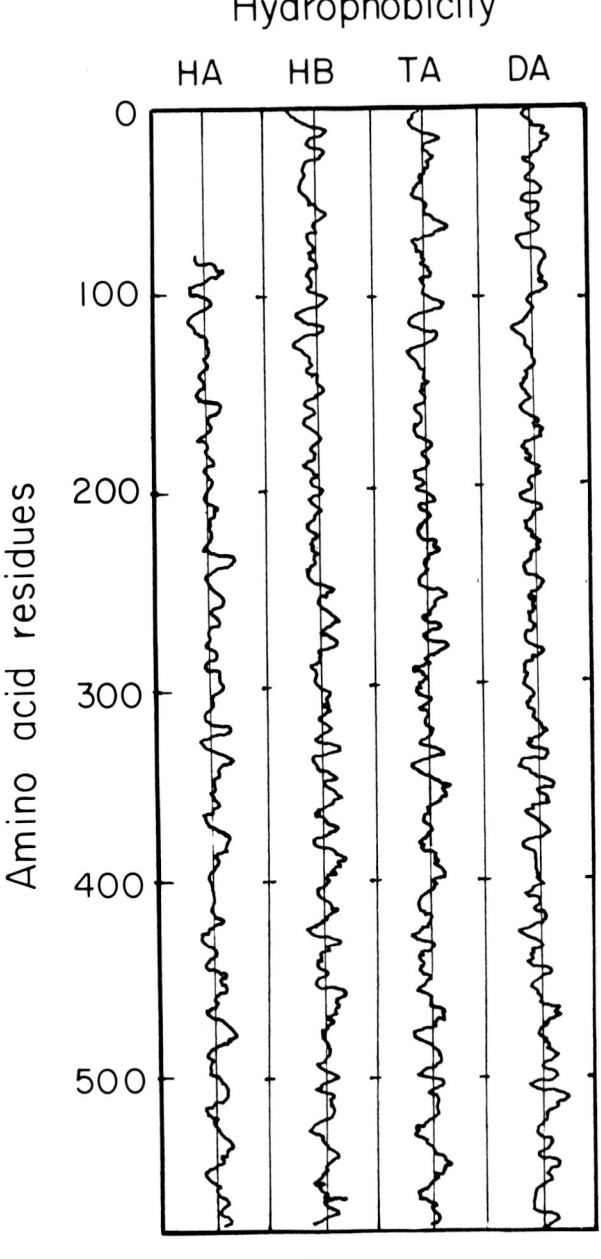

Fig. 3

In order to search for specific conserved domains in human AChE that could be involved in its hydrolytic activity, the above data were combined with generally accepted concepts on the catalytic functioning of carboxylesterases and of serine proteases. Several serine proteases have recently been subjected to site directed mutagenesis (see, for example, Craik et al., 1987). In corroboration of previous enzymology studies, these recent experiments have demonstrated beyond doubt that three key residues are involved in the charge relay mechanism of serine proteases, donating protons to a peptide bond which is subsequently hydrolyzed. These include the active site serine, a basic histidine residue and an acidic aspartate residue. In most of the serine proteases, the three key residues appear in the order His-Asp-Ser with an average distance of 43 and 91 residues between the His and the Asp and between the Asp and the Ser, respectively (Dayhoff et al., 1983). Each of these key residues is embedded in highly conserved peptides, 8-18 amino acids in length.

The sequence similarities in the peptides surroudning the reactive site serine^{198} are cited in Figure 1. Careful analysis revealed an invariant aspartate at position 170, that is also surrounded by a highly conserved domain, as expected from residues playing important roles in hydrolytic activity. In spite of the non-conserved distance between this Asp^{170} and Ser^{198}, these two residues appear to be very good candidates for the putative key functions in the charge relay system.

Figure 3. Comparative hydrophobicity patterns of members of the ChE family. The predictions of the hydrophobic and hydrophilic regions of human AChE and BuChE, Torpedo AChE and Drosophila AChE are presented, based on the algorithm of Hopp and Woods (1981). The dotted vertical baseline in each box represents a hydrophylicity value of -o-; increasing hydrophylicity is in the right-hand direction and increased hydrophobicity is in the left-hand direction.

The high PH dependence of the catalytic activity of ChEs (Rosenberry, 1975) and their sensitivity to chemical agents that modify imidazole groups (Roskoski. 1974) suggest that a histidine residue is also involved in the charge relay mechanism of ChEs. However, there is no conserved histidine on the amino-terminal site of the reactive serine. On the other hand, a highly conserved peptide including an arginine residue can be found around position 147. Arginine replaces histidine in the charge-relay system of phosphodiesterase ST (Nakamura et al., 1982), suggesting this residue as a substitute for the conservative His.

An alternative possibility suggests that histidine residues in other positions take part in the charge-relay system of ChEs. Indeed, highly conserved peptides that include histidine residues may be found at positions 423 and 438. Both were suggested by Sikorav et al. (1987) to take part in the charge relay mechanism. According to this proposal, the basic histidine residue would therefore be located on the carboxy-side of the reactive serine. An example for reversed positions of the Asp and His residues relative to the Ser may be found in another serine protease, subtilisin (Carter and Wells, 1987).

Site-directed mutagenesis studies will be required to distinguish between the above discussed possibilities for putative residues in human AChE. However, in all cases, ChEs are indicated to have a different charge relay system from that of serine proteases, differeing in the identity of the basic residue, in its distance from the reactive Ser or in its relative position on the primary sequence. The possible combinations for the charge relay key residues of ChEs, including the peptide similarities, are presented in Figure 4.

Figure 4. Putative residue combinations for the charge-relay system of the ChE catalytic sites.

A. Arg^{147} - Asp^{170} - Ser^{198} (23 - 28)

B. His^{423} - Asp^{170} - Ser^{198} (253 - 28)

C. His^{438} - Asp^{170} - Ser^{198} (268 - 28)

Figure 4D. Sequence Similarities

	Arg^{147} region	Asp^{170} region	His^{423} region	His^{438} region
H. AChE	yRvgafgflal	nvgllDqrlal	Hrastlswmgv	pHgyeieftfg
H. BuChE	yRvgalgflal	nmglfDqqlal	Hrssklpwmgv	mHgyeiefvfg
T. AChE	yRvgafgflal	nvgllDqrmal	Hrasnlvwmgv	iHgyeiefvfg
D. AChE	yRvgafgflhl	nvglwDqalai	Hrtstslwmgv	lHgdeieyffg

Legends to Figure 4.

A-C. Possible combinations for the key residues in the charge-relay system of human cholinesterases are presented. Residues are numbered according to their appearance in the mature human BuChE protein. The no. of residues between the 1st and 2nd pairs of amino acids are marked in parentheses for each combination.

D. Sequences were aligned as previously detailed (Prody et al., 1987; Soreq and Gnatt, 1987) for human AChE (H. AChE), human BuChE (H. BuChE), Torpedo AChE (T. AChE) and Drosophila AChE (D. AChE). The position of each putative catalytic site residue within the surrounding 11 amino acid sequence is shown above each region. Residue numbering begins with 1 as the first amino acid of the mature human BuChE protein (Lockridge et al., 1987), since the human AChE sequence is deduced from cDNA data only. Dissimilar residues are underlined and the putative catalytic residues are shown in bold letters.

DISCUSSION

Cholinesterases have a catalytic mechanism similar to that known for the serine proteases (reviewed by Rosenberry, 1975). In line with the widely accepted notions that functional similarity reflects common ancestral genes and that multigene families have developed by gene duplication and subsequent divergence during evolution (Ohno. 1970).

Augustinsson (1968) suggested that ChEs are phylogenetically related to the large family of serine proteases and may be defined as members in the multigene family of serine hydrolases. The 3 out of 8 match in the concensus sequences of the catalytic sites of carboxylesterases and serine proteases, including the invariant serine residue, further suggested a common origin to these two families (Neurath, 1984). Recent molecular cloning and DNA sequencing studies confirmed the phylogenetic relationships within the gene families of serine proteases (Rogers, 1985) and of carboxylesterases (Hedrich and Von Deimling, 1987), but left the question of their inter-relationships open.

Profile analysis of the human AChE amino acid sequence showed no similarity to any of the ≥4500 protein sequences in the European Molecular Biology Laboratory (EMBL) protein data base, with the exceptions of Torpedo and Drosophila AChEs, human BuChE and bovine thyroglobuline. Human AChE did not specifically show any resemblance of serine protease sequences. In view of the sequence information deduced from cloned DNAs for four different ChEs, and based on the above discussed arguments, it now appears that human AChE joins the other ChE species to form a limited minigene family that belongs to the larger family of carboxylesterases type B but appears to be distinct from the other serine proteases. Our analysis therefore extends and supports the recent conclusion of Richmond and colleagues (Myers et al., 1988) in suggesting that ChEs cannot be included in a serine hydrolase multigene family.

Within the ChE family, the high sequence similarities between human AChE and BuChE imply that variations and conservations in the primary amino acid sequence of ChEs may be implicated with distinct differences in the substrate specificity and sensitivity to selective inhibitors that were observed for particular types of ChEs. Detailed analysis of these sequences by site-directed mutagenesis and expression of the modified genes in heterologous systems may therefore reveal the key residues in the charge-relay system of ChEs and lead to the development of improved therapeutic drugs against OP intoxication.

ACKNOLWEDGEMENT

Supported by USARMED contract DAMD 17-87-C-7169 (to H.S.), and by a post-doctoral fellowship from the Muscular Dystrophy Association of America (to C.A.P.).

REFERENCES

Aldridge, WN, Reiner, E (1972). Enzyme inhibitors as substates. North Holland, Amsterdam.

Augustinsson, KB (1968). The evolution of esterases in vertebrates. in: NV Their and J. Roche, eds. Homologous enzymes and biochemical evolution. Gordon and Breach, New York. pp. 299-311.

Austin, L, Berry, WK (1953). Two selective inhibitors of cholinesterase. Biochem J. 54: 694-700.

Carter, P., Wells, JA (1987). Engineering enzyme specificity by "substrate-assisted catalysis". Science 237: 394-399.

Craik, CS, Roczniak, S, Largman, C, Rutter, WJ (1987). The catalytic role of active site aspartic acid in serine protease. Science 237: 909-913.

Dayhoff, MO, Barker, WC, Hunt, LT (1983). Establishing homologies in protein sequences. Methods Enzymol. 91: 524-545.

Hall, LM, Spierer, P (1986). The Ace locus of Drosophila melanogaster: Structural gene for acetylcholinesterase with an unusual 5' leader. EMBO J 5: 2949-2954.

Hedrich, HJ, Von Deimling, O (1987). Re-evaluation of LGV of the rat and assignment of 12 carboxylesterases to two gene clusters. J. Hered. 78: 92-96.

Heymann, E (1980). Carboxylesterases and amidases. in: W. Jackoby, ed. Enzymatic basis of detoxification. Vol. 2. Acad. Press, N.Y., pp. 291-323.

Hopp, TP, Woods, KR (1981). Prediction of protein antigenic determinants from amino acid sequences. Proc. Natl. Acad. Sci. USA 78: 3824-3828.

Koelle, GB (1972). Anticholinesterase agents. in: Goodman, LS and Gilman, A. Eds. McMillan, NY, pp. 445-466.

Lockridge, O, LaDu, BN (1986). Amino acid sequence of the active site of human serum cholinesterase from usual, atypical and atypical-silent genotypes. Biochem. Genet. 24: 485-498.

Lockridge, O, Bartels, CF, Vaughan, TA, Wong, CK, Norton, SE, Johnson, LL (1987). Complete amino acid sequence of human serum cholinesterase. J. Biol. Chem. 262: 549-557.

Maizel, JV, Lenk, RR (1981) Enhanced graphic matrix analysis of nucleic acid and protein sequences. Proc. Natl. Acad. Sci. USA 78: 7665-7669.

Massoulie, J, Bon, S (1982). The molecular forms of cholinesterase and acetylcholinesterase in vertebrates. Ann. Rev. Neurosci. 3: 57-106.

McTiernan, C, Adkins, S, Chattonet, A, Vaughan, TA, Bartles, CF, Kott, M, Rosenberry, TL, LaDu, BN, Lockridge, O (1987) Proc. Natl. Acad. Sci. USA 84: 6682-6686.

Myers, M, Richmond, RC, Oakeshott, JG (1988). On the origin of esterases. Mol. Biol. Evol. 5, in press.

Nakamura, KT, Iwahashi, K, Yamamoto, Y, Litaka, Y, Yoshida, N, Mitsui, Y (1982). Crystal structure of a microbial ribonuclease, RNase St. Nature 299: 564-566.

Neurath, H (1984). Evolution of proteolytic enzymes. Science 224: 350-357.

Ohno, S (1970). Evolution by gene duplication. Springer, New York.

Prody, C, Zevin-Sonkin, D, Gnatt, A, Koch, R, Zisling, R, Goldberg, O, Soreq, H (1986). Use of synthetic oligodeoxynucleotide probes for the isolation of a human cholinesterase cDNA clone. J. Neurosci. Res. 16: 25-35.

Prody, C, Gnatt, A, Zevin-Sonkin, D, Goldberg, O, Soreq, H (1987). Isolation and characterization of full-length cDNA clones coding for cholinesterase from fetal human tissues. Proc. Natl. Acad. Sci. USA 84: 3555-3559.

Rosenberry, TL (1975) Acetylcholinesterase. Adv. Enzymol. 43: 103-218.

Roskoski, RJR (1974). Choline acetyltransferase and acetylcholinesterase: Evidence for essential histidine residues. Biochem. 13: 5141-5144.

Rogers, J (1985). Exon shuffling and intron insertion in serine protease genes. Nature 315: 458-459.

Schumacher, M, Camp, S, Maulet, Y, Newton, M, MacPhee-Quigley, S, Taylor, P (1986). Primary structure of Torpedo californica acetylcholinesterase deduced from its cDNA sequence. Nature 319: 407-409.

Sikorav, JL, Krejci, E, Massoulie, J (1987). cDNA sequences of Torpedo marmorata acetylcholinesterase: Primary structure of the precursor of a catalytic subunit; existence of multiple 5'-untranslated regions. EMBO J. 6: 1865-1873.

Silman, I, Futerman, TH (1987). Modes of attachment of acetylcholinesterase to the surface membrane. Eur. J. Biochem. 170: 11-22.

Soreq, H, Gnatt, A (1987). Molecular biological search for human genes encoding cholinesterases. Molecular Neurobiol. 1: 47-80.

Zakut, H, Matzkel, A, Schejter, E, Avni, A, Soreq, H (1985). Polymorphism of acetylcholinesterase in discrete regions of the developing fetal human brain. J. Neurochem. 45: 382-389.

COMPARISON OF THE INTERACTION OF THE HISTAMINE H_2-ANTAGONISTS HISTAMINE AND DIMAPRIT

E.E.J. Haaksma, G.M. Donné-Op den Kelder, H. Timmerman.
Department of Pharmacochemistry, Vrije Universiteit, De Boelelaan 1083, 1081 HV Amsterdam, The Netherlands.
P. Vernooijs, W. Ravenek.
Department of Quantum Chemistry, Vrije Universiteit, De Boelelaan 1083, 1081 HV Amsterdam, The Netherlands.

INTRODUCTION

Since the definition of the histamine H_2-receptor by Ash & Schild (1966) and by Black et al.(1972) various studies have been dedicated to the elucidation of the mode of action of histamine H_2-agonists and antagonists. The first analysis concerning the mode of interaction of histamine with the histamine H_2-receptor was done by the using various histamine analogues (Durant et al. 1975). These 4(5) substituted histamine derivatives in combination with the heterocyclic analogues showed that a prerequisite for histamine H_2-receptor activation is a 1,3 prototropic tautomeric system.This in contradiction to the histamine H_1-receptor. For both types of receptors the N- t form is the active form (fig. 1.). Later studies investigated the effects of neutralization of histamine (Richards et al. 1979, Weinstein et al. 1976). This was based upon the assumption that histamine binds in the monocation form to most probably a negatively charged group at the receptor and is converted into the neutral form. The effect of neutralization on the tautomeric preference of the imidazole ring appeared to be stabilization of the N-p form. Was the N-t form the preferred tautomeric species of charged histamine,

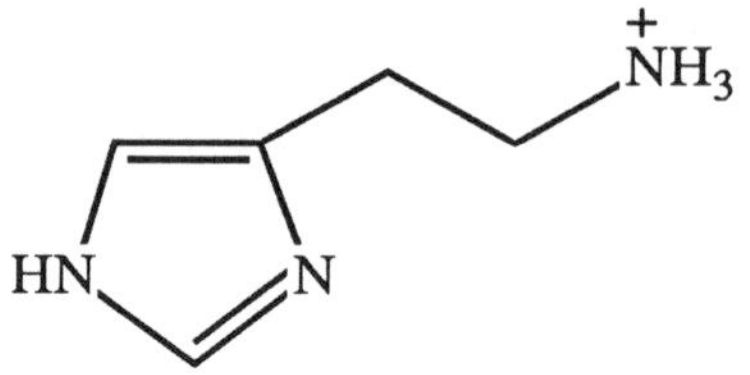

Figure 1. Molecular structure of histamine.

neutralization histamine makes the N-p form the most stable one. It was also shown that it did not differ whether calculations were performed on neutralized or on neutral histamine. These studies have been extended to other compounds (Reggio et al. 1986) and have finally resulted in a theoretical model describing a proton transfer (Weinstein et al. 1986).

One important conclusion can be drawn from these studies. The molecular electrostatic potential in the plane of the imidazole ring is a good parameter for studying effects related to H_2-agonism. So, if a compound is designed to have H_2-agonistic activity, it must have molecular electrostatic potentials comparable to that of histamine.

Dimaprit was introduced in 1977 as a specific H_2-agonist with 50 - 70 % of the activity of histamine (Durant et al. 1977, Parsons et al. 1977). Dimaprit contains features essential for H_2-agonistic activity, i.e. a charged amino group and a thioureum moiety capable of tautomerism. Like histamine, dimaprit occurs in various ionic forms in solution. Based on comparison with histamine, the monocationic form of dimaprit is thought to be the active form (fig. 2.). For the thioureum moiety, however, no calculations concerning a possible proton transfer mechanism have been performed yet. This is an interesting subject as two possible ways of interaction have been suggested (Durant et al. 1977, Green et al. 1978). One in which one of the two nitrogen atoms of the thioureum moiety acts as proton acceptor in the proton transfer system and one in which the sulphur atom acts as the proton acceptor. We tried to study these two models of interaction by using computer graphics,

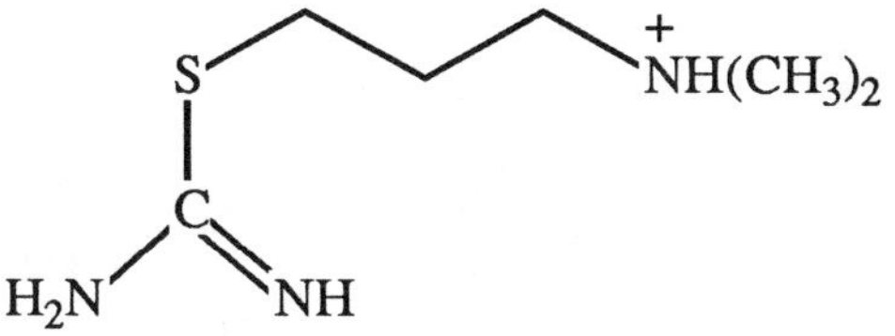

Figure 2. Molecular stucture of dimaprit.

semiempirical and non-empirical quantum chemical calculations. Finally a survey of the preliminary results are presented.

METHOD

First the methylthioureum part of dimaprit was optimized using the quantum chemical Hartree-Fock-Slater Xa (HFS) procedure (Baerends at al. 1973). This quantum chemical procedure was choosen instead of MNDO as it is known that the latter program does not take into account delocalization effects. Delocalization is supposed to effect the relative positions of the hydrogen atoms connected to both of the nitrogens of the thioureum group. As one of these protons is considered to play a key role in the proton relay system its geometry must be accurately known.

To reduce the amount of computer time the less crucial end part of dimaprit, the propyl dimethylamino group, was optimized using MNDO. After optimizing the thioureum and the propyldimethylamino group, the two fragments were connected and the whole subsequently used in a fitting procedure. Optimizing the two fragments separately and with different methods is allowed as in the fitting procedure the accurately determined thioureum moiety was fixed and the propyldimethylamino part was allowed to rotate internally. Fitting will change the initial conformation so that an accurate estimation of the geometry of the side chain of dimaprit will be sufficient. Afterwards only a restricted geometry optimization is allowed (one that does not change the relative position of the functional groups).

Once the geometry of dimaprit was established it was fitted to histamine. The conformation of histamine was based on calculations performed by Weinstein et. al. (1976). This conformation resembles the X-ray data on histamine presenting the fully extended conformation ($t_1 = 90°$, $t_2 = 180°$). For the fitting procedure the computer graphics program CHEM-X (Chem-x) was used. A stepwise method was applied. First a rigid fit was performed in which the geometry of dimaprit was kept frozen. The thioureum part of dimaprit was fitted to the imidazole ring of histamine. Two types of fits were considered. One in which a sulphur atom acts as a hydrogen bond acceptor (further referred to as S-fit) and one in which a nitrogen atom fulfills this type of interaction (N-fit). In both types of fits the amino type nitrogen acts as the proton donor. In a second step of the fitting procedure a flexible fit was performed in which the side chain of dimaprit was allowed to adopt different conformations in order to superimpose the charged amino groups of dimprit and histamine. Finally the semiempirical quantum chemical procedure MNDO was used to further optimize the fitted structure of dimaprit. During this optimization the relative positions of the thioureum moiety and the charged nitro group were fixed.

Once the two conformations of dimaprit in the S- and the N-fit were derived, their molecular electrostatic potentials were calculated. Also for histamine the potentials were calculated. The plane of the imidazole ring of histamine was chosen as the reference plane for calculating the electrostatic potential. Calculations were carried out for the protonated species as this is the species supposed to be recognized by the histamine H_2-receptor.

Also the energies of both dimaprit conformations were calculated with the HFS program. As this program can, due to the assumptions on which it is based upon, only calculate the bonding energy (the statistical total energy) the results will be expressed as relative values.

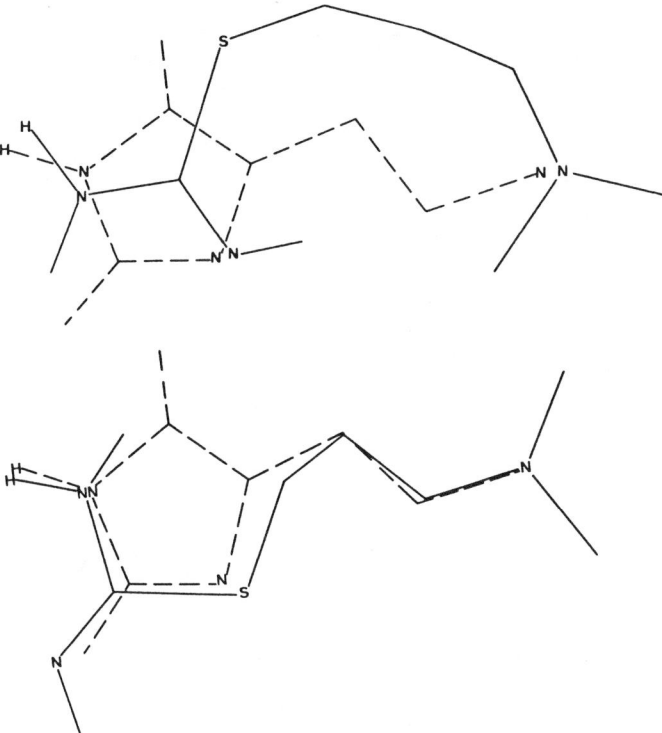

Figure 3. (a) N-fit of dimaprit to histamine (b) S-fit of dimaprit to histamine. Drawn lines: dimaprit, dashed lines: histamine.

DISCUSSION

Both fits are displayed in figure 3. These two figures reveal that both fits, from a geometrical point of view, are feasible. The closer resemblance of dimaprit to histamine in the S-fit points points to a higher probability of this conformation of dimaprit when bound to the H_2-receptor. Not only the side chain fits more closely, also, and more important, the position of the N-H acting as the hydrogen bond donor is more in harmony with the N-H of the imidazole ring of histamine. The distance between the functional groups of histamine and dimaprit are ≤ 0.38 Å for the N-fit and ≤ 0.31 Å for the S-fit.

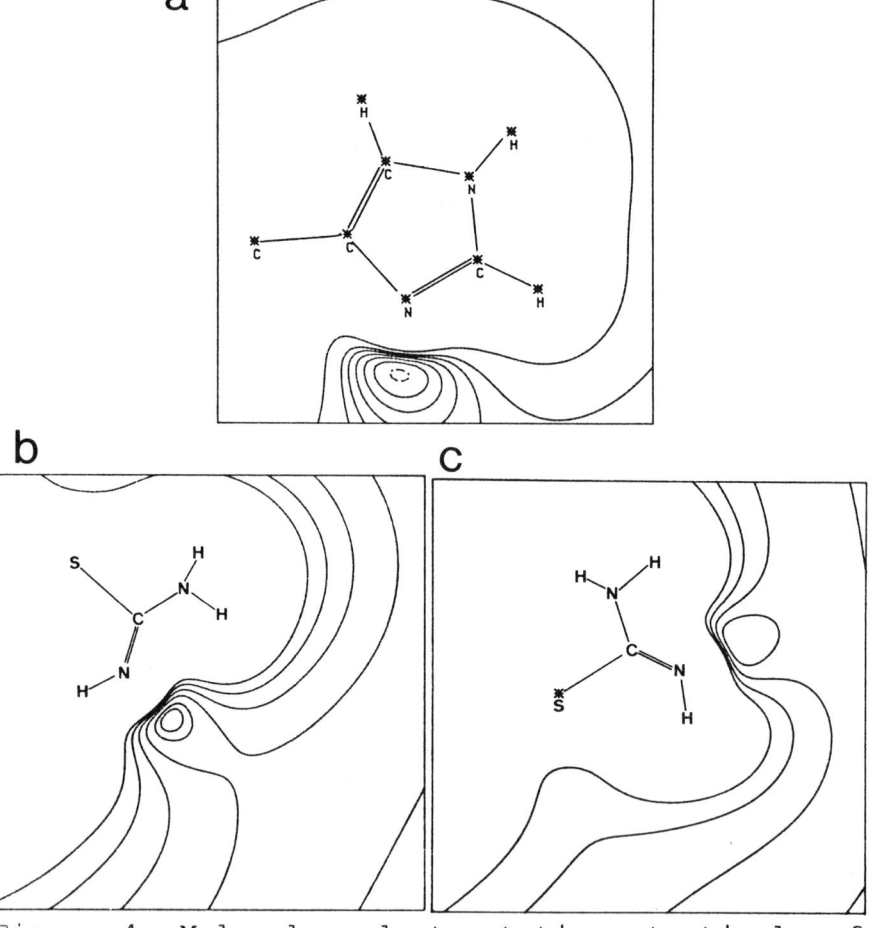

Figure 4. Molecular electrostatic potentianls of the charged forms of (a) histamine (b) dimaprit N-fit (c) dimaprit S-fit. Contour values for (a): 0, 10, 20, 30, 40, 50, 75 kcal/mole, drawn lines: positive, dash-dot: zero. Contour values for (b),(c): 20, 30, 40, 50, 60 kcal/mole.

A further distinction can be made by comparing the molecular electrostatic potentials. In figure 4 the potential map of histamine and the two conformers of dimaprit are shown. Histamine has a clear deep minimum near the nitrogen atom acting as the H-bond acceptor. A minimum in the electrostatic potential is also found for the N-fit of dimaprit, close to the -N= atom but its location is strikingly different from the position of the histamine minimum.

The S-fit of dimaprit does not show a clear minimum in the region near the sulphur atom, the supposed H-acceptor, although a remarkable deviation in the molecular electrostatic potential in this region is perceptible. A minimum in the electrostatic potential needs to be present near the sulphur as only in that case it is able to act as a proton acceptor. Therefore, the electrostatic potential of the neutral form of dimaprit in the S-fit was calculated to investigate whether or not this minimum would become more pronounced. This handling is justified as histamine after binding to the receptor is thought to be neutralized. The neutral form of dimaprit shows a sharp minimum near the sulphur atom. The relative position of this minimum with respect to the sulphur atom closely resembles the position of the electrostatic minimum of the imidazole ring of histamine relative to its nitrogen atom (fig. 5). These results validate the S-fit.

This is further supported by the fit of nordimaprit which is an analogue of dimaprit (Durant et al. 1977). This compound in which the length of the side chain has been shortened by one methylene group is completely inactive. For this compound only a N-fit is possible as a S-fit makes it impossible to match the charged dimethylamino group with the charged amino group of histamine (fig. 6).

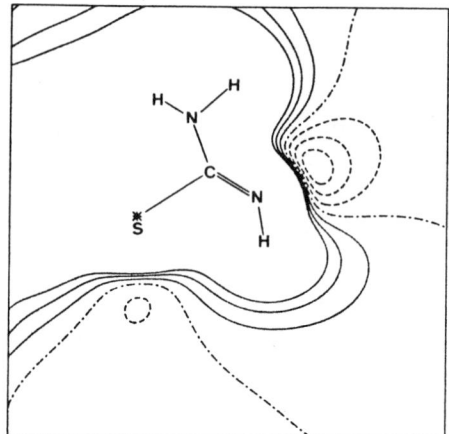

Figure 5. Molecular eletrostatic potential of the neutral form of the S-fit of dimaprit. Contour values 0, ±10, ±20, ±30 kcal/mole, drawn lines: positive, dashed lines: negative, dash-dot line: zero.

As in this compound the thioureum moiety has almost the same relative position to the imidazole ring of histamine as in the N-fit of dimaprit (fig. 5) and is inactive, this also supports the validity the S-fit of dimaprit.

The conclusion which can be drawn from these preliminary results is that the sulphur atom in dimaprit presumably acts as the proton acceptor in the proton relay system needed for activation of the histamine H_2-receptor.

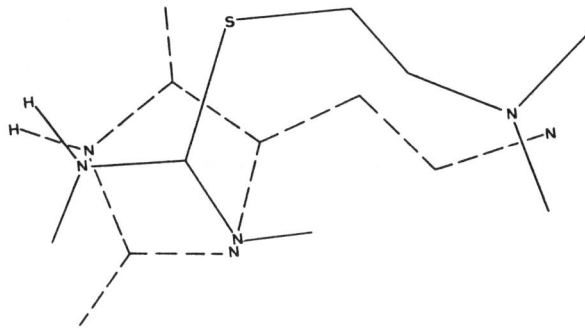

Figure 6. The fit of nordimaprit (drawn lines) to histamine (dashed lines).

REFERENCES

Ash ASF, Schild HP (1966). Receptors mediating some actions of histamine. Br J Pharmacol Chemother 27: 427-439.

Baerends EJ, Ellis DE, Ros P (1973). Self-consistent molecular Hartree-Fock-Slater calculations. I The computational procedure. Chem Phys 2: 41-51.

Black JW, Duncan WAM, Durant CJ, Ganellin CR, Parsons EM (1972). Definition and antagonism of histamine H_2-receptors. Nature 236: 385-390.

Chem-X, developed and distributed by Chemical Design Ltd., Oxford, England.

Durant GJ, Ganellin CR, Parsons ME (1975). Chemical differentiation of histamine H_1- and H_2-receptor agonists. J Med Chem 18: 905-909.

Durant GJ, Ganellin CR, Parsons ME (1977). Dimaprit - [S-[3-(N,N-dimethylamino)propyl]-isothiourea] - A highly specific histamine H_2-receptor agonist. Part 2. Structure - Activity considerations. Agents & Actions 7: 39-43.

Green JP, Johnson CL, Weinstein H (1978). In "Psychopharmacology: A Generation of progress", Ripton M.A., Dimascio A., Killam K.F.(eds.); Raven, New York, p. 319.

Parsons ME, Owen DAA, Ganellin CR, Durant GJ (1977). Dimaprit [S-[3-(N,N-dimethylamino)-propyl]-isothiourea] - A highly specific histamine H_2-receptor agonist. Part 1.

Pharmacology. Agents & Actions 7: 31-37.

Reggio P, Topiol S, Weinstein H (1986). Molecular determinants for the agonist activity of 2-methylhistamine and 4-methylhistamine at H_2-receptors, J Med Chem 29: 2412-2413.

Richards WG, Wallis J, Ganellin CR (1979). Imidazole geometry and the calculations of automer preference for histamine. Eur J Med Chem 14: 9-12.

Weinstein H, Chou D, Johnson CL, Kang S, Green JP (1976). Tautomerism and the receptor action of histamine: a mechanistic model. Mol Pharmacol 12: 738-745.

Weinstein H, Mazurek AP, Osman R, Topiol S (1986). Theoretical studies on the activation mechanism of the histamine H_2-receptor: The proton transfer between histamine and a receptor model. Mol Pharmacol 29: 28-33.

SECTION IV. RECEPTOR-DIRECTED DRUG DESIGN

QUANTUM MECHANICAL SCF/CI STUDIES AS PROBES OF MACRO-
MOLECULAR STRUCTURE: METHODOLOGICAL ASPECTS OF SPECTRAL
COMPARISONS

James D. Petke, Gerald M. Maggiora and Ralph E. Christoffersen

The Upjohn Company, Kalamazoo, Michigan 49001

INTRODUCTION

On the one hand, progress toward understanding and predicting geometric structural features of macromolecules has occurred rapidly in recent years, both from a theoretical and experimental point of view (Dean, 1987; McCamman and Harvey, 1987; Saenger, 1980; Schulz and Schirmer, 1980). Indeed, considerable progress has been made in the characterization of primary, secondary and tertiary protein structure (Schulz and Schirmer, 1980).

On the other hand, while experimentally observed electronic structural features of macromolecules, (e.g., absorption, CD) are rich in detail and potential insight (Callis, 1983), considerably less progress has been made in providing an appropriate theoretical basis for understanding and predicting these spectral features. Indeed, it was noted recently (Robin, 1985) that: "The prognosis for understanding the vacuum-UV spectra of biological materials is not an optimistic one. ... Though there is considerable activity in the vacuum-UV spectroscopy of biological materials, it is almost totally in the area of investigating molecular conformation rather than electronic structure. From the molecular orbital point of view, there is little that one can say regarding such experiments."

Given this situation, a series of studies has been undertaken, designed to adapt or develop theoretical methodologies to allow their systematic application to the study of excited state structure and the visible and spectra of macromolecules. Because of their central role in biology, special attention has been given to proteins and nucleic acids.

In order to provide such methodologies and gain the insight desired, it is believed to be particularly important to create "benchmark" studies that can be used both for insight and for later comparisons with other methodologies. In this report, we present several methodological considerations that have been found to be important when relating calculated spectra from benchmark studies to observed electronic spectra, using cytosine as an example.

METHODOLOGY

The basic methodology employs an <u>ab initio</u> Hartree-Fock self-consistent-field (SCF) quantum mechanical approach on the ground state, followed by multi-reference configuration interaction (CI) studies on low-lying ($\pi \to \pi^*$) and ($n \to \pi^*$) singlet excited states. More specifically, the studies begin with determination of a closed shell molecular orbital (MO) wavefunction

$$\Phi_0(1, 2, ..., 2N) = N_0 \, det\{\phi_1(1)\,\phi_1(2)\,...\,\phi_N(2N)\}, \qquad (1)$$

where the N molecular orbitals (ϕ_i) are expanded in terms of a suitable set of basis functions (χ_j), i.e.,

$$\phi_i = \sum_j c_{ji} \chi_j \qquad (2)$$

and the MO coefficients (c_{ji}) are determined variationally in the SCF process.

The MOs are then used to construct appropriate configurations for CI studies based on the procedure developed by Whitten (Whitten and Hackmeyer, 1969). Namely, the CI wavefunction for state l is expressed as

$$\Psi_l = \sum_k d_{kl} \Phi_k ,\qquad(3)$$

where d_{kl} are CI coefficients and Φ_k are configurations generated by performing single and double excitations from a set of reference configurations, and included in the expansions based on satisfaction of a second-order perturbation theory criterion (Whitten and Hackmeyer, 1969).

In studies of purines and pyrimidines, approximately 5000 configurations were typically utilized for each state, and 40-50 MOs were allowed variable occupancy in the CI calculations. Such studies allowed examination of the 10-15 lowest $\pi \to \pi^*$ and $n \to \pi^*$ states and spectral features associated with them.

In characterizing spectral features, transition energies as well as other properties were examined, including the magnitude and direction of the transition dipole

$$\mu_{kl} = \int \Psi_k^* \, r \, \Psi_l \, d\tau ,\qquad(4)$$

and the oscillator strength:

$$f_{ij} = K \nu_{ij} \left| \mu_{ij} \right|^2 ,\qquad(5)$$

where K is a constant and ν_{ij} is the transition frequency.

Basis functions utilized in the studies were constructed to provide a flexible representation of the pi-system, sufficient for the description of both valence- and Rydberg-type states. Specifically, for each non-hydrogen atom a (4s,2p) double zeta contracted basis was constructed from a (9s,5p) basis, and a (2s) contraction of a (4s) basis was used for hydrogens. In each case, the Dunning contraction (Dunning, 1970) of the Huzinaga (Huzinaga, 1965) uncontracted basis sets was used.

In addition, polarization functions were added at all nuclei to provide additional flexibility in the pi-system. For non-hydrogen nuclei, two out-of-plane d_π-type orbitals (with orbital exponents of 0.8) were added, while for hydrogens a $2p_\pi$ orbital was added with an orbital exponent of 1.0.

Finally, since there is evidence of Rydberg character in certain low-lying $\pi \rightarrow \pi^*$ states of molecules of similar size (Hay and Shavitt, 1974; Matos et al, 1987), diffuse p_π-type functions were also included in the basis. In particular, a pair of p_π functions were added to each non-hydrogen nucleus, with orbital exponents whose values are 30% and 10% of the exponent of the longest range p-type Gaussian in the double zeta basis of the atom to which the diffuse functions are added.

For the specific case of cytosine to be discussed here, the above functions constitute a 127 function basis set, from which the SCF wavefunction was constructed. The CI wavefunction was constructed utilizing 45 MOs with variable occupancy, and a core of 14 filled MOs with fixed occupancy. This resulted in a CI wavefunction possessing 4472 configurations, that allowed examination of 14 low-lying $\pi \rightarrow \pi^*$ states.

It should be noted that not only are these the first ab initio CI studies of spectral features of cytosine, but the basis set and CI study are believed to be flexible enough to provide an appropriate "benchmark" for interpretation of experimental spectra and comparison of semi-empirical and other theoretical approaches to spectral calculation.

RESULTS AND DISCUSSION

Data from the SCF/CI studies on the lowest 14 $\pi \rightarrow \pi^*$ states of cytosine are given in Table 1. Included in the data are calculated transition energies, oscillator strengths and transition moment directions.

Table 1. Calculated Spectral Information for Cytosine

State	ΔE (cm^{-1})	f, θ*
1A	---	---
2A	45,105	0.072 (64)
3A	54,617	0.011 (-46)
4A	56,300	0.266 (-5)
5A	59,452	0.079 (-34)
6A	61,792	0.029 (-47)
7A	63,519	0.003 (-85)
8A	65,253	0.012 (-49)
9A	66,067	0.003 (54)
10A	67,166	0.100 (-39)
11A	67,367	0.091 (-33)
12A	71,960	0.244 (-13)
13A	73,812	0.608 (-47)
14A	77,583	0.019 (70)

*Oscillator strengths were calculated using Eq. 4; transition moment angles θ depicted in Figure 1, are given in parentheses.

For comparison, Figure 1 shows the experimentally observed absorption spectrum for the cytosine nucleotide in water (Zaloudek et al, 1985; Sprecher and Johnson, 1977). This spectrum has been described in terms of two maxima and two inflections, thought to be interpretable in terms of four transitions. Transition moment directions were also assigned, and the assigned transition energies and possible transition moments are summarized in Table 2.

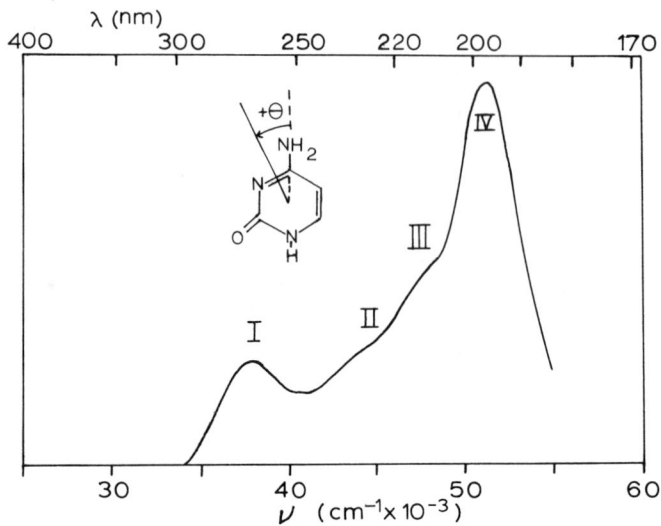

Figure 1. Experimental absorption spectrum of cytosine.

Table 2. Interpretation of Cytosine Experimental Spectra Features

Transition	ΔE (cm^{-1})	Transition Moment Direction (θ)*
I	37,594	+6 (+54)
II	43,478	-46
III	47,170	+76 (-17)
IV	50,761	-27 or +86

*Two possible values arise from experiments. When possible, the preferred value is listed first with the other value in parentheses. The angle θ is depicted in Figure 1.

In order to relate calculated to observed spectra, a process has been created for photosynthetic and related systems (Petke et al, 1978) that can be applied to the current case. In particular, the basis set and CI wavefunctions, while large and flexible, are nevertheless limited in their size and scope. As a result, absolute energies calculated for each state will be substantially above the exact energies for that state. Use of ground state MOs for excited state descriptions also contribute to errors in calculated absolute energies. However, the comparisions of interest are with transition energies, which relate to differences in state energies. Since the basis set and CI wavefunctions have been designed to give relatively balanced descriptions of both the ground and low-lying excited states, it is expected that a direct relationship between the calculated CI states and observed experimental transition energies is likely. The simplest of such relationships is a linear relation between calculated and observed transition energies, and such a relationship has been found in other studies (5, Petke et al, 1978).

For the case of cytosine, creating a linear relation gives rise to two possibilities. In particular, examination of the data in Tables 1 and 2 reveals that the 13A transition corresponds well to Transition IV observed experimentally. However, Transition I in the experimental spectra can be correlated with either the 2A or 4A calculated transition. Both of these possibilities have been examined, and the data are presented in Table 3 along with the results obtained from RPA calculations (Jensen et al, 1988) that have been scaled in a similar manner. To illustrate these results pictorially, Figures 2 and 3 show how the (2A,13A) and (4A,13A) alignments compare with the observed spectra.

Table 3. Alternative Spectra Assignment for Cytosine

State	(2A, 13A) Alignment (cm-1)	(4A, 13A) Alignment (cm-1)	f, (Θ)	RPA Results* (cm-1)	f, (Θ)
2A	37,594	29,177	.07 (64)	37,594	0.11 (+50)
3A	41,957	36,328			
4A	42,729	37,594	.266 (-5)	48,127	0.434 (-18)
5A	44,175	39,964	.079 (-34)	48,735	0.09 (-52)
6A	45,248	41,723			
7A	46,040	43,022			
8A	46,836	44,325			
9A	47,209	44,937			
10A	47,713	45,764			
11A	47,805	45,915			
12A	49,912	49,368			
13A	50,761	50,761	.108 (-47)	50,761	0.4 (-44)
14A	52,491	53,596			

*The transition energies have been scaled to match the lowest and highest values with experimental abosrption maxima I and IV.

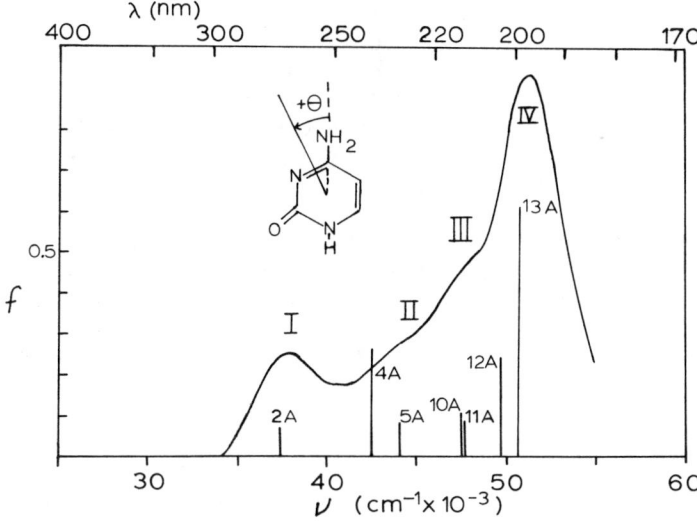

Figure 2. (2A,13A) assignment of calculated transitions for cytosine f is the calculated oscillator strength.

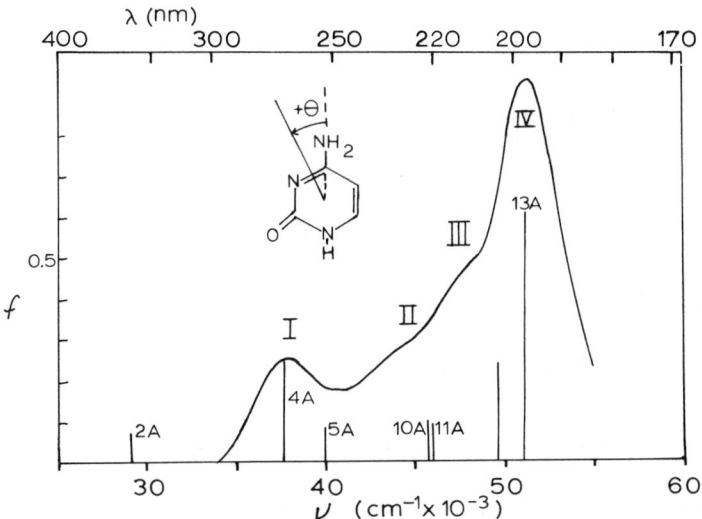

Figure 3. (4A,13A) assignment of calculated transitions for cytosine f is the calculated oscillator strength.

Choice of the appropriate alignment can be accomplished by noting several points. First, if the (4A,13A) alignment is chosen, the 2A transition appears at very low energy (29177 cm^{-1}), for which there is no experimental evidence. On the other hand, the (2A,13A) alignment provides state locations in good agreement with the observed spectra, and the 2A transition moment (+64) agrees well with the transition moment measured for Transition I (+54). Furthermore, the scaled RPA results are in excellent agreement with the (2A,13A) alignment in terms of oscillator strengths and transition moment directions, even though the number of states found in the RPA studies is substantially smaller than found in the current CI studies.

The only substantive discrepancy in the (2A,13A) alignment is the calculated oscillator strength for the 2A transition. However, this state is composed of multiple (6-7) major "parent" configurations (instead of 2-3 in most other state descriptions), which is expected to give rise to greater uncertainty in calulated properties such as the oscillator strength. Given that situation and the excellent agreement with other spectral features, the (2A,13A) alignment is believed to be appropriate.

Thus, it is seen that rich detail is contained in the absoption spectra of cytosine, which can be explained and understood directly from the SCF/CI studies. Many more states are present than have been inferred previously, although their location and nature are quite consistent with the observed spectra. Application of such CI studies to spectral features of other DNA bases can be expected to provide similar results, and such studies are under way.

REFERENCES

Callis PR (1983). Electronic states and luminescence of nucleic acid systems. Ann. Rev Phys Chem 34:329-357.
Dean PM (1987). "Molecular Foundations of Drug-Receptor Intraction." Cambridge: Cambridge University Press.
Dunning TH (1970). Gaussian basis functions for use in molecular calculations. I. Contraction of (9s,5p) atomic basis sets for the first-row atoms. J Chem Phys 53:2823-2833.

Hay PJ, Shavitt I (1974). Ab initio configurations interaction studies of the π-electron states of benzene. J Chem Phys 60:2865-2877.

Huzinaga S (1965). Gaussian-type functions for polyatomic systems. I. J Chem Phys 42:1293-1302.

Jensen H-JA, Koch H, Jorgensen P (1988). Direct iterative RPA calculations. Applications to ethylene, benzene, and cytosine. Chem Phys 119:297-306.

Matos JMO, Roos BO, Malmquist P-A (1987). A CASSCF-CCI study of the valence and lower excited states of the benzene molecule. J Chem Phys 86:1458-1466.

McCammon AJ, Harvey SC (1987). "Dynamics of Proteins and Nucleic Acids." Cambridge: Cambridge University Press.

Petke, JD, Maggiora GM, Shipman LL, Christoffersen RE (1978). Stereo-electronic properties of photosynthetic and related systems. III. Ab initio configuration interaction calculations on the ground and lower excited singlet and triplet states of magnesium porphine and phorphine. J Mol Spectrosc 71:64-84.

Robin MB (1985). "Higher Excited States of Polyatomic Molecules III." New York: Academic Press, pp 395-397.

Saenger W (1980). "Principle of Nucleic Acid Structure." New York: Springer-Verlag.

Schulz GE, Schirmer RH (1979). "Principles of Protein Structure." New York: Springer-Verlag.

Sprecher CA, Johnson WC (1977). Circular dichroism of the nucleic acid monomers. Biopolymers 16:2243-2264.

Whitten JL, Hackmeyer M (1969). Configuration interaction studies of ground and excited states of polyatomic molecules. I. The CI formulation and studies of formaldehyde. J Chem Phys 51:5584-5596.

Zaloudek F, Novros JS, Clark LB (1985). The electronic spectrum of cytosine. J Am Chem Soc 107:7344-7351.

COMPARATIVE MODELING OF PROTEINS IN THE COMPLEMENT PATHWAY

Jonathan Greer, Karl W. Mollison, George W. Carter and Erik R. P. Zuiderweg

Pharmaceutical Products Division, Abbott Laboratories, Abbott Park, Illinois 60064

INTRODUCTION

The complement system is one of the major host defense mechanisms. Byproducts of complement activation, the anaphylatoxins, display a variety of biological actions including smooth muscle contraction, vasodilation, vascular permeability, and immunoregulation. Members of this family include C3a, C4a and C5a. One of these proteins, C5a, is a powerful inflammatory mediator derived from complement acting to recruit leukocytes and stimulate them to release tissue damaging enzymes. Thus, there is considerable interest in the study of these proteins and, in particular, in the development of an antagonist to C5a activity as a new class of anti-inflammatory agent.

The anaphylatoxins are proteins consisting of approximately 75 amino acids (Fig. 1). The sequences clearly indicate that all three poteins are homologous. Biological studies have shown that removal of one or more of the C-terminal residues of C3a destroys biological activity (Hugli, 1981) and that a peptide which contains the C-terminal 21 amino acids of C3a has full activity (Lu et al., 1984; Caporale et al., 1980). C5a also shows significantly reduced activity when one or more C-terminal residues are removed (Chenoweth and Hugli, 1980); however, in this case, little if any activity is obtained with C-terminal peptides (Chenoweth et al., 1979; Khan et al., 1985). Consequently, it is possible that C5a has additional sites that bind with the C5a receptor beyond the C-terminal region. In order to explore this question and to develop a structural basis to aid in performing site-specific mutagenesis experiments, we undertook to model the three-dimensional structure of C5a.

```
                 5         10        15        20        25        30        35
C3aH    S V Q L T E K R M N K V G K Y - P K E L R K C C E D G M R Q N P M R F S C E R
                 5         10        15        20        25        30        35
C5aH        T L Q K K I E E I A A K Y K H S V V K K C C Y D G A C V N N D E - T C E Q
C5aB        M L K K K I E E E A A K Y R N A W V K K C C Y D G A H R N D D E - T C E E
C5aP        M L Q K K I E E E A A K Y K Y A M L K K C C Y D G A Y R N D D E - T C E E
C5aR  D L Q L L H Q K V E E Q A A K Y K H R V P K K C C Y D G A R E N K Y E - T C E Q
C5aM  N L H L L R Q K I E E Q A A K Y K H S V P K K C C Y D G A R V N F Y E - T C E E

         40        45        50        55        60        65        70        75
C3aH  R T R F I S L G E A C K K V F L D C C N Y I T E L R R Q H A R A S H L G L A R
         40        45        50        55        60        65        70        74
C5aH  R A A R I S L G P R C I K A F T E C C V V A S Q L R A N I S H K D M Q L G R
C5aB  R A A R I A I G P E C I K A F K S C C A I A S Q F R A D E H H K N M Q L G R
C5aP  R A A R I K I G P K C V K A F K D C C Y I A N Q V R A E E S H K N I Q L G R
C5aR  R V A R V T I G P H C I R A F K E C C T I A D H I R K N E S H K G M L L G R
C5aM  R V A R V T I G P L C I R A F N E C C T I A N K I R K E S P H K P V Q L G R
```

Fig. 1. Amino acid sequences of human C3a (Hugli, 1975) and of human (Fernandez and Hugli, 1978), bovine (Gennaro et al., 1986), porcine (Gerard and Hugli, 1978; Zimmerman and Vogt, 1984), rat (Cui et al., 1985), and mouse (Wetsel et al., 1985) C5a's.

STRUCTURES OF THE ANAPHYLATOXINS

C3a Crystal Structure

An important milestone in the study of the anaphylatoxins was the determination of the crystal three-dimensional structure of human C3a by Huber and coworkers in 1980 (Paques et al., 1980; Huber et al., 1980). This structure could be used as a basis for construction of a model of C5a employing comparative modeling methods (Greer, 1981). The details of the C3a structure, as reported, suggest that comparative modeling based upon this structure should be considered in three parts. The central portion of the molecule consists of three α-helices with turns between them. This region forms a good basis for comparative modeling. On the other hand, the first 12 residues of the molecule are disordered in the crystal and thus cannot be used for modeling. Finally, the last five residues of the molecule are non-helical yet have a defined conformation in the crystal due to intermolecular interactions peculiar to the crystal. The conformation of this C-terminal region in solution is expected to be different and probably disordered in solution (Huber et al., 1980).

C5a Model Structure

A critical step in the modeling of a new structure is the alignment of its amino acid sequence with that of the known structure. The method used to align sequences has previously been described (Greer, 1981) and involves localization of the structurally conserved regions of the molecule to stretches of strong sequence homology which are characteristic of all members of that family. The sequence alignment used in this study is shown in Fig. 1. The structure of human C5a was modeled from that of human C3a (Greer, 1985).

Using this sequence alignment, the structure of the three helical regions was taken from that of C3a with the appropriate "mutation" of the side chains to match the C5a sequence. The turns were more difficult to construct. The initial turn at residues 13 to 15 has an additional residue in C5a relative to human C3a. A starting conformation was chosen for this turn which eliminated bad contacts and overlaps. The second turn, residues 26 to 32, has a relative deletion and was built with a gap that was closed by limited energy minimization using the program DISCOVER (Biosym Technologies, 1987). The last turn, residues 40 to 45, has the same number of residues in C3a and C5a; however, the introduction of a proline residue at position 45 in C5a has a significant influence on the conformation. No attempt was made to perform a complete conformational search on any of these loops in order to find better conformations because of the major computational effort involved (Bruccoleri and Karplus, 1987; Moult and James, 1986; Fine et al., 1986). Instead we decided to wait for experimental structure data to be obtained by NMR analysis in order to further refine these loops.

The resulting structure showed several interesting features both from a modeling and a molecular functional perspective. While the sequence identity between human C3a and C5a is not particularly high, ~36%, the model structure of C5a shows that virtually all of these conserved residues lie in the core of the molecule and form the structural basis of the molecule, see Fig. 2. On the other hand, almost all of the residues that make up the outside of C3a and C5a are very different from each other, both in size and chemical character. As a result, "nature" has constructed a pair of proteins with virtually the same internal structural framework, yet from the outside they look like completely different proteins in shape and properties.

The disordered N-terminus in C3a does not provide a structure that can be used to build this part of C5a. However, examination of the sequences of C5a in this region, Fig. 1, suggested that these residues may form an amphipathic helix with Leu 2 and Ile 6 buried against the rest of the molecule and the very polar Thr 1, Gln 3, Lys 4, Lys 5, Glu 7, and Glu 8 pointing out into the solvent. Several considerations were used to try to identify a possible docking site for this proposed α-helix on the rest of the molecule. The disposition on the model structure of the sites of interspecies sequence variation was examined using the two C5a sequences which were known at that time: human and porcine (Fig. 1). When the sequence variable positions were analyzed, (see Fig. 3 of Greer, 1985), it became clear that there was a unique contiguous patch of hydrophobic residues, including Cys 22, Tyr 23, Gly 25, Ala 26, Cys 54, Cys 55, Val 57, Ala 58, and Leu 61, which was conserved in

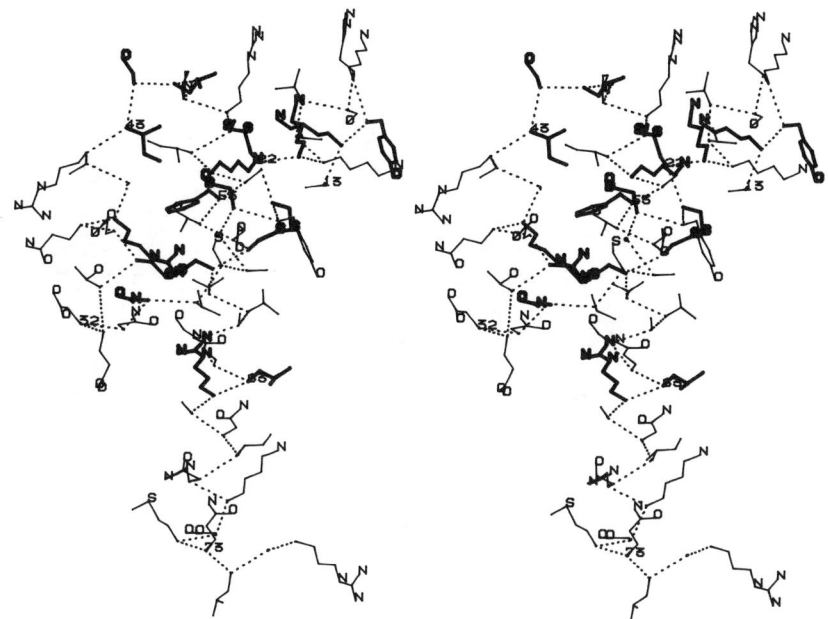

Fig. 2. Cα plot of C5a (dotted lines) showing the side chains that are conserved with C3a (thick lines) and those that differ from C3a (thin lines).

these two C5a sequences, and as we now know, in all the currently available C5a sequences. This region could potentially form a docking site for the proposed N-terminal α-helix. The helix was docked manually using an Evans and Sutherland MPS graphics system and it was found to fit very snugly against the first and third helices of the central portion of the molecule. In this conformation, it was only necessary to partially unwind the last half turn of the helix to join this N-terminal fragment onto the rest of the molecule at residue 11. The resulting structure is shown in Fig. 3. It demonstrates that this additional helix makes the overall structure more globular and symmetric. The structure now consists of four α-helices which are arranged in the typical coiled-coil of helices expected for a structure of this type and which are observed in the experimentally determined structures of several other four-helix proteins (Stenkamp et al., 1983).

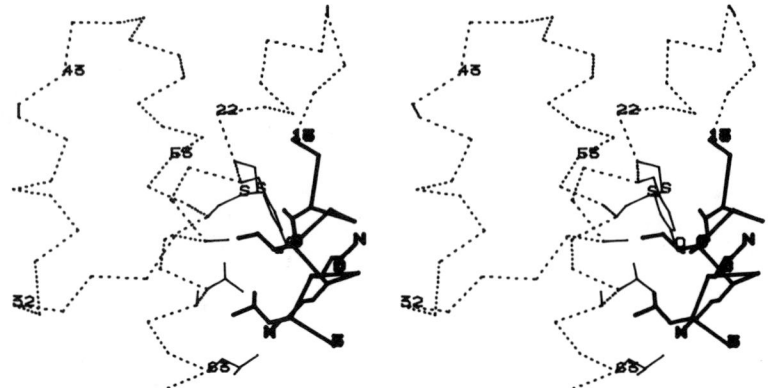

Fig. 3. N-terminal α-helix (thick lines) with contact residues between this helix (thick lines) and those on the central core of the molecule (thin lines). Cα plot of C5a is also shown (dotted lines).

Reexamination of the C3a N-terminus

The modeling studies described above on C5a (Greer, 1985) and similar modeling studies on C4a (Greer, 1986) have suggested that both C5a and C4a begin with an N-terminal α-helix rather than a disordered conformation as reported in the parent C3a crystal structure (Huber et al., 1980). This leads inevitably to the question of whether such a helix might exist in C3a in solution if not in the crystal.

When the sequence at the N-terminus of C3a is examined (Fig. 1), at first glance the pattern of hydrophobic residues separated by two or three hydrophilic ones so characteristic of an amphipathic helix does not occur in this protein. The previously hydrophobic Ile (or Val) residue at position 6 of C5a (Table 1) is now an Arg. Nevertheless, when we constructed an N-terminal α-helix and docked it as was proposed for C5a and C4a, then only simple rotations of the side chain χ angles were required to move the guanidinium group into solvent leaving the aliphatic balance of the long Arg side chain to participate in forming the hydrophobic contact. Thus, it is quite possible that a helix may exist for C3a, as well. Proposed helix contacts for C3a are compared with those for C5a and C4a in Table 1.

Table 1. N-terminal Helix Contacts

No.	C3a	C5a	C4a
N-terminal Helix:			
4[a]	Leu	Leu	Phe
8	Arg	Ile/Val	Ile/Val
9	Met	-- [b]	-- [b]
Central Portion:			
23	Cys	Cys	Cys
27	Met	Ala	Val/Leu/Met
56	Cys	Cys	Cys
59	Tyr/His	Val/Ile	Phe
60	Ile	Ala	Ala
63	Leu	Leu/Val/Ile/Phe	Leu/Ile

[a] Residue numbers listed are those of C3a. The corresponding numbers for C5a can be determined from Fig. 1.
[b] The residues at this position are smaller than Met and therefore do not form part of the contact.

If a helix could exist in C3a, why does it not appear in the crystal structure? Examination of the space group of the C3a crystal shows that there is no room for the helix to fit into the molecular packing of these crystals. Major collisions would occur with neighboring molecules in the lattice. Thus, if an equilibrium exists between C3a molecules with the helix formed and docked or disordered, the crystal lattice constraints would select for molecules with the helix disordered. In order to test this hypothesis, structural studies, by NMR or other spectroscopic methods, should be performed on C3a in solution (Nettesheim et al., 1988).

Comparison of Model and NMR Structures for C5a

Using modern 2D NMR methods (Wuthrich et al., 1982), a three-dimensional structure (Zuiderweg et al., 1988b) was determined for human C5a derived from recombinant DNA sources (Mandecki et al., 1985; Mollison et al., 1987). The details of the structure determination have been reported elsewhere (Zuiderweg et al., 1988a). Based upon the NMR analysis, a family of structures was generated which has the property that they all fit the experimental NMR data, Fig. 4(a). As can be seen, these structures are very similar, varying mostly in the external turns or loops, probably due to the dynamic nature of their conformation in solution.

Superimposed upon the family of NMR structures is the model C5a structure originally proposed (Greer, 1985) and described above. It is clear that the overall fit of the model to the experimental structures is excellent. The helices of the central portion of the molecule coincide very nicely. Most exciting is the direct experimental evidence confirming the prediction that the N-terminus of C5a is indeed α-helical and that this helix docks onto the central portion of the molecule almost exactly as predicted in the modeling. In fact, the NMR results show that this helix is one-half turn longer than in the model structure.

The deviations between the model and the NMR structures, Fig. 4(b), show graphically that the helices match quite well. The loops between the structures though differ much more. Despite the variation in the loops amongst the ensemble of NMR structures, the model structure differs significantly from the ensemble as a whole. This is due in part to the fact that no effort was made to refine these loops, but also reflects the great challenge that folding these parts of the structure continues to present to the field of comparative modeling of proteins.

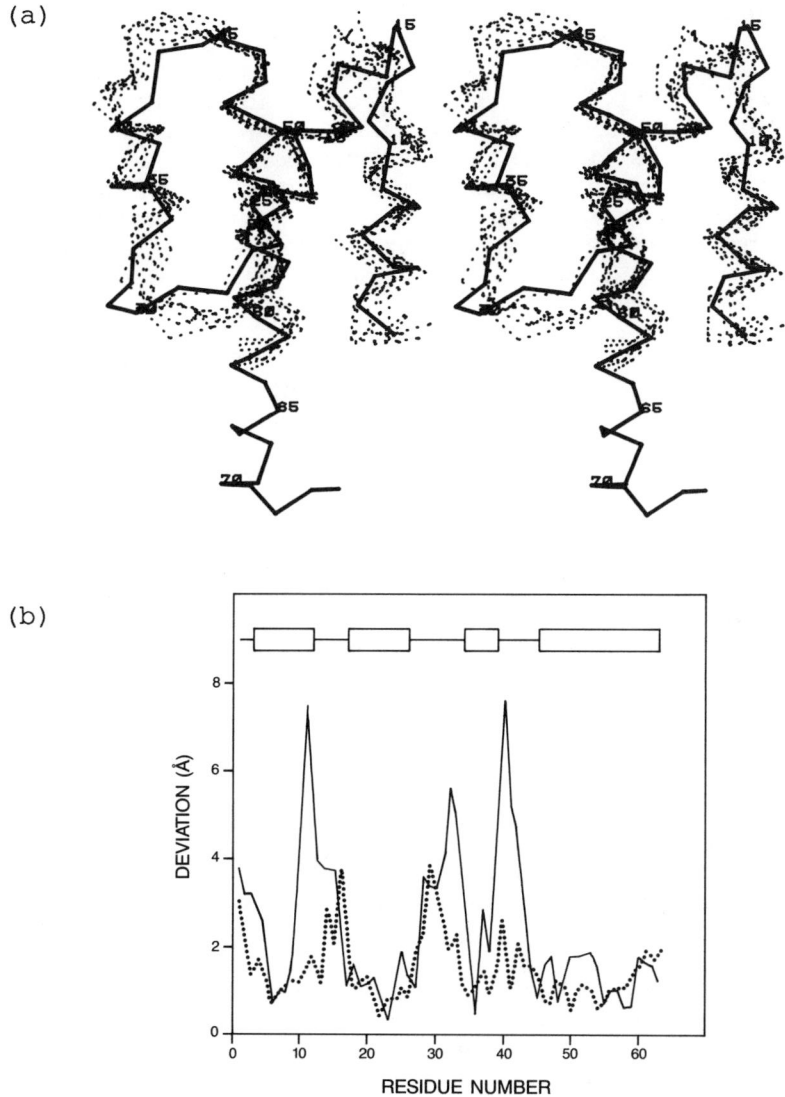

Fig. 4. (a) Cα plots of the seven NMR generated structures (dotted lines) with the model structure superimposed (solid lines). (b). The maximum deviation of the NMR structures (dotted lines) and the deviation of the model (solid lines) from the average of the NMR structures is plotted along the structure. The boxes indicate the α-helices.

There was one other area where a major deviation was observed. The NMR results indicate that the stem of the C5a molecule extending beyond the core is not helical as predicted in the model from the C3a crystal structure. The NMR study shows that the stretch from residues 62 to 66 is increasingly less helical and beyond 66 appears to be completely a random coil. This result is important since the immediate C-terminal residues have been implicated in receptor binding and biological function (Hugli, 1981).

The amino acid sequences for C5a's (Fig. 1) include that for mouse which was not available when the original modeling work was performed on C5a (Greer, 1985). Examination of the sequence of mouse C5a shows that two prolines appear at residues 66 and 69 where the model predicted α-helix based upon the C3a crystal structure. It is exceedingly unlikely that mouse C5a is helical in this region. If the mouse sequence had been available when C5a was initially modeled, then we would have questioned whether residues 66 to 69 should be designated as helix in the C5a model.

CONCLUSIONS

The experimental structure determination for C5a by NMR methods has demonstrated that the model structure is largely accurate. The correct positioning of the helices, especially that of the predicted N-terminal α-helix, as well as the side chain conformations, has shown that careful comparative model building can produce a structure which is close enough to the true structure to be useful for designing experiments and for helping to interpret the results.

One of the major goals of the study of C5a is to determine which parts of the molecule interact with its specific receptor on polymorphonuclear leukocytes. The model structure of C5a can help in this study in several ways. It has served as the basis for selecting side chains of C5a to be modified by site-specific mutagenesis or by chemical or enzymatic modification. The model structure allows us to select residues which lie on the surface of the molecule and thus avoid obtaining inactive C5a mutants because of global structural changes. Knowledge of the structure also permits the systematic quartering of the surface of the molecule with the minimum number of mutants.

The three-dimensional structure can also help more directly in identifying sites for modification. Using the sequence variation between species for C5a's (Fig. 1), one can map the location of the side chains that remain the same in all species and those which vary in one or more species on the three-dimensional structure. This permits the localization of patches of the surface which are conserved in all C5a's and stretches which are variable (Greer, 1986). It is likely that the receptor binding site or sites on C5a will be conserved between species. Such considerations are currently being used to explore the possible receptor binding sites on the C5a molecule.

REFERENCES

Bruccoleri RE, Karplus M (1987). Prediction of the folding of short polypeptide segments by uniform conformational sampling. Biopolymers 26:137-168.
Caporale LH, Tippett PS, Erickson BW, Hugli TE (1980). The active site of C3a anaphylatoxin. J Biol Chem 255:10758-10763.
Chenoweth DE, Erickson BW, Hugli TE (1979). Human C5a-related synthetic peptides as neutrophil chemotactic factors. Biochem Biophys Res Commun 86:227-234.
Chenoweth, DE and Hugli, TE (1980). Human C5a and C5a analogs as probes of the neutrophil C5a receptor. Mol Immunol 17:151-161.
Cui L-X, Ferreri K, Hugli TE (1985). Characterization of rat C5a, a uniquely active spasmogen. Complement 2:18-19.
Fernandez HN, Hugli TE (1978). Primary structure analysis of the polypeptide portion of human C5a

anaphylatoxin: Polypeptide sequence determination and assignment of the oligosaccharide attachment site in C5a. J Biol Chem 253:6955-6964.

Fine RM, Wang H, Shenkin PS, Yarmush DL, Levinthal C (1986). Predicting antibody hypervariable loop conformations. II. Minimization and molecular dynamics studies of MCPC603 from many randomly generated loop conformations. Proteins 1:342-362.

Gennaro R, Simonic T, Negri A, Mottola C, Secchi C, Ronchi S, Romeo D (1986). C5a fragment of bovine complement. Purification, bioassays, amino-acid sequence and other structural studies. Eur J Biochem 155:77-86.

Gerard C, Hugli TE (1978). Amino acid sequence of the anaphylatoxin from the fifth component of porcine complement. J Biol Chem 255:4710-4715.

Greer J (1981). Comparative model building of the mammalian serine proteases. J Mol Biol 153:1027-1042.

Greer J (1985). Model Structure for the Inflammatory Protein C5a. Science 228:1055-1060.

Greer J (1986). Comparative Structural Anatomy of the Complement Anaphylatoxin Proteins: C3a, C4a, and C5a. Enzyme 36:150-163.

Hoeprich PD, Hugli TE (1985). Synthesis of COOH-terminal peptides from human C3a. Fed Proc 44:1185.

Huber R, Scholze H, Paques EP, Deisenhofer J (1980). Crystal structure analysis and molecular model of human C3a anaphylatoxin. Hoppe Seyler's Z Physiol Chem 361:1389-1399.

Hugli TE (1975). Human anaphylatoxin (C3a) from the third component of complement. Primary structure. J Biol Chem 250:8293-8301.

Hugli TE (1981). The structural basis for anaphylatoxins and chemotactic factors. CRC Crit Rev Immunol 1:321-366.

Khan SA, Erickson BW, Kawahara MS, Hugli TE (1985). A synthetic analogue of huamn C5a with spasmogenic activity. Complement 2:42.

Lu Z-X, Fok K-F, Erickson BW, Hugli TE (1984). Conformational analysis of COOH-terminal segments of human C3a: Evidence of ordered conformation in an active 21-residue peptide. J Biol Chem 259:7367-7370.

Mandecki W, Mollison KW, Bolling TJ, Powell BS, Carter GW, Fox JL (1985). Chemical synthesis of a gene encoding the human complement fragment C5a and its expression in Escherichia coli. Proc Natl Acad Sci USA 82:3543-3547.

Mollison, KW, Fey, TA, Krause, RA, Mandecki, W, Fox, JL, Carter, GW (1987). High level C5a gene expression and recovery of recombinant C5a from Escherichia coli. Agents and Actions 21:366-370.

Moult J, James MNG (1986). An aligorithm for determining the conformation of polypeptide segments in proteins by systematic search. Proteins 1:146-163.

Nettesheim DG, Edalji RP, Mollison KW, Greer J, Zuiderweg ERP (1988). Secondary structure of C3a anaphylatoxin in solution as determined by NMR spectroscopy. Proc Natl Acad Sci USA submitted for publication.

Paques EP, Scholze H, Huber R (1980). Purification and crystallization of human anaphylatoxin, C3a. Hoppe Seyler's Z Physiol Chem 361:977-980.

Stenkamp RE, Sieker LC, Jensen LH (1983). Adjustment of constraints in the refinement of methemerythrin and azidomethemerythrin at 2.0 Å resolution. Acta Crystallogr B39:697-703.

Wetsel RA, Ogata RT, Tack BF (1985). Isolation and sequence of a murine cDNA clone encoding the fifth complement component. Fed Proc 44:987.

Wuthrich K, Wider G, Wagner G, Braun W (1982). Sequential Resonance assignments as a basis for the determination of spatial protein structures by high resolution proton nuclear magnetic resonance spectroscopy. J Mol Biol 155:311-319.

Zimmermann B, Vogt W (1984). Amino-acid sequence and disulfide linkages of the anaphylatoxin, Des-Arg-C5a, from porcine serum. Hoppe Seyler's Z Physiol Chem 365:151-158.

Zuiderweg ERP, Mollison KW, Henkin J, Carter GW (1988a). Sequence Specific Assignments in the ^{1}H NMR spectrum of the human inflammatory protein C5a. Biochemistry, in press.

Zuiderweg ERP, Mollison KW, Carter GW (1988b). Solution structure of C5a by 2D NMR methods. Manuscript submitted.

A MOLECULAR THEORETICAL MODEL OF RECOGNITION AND ACTIVATION AT A 5-HT RECEPTOR

Gustavo A. Mercier, Roman Osman and Harel Weinstein

Departments of Physiology and Biophysics and of Pharmacology, Mount Sinai School of Medicine, CUNY, New York, NY 10029, USA

INTRODUCTION

The formulation of a molecular mechanism of receptor activation is hampered by the scarcity of three dimensional structural data on receptors. This obstacle prevents, at the present time, the formulation of both the specific molecular events that are triggered by the drug in a localized area of the receptor and the ensuing change in the overall conformation in the receptor protein. The latter is likely to be responsible for the interaction between the receptor and effector proteins (e.g., a G-protein) which leads eventually to the observed cellular response. While the drug-receptor interaction and the triggering events can be modeled with heuristic methods, the conformational change in the overall receptor-protein must await the elucidation of the three-dimensional structure of the receptor protein. The available structural information for membrane bound proteins is limited to the high resolution structure of the photosynthetic reaction center of R. Sphaeroides (Allen et al., 1987), and the very low resolution structures of the photosensitive protein bacteriorhodopsin (Engelman et al., 1982 and references therein) and of the nicotinic receptor (Stevens, 1985 and references therein). The structural information for other receptors is limited to the amino acid sequences obtained by cloning procedures (Dixon et al., 1986; Gocayne et al., 1987; Grenningloh et al., 1987; Hulme and Birdsall, 1986; Kobilka et al., 1987; Kubo et al., 1986; Marx, 1987; Schofield et al., 1987; Stevens, 1987; Trowbridge, 1987).

In the past we took a heuristic approach in the attempt to identify the molecular determinants for receptor recognition

(Weinstein, et al., 1980, 1987 and references therein) and activation (Osman et al., 1985; Osman et al.,1987). This approach was based on the identification of molecular determinants for recognition from the analysis of the molecular properties of the ligands that are recognized by the receptor. The molecular model for activation was based on the changes induced in a receptor model, that possessed the discriminant recognition characteristics, by the interaction with those ligands. The major molecular property related to recognition at receptors for serotonin (5-hydroxytryptamine; 5-HT) was identified as the directional character of the molecular electrostatic protential above the indole portion of 5-HT (Weinstein, et al., 1981). This directionality implied a complementary charge distribution in the receptor, thus forming the basis for selective recognition. The observed directionalities in the molecular electrostatic potential of various ligands that are recognized by a 5-HT receptor gave rise to a hypothesis that linked recognition with measured affinities of these ligands to a 5-HT receptor (Weinstein, et al., 1987). To test the directionality hypothesis, an imidazolium cation was chosen because of its known interactions with indole (Shinitzky and Katchalski, 1968), and because its charge distribution exhibits the implied complementary property required for interaction with the molecular electrostatic potential of the indole portion of 5-HT (Weinstein, et al., 1978). Indeed, imidazolium showed the expected preference with regard to the orientational directionality, which was based primarily on the electrostatic nature of its interaction with the corresponding ligands (Osman, et al., 1980; Weinstein et al., 1981, 1987).

The analysis of the changes produced in the electron density distribution of imidazolium, the heuristic receptor model, by its interaction with 5-HT indicated that the major rearrangement is around the N-H bond. The change in proton affinity of the nitrogen suggested that the interaction may induce a proton transfer from the imidazolium cation to a potential proton acceptor (Osman, et al., 1985, 1987). The construction of the proton transfer model (PTM) followed the heuristic approach in that it conserved the elements of recognition which characterized the imidazolium cation and extended the receptor model to contain a responsive function in the form of a proton transfer from imidazolium to an ammonia molecule which could function as the proton acceptor.

Using this heuristic model our studies showed that in the absence of 5-HT the process of proton transfer from imidazolium to ammonia has a high barrier and a minimal driving force. When 5-HT approaches the PTM, the interaction lowers the barrier and generates an appreciable driving force for the proton transfer. These studies pointed to the molecular determinants for recognition and the primary steps in a plausible molecular mechanism of the activation of a 5-HT receptor (Osman, et al., 1985, 1987). It is premature to conclude from such studies that the 5-HT receptor indeed contains a proton transfer system composed of an imidazolium (e.g., from histidine) and an ammonia (e.g., from lysine). However, proton transfer systems are ubiquitous in proteins and often serve as the responsive - catalytic components in enzymes (Fersht, 1985). Inasmuch as the proton transfer process takes place in a protein, the energetics of such a process will be affected by its environment. Several works attempted to understand the effect of the protein environment on the proton transfer process (Allen, 1981; Tapia, et al., 1985; Thole and van Duijnen, 1983; Umeyama, et al., 1984; Warshel and Russel, 1986) and demonstrated that the primary effect is electrostatic in nature.

In order to gain insight into the nature of the electrostatic effect that the protein environment may have on a putative proton transfer process in a receptor whose structure is not known, we have adopted a heuristic approach in which we evaluate the effect of a protein of known structure on a proton transfer process. In this approach we simulated a proton transfer inside a known protein, actinidin, and evaluated the effects of the protein in terms of the secondary and tertiary structure of the protein. Actinidin was selected because it contains an appropriate juxtaposition of groups that can participate in a proton transfer process and its structure is well characterized in terms of its secondary and tertiary structural components. Such components are identified in all proteins including neurotransmitter receptors. Based on the premise that the structural elements which are conserved across different proteins and have similar properties, also conserve their function, the heuristic approach we take represents a new strategy in the study of structure-function relationships in receptors.

COMPUTATIONAL DETAILS

From among the proteins in the Brookhaven Protein Data Bank (Bernstein et al. 1977), actinidin (E.C. 3.4.22.14) was identified as a protein that fulfills the requirement of containing a putative proton transfer model. The proton transfer system consists of a His 162 hydrogen bonded to Cys 25. Trp 184 is in close proximity to this system and is aligned in a stacking geometry with respect to the imidazole ring of His 162. These residues are also the main components of the active site of the enzyme (Polgar and Halasz, 1982).

The potential energy curve for proton transfer between ND1 of His 162 and SG of Cys 25 was obtained from <u>ab initio</u> quantum mechanical calculations on a model system in which His 162 was modeled by imidazolium and Cys 25 by methanethiol. Thus, the proton transfer system represents an inactive form of the enzyme that exists at low pH. The relative geometry of the components of the proton transfer system were taken from the crystallographic coordinates of the enzyme. The critical points on the potential energy curve for proton transfer were obtained by a full optimization of the positions of the hydrogen involved in the proton transfer process and the hydrogens bound to the methanethiol; all other internal coordinates were kept frozen. To test the basis set dependence of the calculations, several were used ranging from a minimal STO-3G to an extended 6-21G** basis set. The qualitative nature of the results was found to be basis set independent; we present here results from calculations with the minimal basis set.

The electrostatic effect of the protein environment was calculated as ES = $\sum_j V(R_j) \cdot Q_j$, where $V(R_j)$ is the electrostatic potential generated by the proton transfer system at the position of the point charge Q_j, representing an atom of the protein. The electrostatic potential was calculated from the wave function of the proton transfer system. The point charges, Q_j, were obtained from a dipole conserving population analysis (Thole and Van Duijnen, 1983a) of the wave functions of the amino acids of the protein, obtained from Hartree-Fock calculations with the Mehler-Paul basis set (Mehler and Paul, 1979). The incorporation of the point charges from the environment, Q_j, into the quantum mechanical Hamiltonian gives a small polarization of the wave function that is nearly constant as a function of the proton movement. A useful approximation of the electrostatic interaction energy can be obtained by representing the quantum motif by a collection of point charges obtained from Mulliken population analysis of its wave function. This approximation reproduces quite reliably the results obtained with other methods.

RESULTS AND DISCUSSION

The Effect of the Entire Protein on the Proton Transfer Process

The potential energy curve of the proton transfer in the imidazolium/methanethiol complex calculated in vacuum, shown in Figure 1, reflects the basic difference in the proton affinities of the neutral molecules. In the complex, the form with the proton on the sulfur (M2) is 35.6 Kcal/mole higher than that with the proton on imidazole (M1). Furthermore, the barrier (TS) for the proton transfer from imidazolium to methanethiol is 39.0 Kcal/mole. In the presence of the protein the potential energy curve changes as shown in Figure 1. The curve no longer possesses a double-well character which is typical of hydrogen bonded systems. This is due to a differential effect of the protein on the three extrema of the curve. The electrostatic interaction energy between the proton transfer system in the M1 geometry and the protein is -315.5 Kcal/mol while in geometry M2 the interaction is only -299.7 Kcal/mole and in the TS geometry it is -304.6 Kcal/mole. Consequently, the energy of M2 relative to that of M1 is raised by 15.8 kcal/mole.

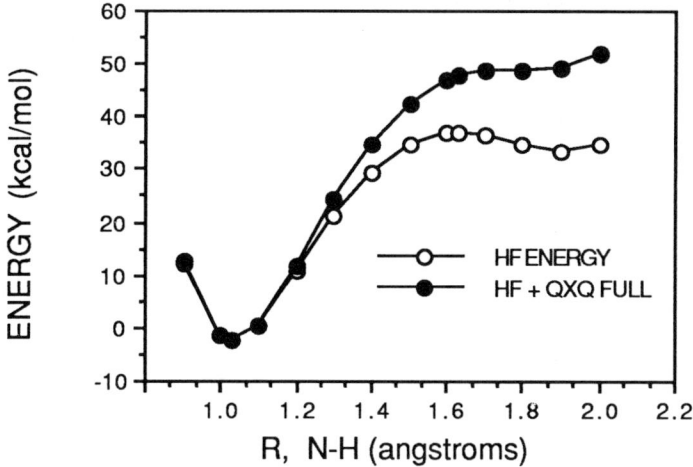

Figure 1. Potential energy curve for proton transfer in the imidazoliun/methanethiol system in vacuum and in the presence of the full protein. Energies are relative to separated imidazolium and methanethiol.

This effect of the protein on the proton transfer system is consistent with the proposed catalytic role of the protein (Thole and van Duijnen, 1983; Kollman and Hayes, 1981), which stabilizes the zwitterionic state in which a proton has been transferred from Cys 25 to His 162, i.e., in a direction opposite to the one considered here. As in the native enzyme, in the proton transfer system the protein prefers the geometry with the proton on imidazole (M1) over that with the proton on the sulfur (M2). However, there is an important difference between the effect of the protein on the proton transfer system and its original effect on the catalytic site. In the enzyme the protein is stabilizing a charge separated, zwitterionic, state which differs very significantly from the neutral state primarily by a large dipole moment generated by the charge separation. In the proton transfer system, on the other hand, the protein exhibits a sensitivity to the position of a proton that moves across a distance of approximately 1Å between two neutral ligands. Because the movement of the proton does not drastically change the dipole of the system, the question is whether this sensitivity is due to the strong inhomogeneity of the potential generated by the protein in the limited space between the proton donor and acceptor, or due to the sensitivity of the protein to the changes in the components of the proton transfer system.

Within the electrostatic approach used here the effect of the protein on the individual components of the proton transfer system can be evaluated; the total effect is the sum of the effects on the individual components. Such a decomposition is shown in Figure 2 as a function of the position of the proton.

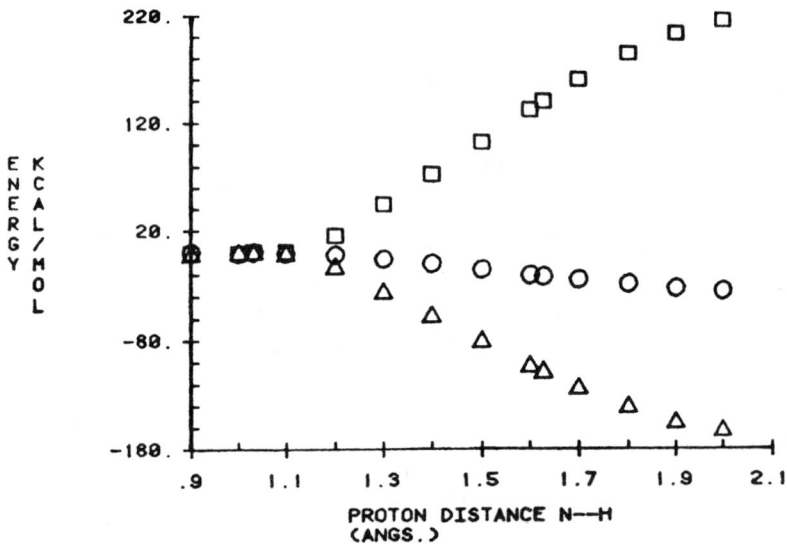

Figure 2. The effect of the protein on the components of the proton transfer system as a function of the position of the proton: (□) imidazole, (△) methanethiol, (○) proton.

The results clearly demonstrate that the electrostatic potential generated by the protein is not strongly inhomogeneous and is not the source of the observed effect. First, the electrostatic interaction energy changes with the moving proton only by 33 kcal/mole over a distance of 1Å. Secondly, the protein stabilizes the proton as it moves from imidazole towards the methanethiol. This effect is opposite to what is expected if the effect of the protein on the movement of the proton were due to an inhomogeneity in the electrostatic potential. In fact, the effect of the protein on the methanethiol and the imidazole is much larger and ultimately determines the total electrostatic effect. As the proton moves from the imidazole towards the methanethiol the protein stabilizes the methanthiol by as much as 154 kcal/mole and destabilizes the imidazole by as much as 202 kcal/mole. Since the changes in the proton donor and the proton acceptor are primarily due to charge redistribution upon the movement of the proton, these results demonstrate that the protein is more sensitive to the change in charge distribution than to the position of the proton.

This conclusion has important implications for the suitability of the proton transfer system as a heuristic model of a receptor. As was previously demonstrated (Osman et al., 1985, 1987) the driving force for the proton transfer in our earlier proton transfer model was due to the charge redistribution induced in imidazolium by the interaction with 5-HT. Similarly, in the present proton transfer system it is likely that the interaction between an incoming ligand (e.g., 5-HT) and the proton transfer system will induce a charge redistribution. The induced charge redistribution will be further stabilized by the protein, thus facilitating the proton transfer and enhancing the effect of the incoming ligand. This observation also emphasizes the importance of the quantum mechanical description of the proton transfer process. It is only at this level that the correct charge redistribution can be observed, and its interaction with the electrostatic potential of the protein correctly calculated. Had this simulation been performed with molecular mechanical methods, which do not allow intramolecular charge redistribution, the entire effect of the protein would have been overlooked.

The Effect of Individual Helices on the Proton Transfer Process

In order to understand the effect of the protein in terms of structural components we have chosen to study the effect of the helices in actinidin on the proton transfer. Helices are ubiquitous secondary structures in proteins and the understanding of their effect as structural units in one system may be transferable to other systems. The transferability is based on the premise that conserved properties encoded in the structure may have similar function in different systems.

Actinidin has six helices of which one, the A1 helix, has been implicated in the catalytic activity of the enzyme (Thole and van Duijnen, 1985). The importance of the helices can be seen in their electrostatic effect on the proton transfer process. The effect of each individual helix is presented in

Figure 3A. Clearly, their effect is not uniform; while helices A1 and A5 increase the barrier to proton transfer (TS-M1) and the energetic difference between the two minima (M2-M1) all the other helices act in the opposite way. Of special interest is the effect of helices A1 and A2 because they make large contributions to the barrier and to the differences between the minima, but act in opposite directions. Helix A1 increases the barrier by 7.0 kcal/mole whereas A2 decreases it by 5.9 kcal/mole. Likewise, A1 increases the difference between the minima by 9.8 kcal/mole and A2 decreases it by 8.2 kcal/mole.

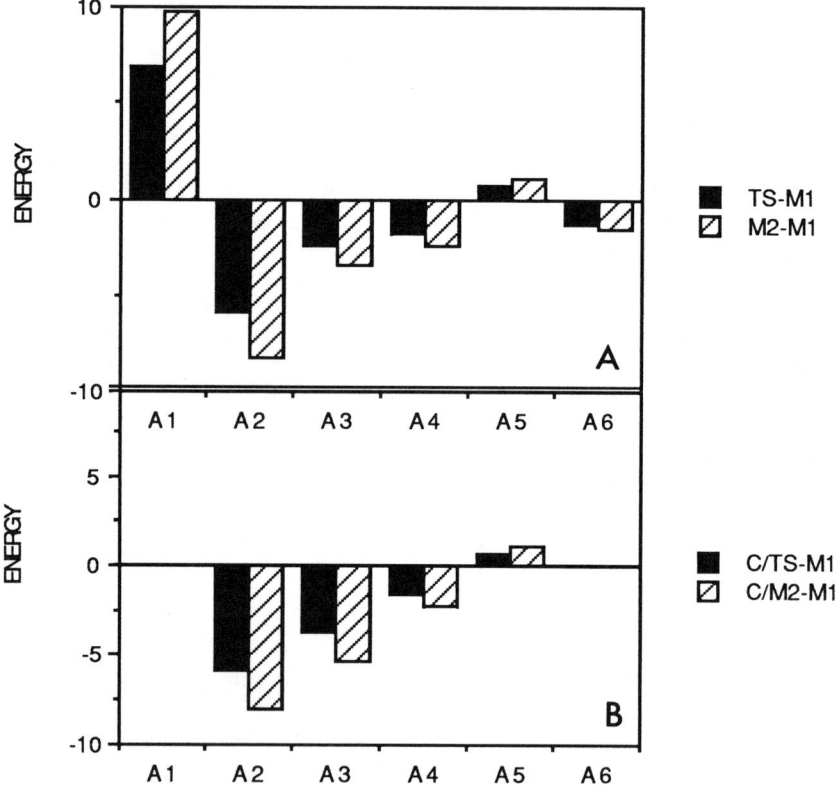

Figure 3. The effect of individual helices on the barrier (TS-M1) and the driving force (M2-M1) for a proton transfer. A. Effect from complete helices. B. Effect from the charged residues of the helices only.

A careful scrutiny of the primary structure of the helices identifies the commonality between helices A2-A5 and their difference from helices A1 and A6. While helices A1 and A6 are neutral, helices A2-A5 contain negatively charged amino acid residues: In helix A2 - Glu 50, Glu 52, Asp 55; in helix A3 - Asp 72 and Asp 80; in helix A4 - Asp 99 and Asp 104; in helix A5 - Glu 121.

The electrostatic effect of each individual helix can be decomposed into the contribution from the net charges and from the rest of the helix. Since the rest of the helix is neutral, the leading term in a multipole expansion of the charge distribution should be the dipole of the helix. The dipole is a special property of the helix and its dependence on structure is relatively simple to predict. Thus, such a decomposition gives us an opportunity to examine the electrostatic contributions from net charges and from the dipole associated with a specific helix.

Inspection of Figure 3B, in which the contributions from the negatively charged residues are plotted, reveals that the electrostatic effects of helices A2-A5 come primarily from the net charges. The effects due to the dipolar component of the electrostatic effect is often small and in the opposite direction from the monopolar effect. In helix A1 however, the only effect is dipolar and its size is very large - five to ten times that of any other helix dipoles. The size of the effect directly correlates with the size of the helix, which also determines the size of the dipole, and with its distance from the proton transfer system. The dipole moment of helix A1, calculated from the point charges, is 21.1 Debye and Cys 25 is part of the A1 helix near the amino terminus.

The detailed analysis of the electrostatic effects of alpha helices on the process of proton transfer reveals a complex interplay between the electrostatic properties of the elements of secondary structure and the primary sequence of which they are composed. Charged groups in a helix have an overwhelming electrostatic effect and the dipolar contribution from the helix dipole is small. However, when the helix is large and positioned proximally to the proton transfer site, its dipolar effect can become significant - especially when charged groups have no contribution.

CONCLUSIONS

The heuristic approach outlined in this work allowed us to examine the effect of the protein environment on a proton transfer process which was designed to model a plausible first step in the activation of a 5-HT receptor (Osman, et al., 1985, 1987). The protein that was chosen, actinidin, has an important effect on the proton transfer process rendering it a good model to study the ligand activated proton transfer as a model for receptor activation. This effect is related to its interaction with the components of the proton transfer system; the protein is much more sensitive to the charge redistribution in the proton donor and the proton acceptor than to the moving proton. Since the interaction of 5-HT with a receptor model induces a charge redistribution in its components, the protein may affect the proton transfer by interacting favorably with the charge polarized groups and thus facilitating the proton transfer.

The decomposition of the electrostatic effect of the protein into contributions from elements of secondary structure served two purposes. First, it revealed the complex interplay between the effects from charged residues and from the neutral alpha-helical components. Secondly, it established a systematic basis for a possible generalization of the effects of well defined elements of secondary structures, e.g. alpha-helices. This generalization should prove important in such heuristic approaches as presented here in which the effect of the protein is constructed from the effects of its elements. Such a constructive approach will depend on the successfull demonstration that elements with conserved structure exhibit similar functionalities in other proteins.

ACKNOWLEGMENTS

We thank Mr. H. Dijkman for many helpful discussions and for technical help. This work was supported by a grant from the National Institute on Drug Abuse (DA-01875). H.W. is a recipient of a Research Scientist Development Award (DA-00060) from the National Institute of Drug Abuse. G.A.M., Jr., is a recepient of a Medical Scientist Training Program Award under the National Research Service Award Program of the National Institute of Health (5-T32-GM7280-80). A generous grant of computer time from the University Computer Center of the City University of New York is gratefully acknowledged. Data analysis and curve fitting were done on the PROPHET computer system - a national computational resource sponsored by NIH through the Division of Research Resources. Computations for the project were also done at the Cornell National Supercomputer Facility, Center for Theory and Simulation in Science and Engineering which is funded, in part, by the National Science Foundation, New York State, and IBM Corporation.

REFERENCES

Allen, L.C. (1981) In Weinstein, H. and Green, J.P. (ed.), Quantum Chemistry in Biomedical Sciences. Annals of the New York Academy of Sciences. The New York Academy of Sciences, New York 367, pp. 383-406.
Allen, J., P. Feher, G., Yeates, T., Komiya, H., and Rees, D.C. (1987) Proc. Natl. Acad. Sci. USA, 84, 6162-6166.
Bernstein, F.C., Koetzle, T.F., Williams, G.J.B., Meyer, E.F., Brice, M.D., Rogers, J.R., Kennary, D., Simanouchi, T., and Tasumi, M. (1977) J. Mol. Biol., 112, 535-542.
Dixon, R.A.F., Kobilka, B.K., Strader, D.J., Benovic, J.L., Dohlman, H.G., Frielle, T., Bolanowski, M.A., Bennett, C.D., Rands, E., Diehl, R.E., Mumford, R.A., Slater, E.E., Sigal, I.S., Caron, M.G., Lefkowitz, R.J., and Strader, C.D. (1986) Nature, 231, 75-79.
Engelman, D.M., Golman, A., and Steitz, T.A. (1982) In Packer, L. (ed.), Methods in Enzymology. Academic Press, New York, 88, pp. 81-88.
Fersht, A. (1985) Enzyme Structure and Mechanism. W.H. Freeman and Co., New York.
Gocayne, J., Robinson, D.A., Fitzgerald, M.G., Chung, F., Kerlavage, A.R., Lentes, K., Lai, J., Wang, C. Fraser, C.M., and Venter, J.C. (1987) Proc. Natl. Acad. Sci. USA, 84, 8296-8300.
Grenningloh, G. Rienitz, A., Schmitt, B., Methfessel, C., Zensen, M., Beyreuther, K., Gundelfinger, E.D., and Betz, H. (1987) Nature, 328, 215-220.
Hulme, E. and Birdsall, N. (1986) Nature, 323, 396-397.
Kobilka, B.K., Matsui, H., Kobilka, T.S., Yang-Feng, T.L., Francke, U., Caron, M.G., Lefkowitz, R.J., and Regan, J.W. (1987) Science 238, 650-656.
Kollman, P.A. and Hayes, D.M. (19810) J. Am. Chem. Soc., 103, 2955-2961.
Kubo, T., Fukuda, K., Mikami, A., Maeda, A., Takahashi, H., Mishina, M., Haga, T., Haga, K., Ichiyama, A., Kangawa, K., Kojima, M., Matsuo, H., Hirose, T., and Numa, S. (1986) Nature, 323, 411-416.
Marx, J.L. (1987) Science, 238, 615-616.
Masu, Y., Nakayama, K., Tamaki, H., Harada, Y., Kuno, M., and Nakanishi, S. (1987) Nature, 329, 836-838.
Mehler, E.L., and Paul, C.H. (1979) Chem. Phys. Let., 63, 145-150.
Osman, R., Topiol, S., Weinstein, H.,and Eilers, J.E. (1980) Chem. Phys. Lett. 73, 399-403
Osman, R., Weinstein, H. Topiol., S., and Rubenstein, L. (1985) Clin . Physiol. Biochem., 3, 80-88.
Osman, R., Topiol, S., Rubenstein, L., and Weinstein, H. (1987) Mol. Pharm., 32, 699-705.

Polgar, L., and Halasz, P. (1982) Biochem. J., 207, 1-10.
Schofield, P.R., Darlson, M.G., Fujita, N., Burt, D.R., Stephenson, F.A., Rodriguez, H., Rhee, L.M. Ramachandran, J., Reale, V., Glencorse, T.A., Seeburg, P.H., and Barnard, E.A. (1987) Nature, 328, 221-227.
Shinitzky, M., and Katchalski, E. (1968) in Molecular Associations in Biology, B. Pullman, ed. Academic Press, New York, pp. 361-376.
Stevens, C.F. (1987) Nature, 328, 198-199.
Stevens, C.F. (1985) TINS, 8, 335-336.
Tapia, O., Stamato, F.M.L.G., and Smyers, Y.G. (1985) J. Mol. Struc . (THEOCHEM), 123, 67-84.
Thole, B.T. and van Duijnen, P.T. (1983a) Biophys. Chem., 18, 53-59.
Thole, B.T. and van Duijnen, P.T. (1983b) Theoret. Chim. Acta (Berl.), 63, 209-201.
Trowbridge, I. (1987) Nature, 327, 461-462.
Umeyama, H., Hirono, S., and Nakagawa, S. (1984) Proc. Natl. Acad. Sci. USA, 81, 6266-6270.
Warshel, A. and Russell, S. (1986) J. Am. Chem. Soc., 102, 6218-6226.
Weinstein, H., Osman, R., Edwards, W.D., and Green, J.P. (1978) Int. J. Quantum Chem. QBS5, 449-461.
Weinstein, H., Osman, R., Topiol, S., and Green, J.P. (1981) Ann. N.Y. Acad. Sci., 367, 434-451.
Weinstein, H., Osman R., and Mazurek, A.P. (1987) In Naray-Szabo, G. and Kalman, S. (eds.), Steric Aspects of Biomolecular Interactions. CRC Press, Boca Raton Florida, pp.199-210.

OPIATE RECEPTOR HETEROGENEITY: RELATIVE LIGAND AFFINITIES AND MOLECULAR DETERMINANTS OF HIGH AFFINITY BINDING AT DIFFERENT OPIATE RECEPTORS

Gilda Loew, Lawrence Toll, John Lawson, Gernot Frenking, and Wilma Polgar.

Molecular Theory Department, Biomedical Research Laboratory, SRI International, Menlo Park, California 94025 U.S.A.

INTRODUCTION

Opiates are a particularly complex class of drugs to understand. They have multiple pharmacological effects, principally analgesia, respiratory depression, euphoria, and high physical dependence liability and appear to bind to multiple receptors to elicit these effects. There are also many different classes of compounds which have opiate effects, ranging from the classical fused ring opiates such as morphine to the 31 amino acid endogenous peptide β-endorphin. Among the main goals of current research in this field is to establish, as clearly as possible, a one-to-one correspondence between classes of opiates, their binding to specific types of opiate receptors, biochemical activation of the receptors, and **in-vivo** responses.

The concept of multiple opiate receptors was first introduced to categorize three qualitatively different **in-vivo** activity profiles in dogs elicited by three closely related fused ring opiates, morphine (σ), and two benzomorphans; ethylketocyclazocine (EKC) (κ); and SKF 10047 (σ) (Martin, 1967; Martin et al., 1976). Subsequently, a number of stereoselective saturable opioid binding sites were discovered by the use of different radiolabeled ligands. These have been named by conventions based partly on the radiolabeled ligands which bind with high affinity to them and partly on the original behavioral distinctions.

The current consensus is that there are at least three major opiate receptor types, μ, δ, and κ. The μ-receptor, first defined by high affinity binding of radiolabeled morphine analogs, now has a variety of diverse, more selective opiates which bind with high affinity to it. These more selective ligands range from the small flexible fentenyls (Leysen, 1983), to peptides such as the enkephalin analog DAGO (Kosterlitz and Patterson, 1981; Gillan and Kosterlitz, 1982) or analogs of the naturally occurring tetrapeptide morphiceptin (Chang et al., 1981; Blanchard et al., 1987). The δ-receptor, first identified by high affinity binding of opioid peptides (Chang and Cuatrecasas, 1979), has no endogenous peptides which bind specifically to it and it has been very difficult to obtain synthetic δ-selective peptides. Recently, however, two types of δ-selective synthetic peptides have been reported, linear (DTLET, DSLET) (Fournie-Zalvski et al., 1981; David et al., 1982) and cyclic (DPDPE, DPLPE) (Mosberg et al, 1983; Akiyama et al., 1985) modifications and extensions of met- and leu-enkephalins. The subtype called κ is a high affinity site for EKC (Kosterlitz and Leslie, 1978; Garzon et al, 1984), and the one to which the naturally occurring peptide dynorphin binds with high affinity (James et al., 1984). While many families of opioids have been reported to bind with high affinity to the κ-receptor, it is only recently, with the advent of the 1,2 arylamino amide compounds by Upjohn (such as U50,488H and U69,593) that highly selective ligands for this receptor have been found (VonVoightlander, 1983; Lahti et al., 1985).

In addition to μ-, δ-, and κ-receptor sites, a number of other receptor types have been suggested and partially characterized by a variety of binding studies. Among these putative receptor types are: μ_1, a common high affinity site for most families of opiates (Wolozin and Pasternak, 1981; Lutz et al., 1985); a benzomorphan receptor (Chang et al., 1981); a "σ opiate receptor" (Mickelson, 1984), subtypes of the κ receptor (Su, 1985), and a β-endorphin specific receptor (Chang et al., 1984). Thus, anywhere from three to seven different opioid binding sites have been identified over the past 15 years by the stereospecific saturable binding behavior of various radiolabeled ligands to brain homogenates.

We report here two types of efforts in our laboratory using both experimental and theoretical methods to continue to probe different aspects of opiate receptor heterogeneity. These include: 1) continued characterization of opioid receptor subtypes in different species, and 2) the identification of common molecular determinants of high affinity binding of different families of opiates to each receptor type.

In the first type of study reported here, we have characterized and compared the nature and number of opioid receptor in two species, rat and guinea pig brain, using six different radiolabeled ligands for **in-vitro** binding studies.

The second goal of the studies presented here, the search for molecular modulators of high affinity at each receptor type, is an ongoing effort in our laboratory. It is a challenging task which must take into account such conflicting properties of these binding sites as: 1) Similarities in requirements for high affinity at each site, 2) The diversity of families of opiates which bind with high affinity to each subtype, and 3) The subtle modulation of affinity at each receptor by small changes within each family of opiates.

In studies made over the years in our laboratory we have identified candidate conformers for high affinity binding to µ-receptors for fused-ring and flexible nonpeptides (Loew et al., 1976; Loew et al., 1976; Cheng, 1986; Loew et al., 1987) as well as peptide opioids and have propposed key commonalities among them (Loew et al., Loew et al., 1986).

We have also recently reported a series of studies characterizing conformational requirements for high affinity binding of opiate peptides to δ-receptors and have identified a candidate conformer for such binding (Keys et al., 1988). This conformer, while allowing similar N-terminal tyrosine contact with receptor sites, is significantly different from our candidate µ-selective peptide conformers.

In contrast to µ and δ high affinity requirements, very little work has been reported to determine molecular properties which modulate affinity at the κ-receptors, even for the recently discovered U compounds. Thus, despite years of effort, the molecular determinants of high affinity opiate binding at each receptor site are far from definitely established and there is much left to understand about the nature of each binding site and the similarities and differences among them.

To continue to probe the extent of similarity of the binding sites of the different receptors and their sensitivity to small local changes, we have determined the relative receptor affinities and selectivites of sets of closely related, "well-known" fused-ring opiates with varying N-substituents. Relatively rigid analogs were chosen since these allow the most definitive deductions to be made, by complimentarity, about the nature of the binding sites of the receptors.

In an extension of this study, we have synthesized new morphine analogs with small variations in known N-substituents and used molecular modeling techniques to further probe the conformational effect of the N-substituent on receptor binding and **in-vivo** activities. Specifically, we have determined the **in vitro, in vivo** and conformational consequences of addition of a methyl group to the αC atom of the N-cyclopropylmethyl substituent of normorphine.

In the last study, we report initial results of an ongoing effort in our laboratory to design κ-selective analogs. Such selective drugs could be helpful in studies to determine the unique biochemical or **in vivo** consequences of binding to κ-receptors. In the work reported here, we describe initial efforts to determine the molecular modulators of μ/κ selectivity for a known class of arylaminoamides, the chiral 1,2 aminoamides, U compounds reported by Upjohn. We also describe how we have used the insights obtained to develop a new, related class of achiral 1,1,aminoamides in our laboratory.

METHODS

EXPERIMENTAL RECEPTOR BINDING STUDIES

All receptor binding studies were performed using basic procedures similar to those described by Pasternak (Pasternak et al., 1975). In the studies to characterize receptors in guinea pig and rat brain homogenate, six radiolabeled ligands were used in self- and cross-competitive binding studies with their unlabeled counterparts. The six ligands used were [^{3}H]D-Ala2-D-

Leu5,enkephalin (DADL); [^{3}H]U69,593; [^{3}H]naloxone; [^{3}H]D-Ser2-Thr6 leu enkephalin (DSLET); [^{3}H]ethylketocyclazocine (EKC); and [^{3}H]D-Ala2-glyol5 enkephalin (DAGO). The resulting 6 x 6 "matrix" of inhibition experiments were analyzed together using a nonlinear least-square regression analysis program, "LIGAND" originally developed by Munson and Rodbard (Munson and Rodbard, 1980) and modified in our laboratory (Toll et al., 1984). Affinities and capacities for various binding models were determined by assuming N-independent binding sites, and varying N. The model chosen was that which gave the best overall fit to all the data using statistical criteria such as minimum mean-square deviation. This procedure allows the determination of a set of self-consistent dissociation constants (K_Ds) for all the ligands at each receptor site as well as self-consistent receptor capacities for each site.

For the binding studies of 12 fused-ring opiate analogs, shown in Table 2, four labeled ligands, DSLET, U69,593, DAGO, and (-)EKC were used. The 12 unlabeled drugs were used to inhibit each [^{3}H]ligand, with binding studies conducted in guinea pig brain homogenate.

For the preliminary binding studies of the arylaminoamides, shown in Table 3 (a,b), three labeled ligands found to be most selective at μ(DAGO); δ(DSLET), and κ(U69,593) were used, and IC_{50}s at the μ- and κ-receptors are reported at this time. In the U family of 2,3 arylaminoamides, a number of analogs were kindly supplied to us by Dr. R. Lahti of Upjohn and their relative receptor affinities at μ- and κ-receptors are reported here for the first time. Preliminary μ- and κ-receptor binding results are also given for five analogs of the new class of achiral 1,1,arylaminoamides we have begun to synthesize in our laboratory.

THEORETICAL

For studies of the conformational properites of the N-substituents of morphine and of the two classes of arylaminoamides, an empirical energy program called MOLMEC was used. This program, developed by Oie, Duchamp, and co-workers, is described in detail elsewhere (Oie et al., 1981). It is based on a seven-term empirical energy expression which includes four bonding terms corresponding to bond stretching, bending, out-of-plane and torsion angle motions as well as electrostatic, and exponential (r^{-6}) nonbonding and hydrogen bonding terms. Net atomic charges in the electrostatic term were obtained from a recent modification of the MNDO method (Dewar and Thiel, 1977) with enhanced capability for describing hydrogen binding, MNDO-H (Goldblum, 1987).

In one study of molecular determinants which modulate μ/κ affinities and selectivities, energy-conformational studies were made of three closely related morphine analogs, N-cyclopropylmethyl morphine (2c, Table 2) and two new diesterioisomers made by the addition of a methyl group to the αC atom of the cyclopropylmethyl substituent. In these studies, nested rotations about the two rotatable torsion angles of the N-substitutents were performed, and low energy conformers were totally optimized.

In a second study of the molecular determinants which modulate μ/κ affinities, extensive energy-conformational studies were made of four analogs of the U compounds, shown in Table 4A (U51,754; U50,488H; U47,700; and U47,109), and three of the new "P" compounds, shown in Table 4B (P1, P2, and P13). Both of these families are quite flexible. The Upjohn U compounds have five rotatable torsion angles, and four relative equatorial/axial positions of the 1-amino and 2-amide substituents on the cyclohexane ring. Our new SRI P compounds (Table 4B) have an additional rotatable bond, τ_2, which is fixed in the U compounds, but have only two possible relative positions of the 1-amino,1-amide substituents: ax/eq and eq/ax.

Similar procedures were used to explore conformational space for these two types of analogs. As a first step, only the aminoamide portion of a parent compound in each series with n=1 (U51754, P13), and n=0 (U47,700, P2) was considered by replacing the benzyl or phenyl moeity by a methyl group. For this fragment, all possible combinations of amino and amide substituents were considered, as well as both chair and boat forms of the cyclohexane ring and 8 sets of torsion angles for the U compounds and 11 for the P compounds. This procedure led to 64 optimized structures for the amino amide fragment of the U compound and 22 for this fragment of the P compounds. In the next step, the aromatic portions were added to the 9 lowest energy structures of the U fragments (ΔE < 9 kcal/mol) and the 11 lowest energy conformers of the P fragments (ΔE < 3 kcal/mol). In this procedure, for the n=0 compounds, a phenyl group was substituted for the methyl group and the 9 U compounds (U47,700) and 11 P compounds (P1) reoptimized. The complete n = 1 compounds, were formed by replacing each hydrogen of the methyl group in turn by a phenyl group leading to 3 possible rotamers for each of the aminoamide structures. The 27 structures corresponding to the complete U51,754 compound and the 33 corresponding to the complete P13 compound were then reoptimized.

The effect of two substituent changes on conformation was also explored. In one study, the U series, the amine nitrogen substituent in the U series was changed from a dimethyl amine (U51,574) to a pyrollidine (U50,488H) and the 13 lowest energy conformers ($\Delta E \leq 9$ kcal/mol) reoptimized. In the second, the methyl group on the amide nitrogen was changed to an H atom for the n = 0 analogs (U47,700, P1) and all the conformers reoptimized.

RESULTS AND DISCUSSION

The results of the study of opiate receptor heterogeneity in rat and guinea pig brain, as determined by six different labeled ligands, are summarized in Table 1. Simultaneous analysis of all the competitive binding data has led to a five-receptor site model in rat brain and a four-receptor site model in guinea pig brain as the best overall fit to all the data. Three well-defined receptor binding sites were found in both species and labeled by currently accepted convention as μ, δ, and κ (Gillan and Kosterlitz, 1982). Specifically, we have labeled the site μ which is high affinity for the selective synthetic enkephalin called DAGO. It is also the site which morphine binds with high affinity as does DADL, naloxone, and EKC. We have labeled as δ the site to which the relatively selective synthetic peptide DSLET binds with highest affinity. DADL also binds with high affinity to this site, consistent with previous studies which find DADL to bind with nearly equal affinity to the sites labeled μ and δ. We have labeled as κ the single high affinity site of U69,593. EKC also binds with high affinity to this site. We now see that, in fact, EKC binds with high affinity to both μ- and κ-receptors and that it is unclear which receptor(s) mediates its behavioral effects.

TABLE 1. Opiate Receptor Heterogeneity Determined by Six Labeled Ligands

A. Rat Brain	K_D (nM)				
	μ_1	μ_2	δ	κ	"Site 4"
DAGO	0.3	2.0	667	5,000	21,000
DSLET	5.5	55.6	2.3	--	2,900
U69,593	385	6,200	10,000	1.3	6,250
DADL	1.1	13.0	3.8	--	1,700
Naloxone	1.9	0.6	35.7	270	20.4
EKC	1.3	1.6	25.0	1.6	130
B_{max} pmol/gm	4.5	7.2	11.2	1.2	73

B. Guinea Pig Brain	K_D (nM)			
	μ	δ	κ	"Site 4"
DAGO	1.4	435	42,000	450
DSLET	20.8	1.4	20,000	--
U69,593	1,300	6,250	3.0	27,000
DADL	4.5	1.0	30,000	--
Naloxone	1.0	30.3	16.1	22
EKC	2.3	3.8	4.3	71
B_{max} pmol/gm	2.9	4.2	8.2	33

In addition to these three well-characterized sites, a fourth binding site was found in both species. This is defined by high to moderate affinity of (-)EKC and a number of other fused-ring opiates, and low affinity of the U69,593 compound. This site has characteristics similar to those reported in the past, which have been variously called a "benzomorphan" (Chang et al., 1981), a κ subtype (Su, 1985), and a σ-opioid site (Michelson and Lahti, 1984). Our studies add to the evidence for the presence of such a site. This site appears to have high capacity in both species. Thus far there are no known selective ligands for this site and it could correspond to more than one residual opioid binding site.

In rat brain, but not in guinea pig brain, a fifth site was found which we label as μ_1, in keeping with the initial definition of such a site as a common, high affinity site for most opioid ligands. This site, first found and described by Pasternak (Wolozin and Pasternak, 1981; Pasternak and Snyder, 1975), has been subsequently reported by us (Toll et al., 1984) and other investigators (Lutz et al, 1985; Cruciani et al., 1987) and there are ongoing efforts to probe its function.

In both species, the common sites appear to have similar affinity profiles but different relative capacities. In rat brain, the μ- and δ-receptors have equal capacities, about ten times that of the κ site, which exists in very small amounts and is therefore difficult to characterize in this species. By contrast, in guinea pig brain, the κ-receptor is the most abundant of the three well-defined sites, and the μ-receptor the least well defined. Because of the presence of abundant κ receptors, we have chosen to continue to probe receptor affinities and selectivity of different classes of opiates using guinea pig brain.

In a recent example of these continued studies, we have examined the receptor binding profiles of well-known fused ring opiates. The results are given in Table 2. The families were the 3-fused-ring benzomorphans (1a-c), 5-ring morphine (2a-c) and oxymorphone (3a-c) series, and the 6-fused-ring oripavine family of which etorphine (4a-c) is a prototypical analog. As shown in Table 2, we have systematically investigated the effect of varying the substituent bound to the amine nitrogen in each class of compounds. N-substituent variation was chosen because a number of previous studies have already demonstrated that it plays a key role in modification of opioid efficacy (Harris, 1974). As seen in Table 2, varying the N-substituent from methyl to allyl to cyclopropyl methyl also results in modulation of affinities at μ, δ, κ, and site 4. The N-methyl analogs, metazocine (1a), morphine (2a), and oxymorphone (3a) are all high affinity, relatively selective, μ-binding analogs. Replacement of the methyl substituent by an allyl or cyclopropylmethyl group in these families, changes which are usually associated with diminished efficacy, result in increased affinity at all receptors but with the greatest increase at κ. Thus the cyclopropylmethyl derivatives, SKF 10047 (1c), CPM-morphine (2c), and naltrexone (3c) bind with high affinity at all

TABLE 2. Modulations of Receptor Affinities (K_D nM) in Fused Ring Opiates by N-Substituent and Ring Variations

1 Benzomorphans **2** Morphine **3** Oxymorphone **4** Etorphines

R=	CH_3	allyl	$CH_2{-}\triangleleft$	CH_3	allyl	$CH_2{-}\triangleleft$	CH_3	allyl	$CH_2{-}\triangleleft$	CH_3	allyl	$CH_2{-}\triangleleft$	
analog	1a	1b	1c	2a	2b	2c	3a	3b	3c	4a	4b	4c	B_{max} pmol/gm
μ	4.5	1.3	0.4	2.3	2.1	0.5	1.4	2.4	0.6	1.3	2.8	2.8	3.0
δ	47	4.7	1.6	100	34	11	67	19	10	1.7	3.1	1.1	2.7
κ	45	0.8	0.2	150	5.9	1.0	280	43	2.0	3.4	4.0	3.2	6.2
"4"	420	26	10	150	5.9	1.0	63	260	14	9.1	71	40	13.8

receptors with approximately equal affinity at μ and κ. By contrast, in the etorphine series, with a large lipophilic group at C7, the N-methyl derivative, etorphine itself, (4a), is a universal ligand, binding with high affinity to all receptors. In this series there is little added effect of N-substituent variation. Across families, all receptor types seem to accommodate 3-, 5-, or 6-fused ring opiates with little distinction.

These deceptively straightforward results provide interesting and provocative insights into the nature of the binding site of the different opiate receptor subtypes. At least for the binding pocket defined by these rather rigid analogs, there seems to be a great deal of similarity among the different receptor proteins. The binding site of the princial receptor subtypes, μ, δ, and κ all appear to interact very favorably with a large lipophilic group at position C7 of etorphine, making the effect of the N-substituent on affinity less important. This finding is consistent with the results of previous studies of a series of oripavines with varying C7-substituents (Loew and Berkowitz, 1979). In these studies it was observed that for compounds with long chains at C7, N-substituents have less effect on efficacy.

The receptor binding studies reported here, taken together with previous structure-activity studies of fused-ring opiates, implicate N-substituent variation as modulator of receptor affinity and selectivity as well as a efficacy. One of the interesting questions that remains is the effect on **in-vivo** activity of this dual modulation. Such studies are hampered by the lack of knowledge of the specific **in-vivo** consequences of drug binding to the different receptor subtypes. However, there is long-standing evidence that one particularly deleterious effect of opiates, physical dependence liability, can be altered by compounds with reduced efficacy (Cowan, 1974). More recent studies suggest that physical dependence is also different in compounds which bind to more than one receptor, particularly those which bind with high affinity to μ- and κ-receptors (Gmerek et al., 1987). Thus, while the search for selective ligands is important in continuing to sort out the functions of each receptor subtype, we have been exploring the hypothesis that clinically useful analgesics with **in-vivo** activity profiles altered from that of morphine might be produced by small alterations of N-substituents in morphine, we have particularly focused on the addition of a methyl group to the αC which binds directly to the protonated amine nitrogen.

Table 3, shows the results of studies of one of these types of analogs, which we have synthesized. The addition of a methyl group to the αC of the cyclopropylmethyl analog of morphine (analog 2c, Table 2) makes the αC a chiral center. The two diestereoisomers, 2cR and 2cS were separated and a crystal structure of one, 2cR, was done to determine absolute configuration. Shown in this table are the results of energy-conformational studies of the three N-substituents, exploring rotations about the two torsion angles, τ_1 and τ_2. The parent compound, which binds with high affinity at the μ- and κ-receptors, was found to have two rotational conformers (A and B) with an energy difference of only 1.4 kcal/mol. One obvious inference of this result is that one rotamer binds with high affinity to the μ-receptor and the other to the κ-receptor. Further studies of the two diesterioisomers of the α-methyl compounds show this explanation not to be the case. Isomer 2cR, in which the B form is favored, retains high affinity at μ and κ. Isomer 2cS, which favors the A form, has diminished affinity at both of these receptors. These "counter-intuitive" results provide evidence that a single conformer (labeled B in Table 3), accessible to both 2c and 2cR, allows high

TABLE 3. Addition of a Methyl Group to αC of N-Cyclopropylmethyl Substituent of Morphine Conformational Profile of N-Substituent

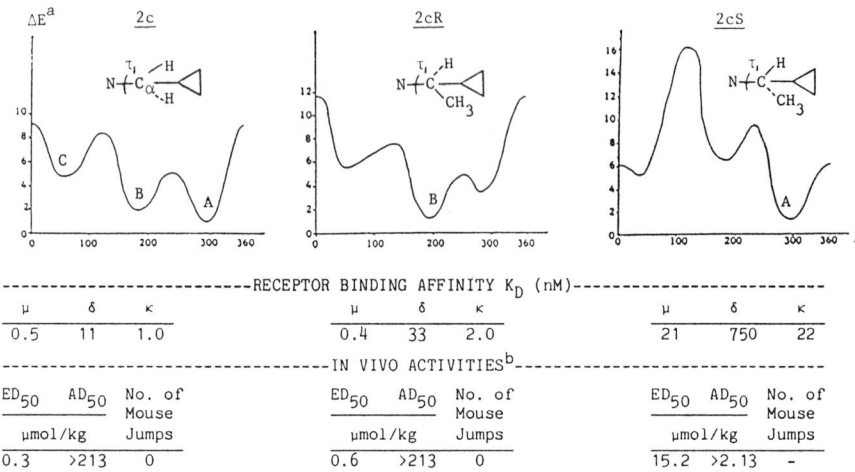

	2c			2cR			2cS	
\- RECEPTOR BINDING AFFINITY K_D (nM) \-								
μ	δ	κ	μ	δ	κ	μ	δ	κ
0.5	11	1.0	0.4	33	2.0	21	750	22
\- IN VIVO ACTIVITIES[b] \-								
ED_{50} μmol/kg	AD_{50}	No. of Mouse Jumps	ED_{50} μmol/kg	AD_{50}	No. of Mouse Jumps	ED_{50} μmol/kg	AD_{50}	No. of Mouse Jumps
0.3	>213	0	0.6	>213	0	15.2	>2.13	-

[a] ΔE in kcal/mol
[b] Morphine ED_{50} = 3.0 μmol/kg; number of mouse jumps = 22

affinity to both the μ- and κ-receptors while the other (A) is not favorable to either. These results add to the evidence indicating very similar μ- and κ-binding sites.

In-vivo behavioral studies of analogs 2c and 2cR indicate that both have potent analgesic activity, ten times morphine, by both subcutaneous and oral administration. They also display low physical dependence liability as measured by the mouse Straub tail response and mouse jump response. This preliminary behavioral profile make them candidates for further study as clinically useful analgesics. These results illustrate that better understanding of the features which modulate μ- and κ-receptor affinities and selectivites can have clinical, as well as mechanistic, significance.

The discovery of the first κ-selective analogs, the 1,2 arylaminoamide U compounds by investigators at Upjohn laboratories has given the study of κ analgesics new impetus.

In the search for κ-selective opiates, many analogs of the U compounds were synthesized, but the relative μ/κ affinities of only three (U50,488H; U62066, and U69593), were reported previously (Lahti et al., 1985). We report here the affinities at μ and κ of these three, and five additional U analogs obtained from Dr. R. Lahti. As can be seen from this small series (Table 4A), there is a very subtle electronic/ steric modulation of μ/κ selectivity in this family of opiates. Addition of a ketal substituent to the cyclohexane ring (U62,066 and U69,593) enhances μ affinity but does not alter κ. Removal of the 3,4 Cl substituents from the phenyl ring greatly diminishes μ affinity, while κ decreases to a lesser extent, leading to the most κ-selective compound found thus far, U69,593. One variation of U50,488H studied is U51,574, in which the pyrrolidine amine substituent is replaced by a dimethyl amine. As seen in Table 4A, this simple substitution greatly diminishes κ-selectivity by diminishing κ-affinity. These results demonstrate the fragile basis for κ selectivity in this family.

TABLE 4. μ/k Receptor Affinities (IC_{50} nM) of Two Classes of Arylamino Amides in Guinea Pig Brain

A. 1,2 Aminoamides (U Compounds, Upjohn)

U	47109	47700	48520	50211	51754	50,488H	62066	69,593
R_1	diMe	diMe	diMe	diMe	diMe	N⟨⟩	N⟨⟩	N⟨⟩
R	H	H	H	H	H	H	O⟨⟩	O⟨⟩
R_2	H	CH_3	CH_3	CH_3	CH_3	CH_3	CH_3	CH_3
n	0	0	0	0	1	1	1	1
R_3	Cl	Cl	H	H	Cl	Cl	Cl	H
R_4	Cl	Cl	Cl	OH	Cl	Cl	Cl	H
Receptor Affinities (IC_{50} (nM))								
μ	150	9	160	>10,000	340	700	210	4,600
κ	275	300	2,000	>10,000	75	2.0	2.5	9.5

B. 1,1 Aminoamides (P Compounds, SRI)

P	1	2	13	2a	13a	13b
R_1	diMe	diMe	diMe	N⟨⟩	diMe	diMe*
R	H	H	H	H	C(O-O)	C(O-O)
R_2	H	H	CH_3	H	CH_3	CH_3
n	0	1	1	1	1	1
Receptor Affinities (IC_{50} (nM))						
μ	25	170	1,000	7,000	90	1,000
κ	130	64	64	64	12	230

*des-CL ∅ analog

Extensive energy conformation studies of the N-dimethyl compound reveal that it and the pyrolidine derivative have the same low energy conformation. Results of these studies indicated that the chair form of the cyclohexane ring is significantly more stable than the boat form ($\Delta E \geq 6$ kcal/mol) for all conformers studied. In addition, of the four isomeric arrangements of the 1-amine/2-amide substituents of the cyclohexane ring, the eq/eq configuration was the lowest energy ($\Delta E \leq 7.5$ kcal/mol). Torsion angles calculated for the lowest energy eq/eq conformers of U51,574 and 50,488H are given in Table 5. Their similar conformations indicate that modulation of μ/κ selectivity by the local effects of the N-substituent itself, in a manner reminiscent of that observed for the rigid fused-ring opiates. Thus, there do not appear to be intrinsic features of the 1,2 arylaminoamide structure that confer high κ-selectivity to this family of compounds. Rather, subtle changes in electronic or steric interactions modulate μ- and κ-affinities.

Our binding studies reveal the dramatic conclusion that the 1,2,arylaminoamides become μ-selective or nonselective

TABLE 5. Calculated Torsion Angle Value for Low Energy Conformers of U and P Compounds

	n = 1				n = 0[b]		
	U51,754	U50,488H	P_{13}		U47,700		P_1
τ_1	- 43	- 46	- 42	179	- 43	- 41	174
τ_2[a]	45	46	46	61	46	45	72
τ_3	51	50	- 93	84	50	-105	89
τ_4	-160	-165	160	-179	-173	155	179
τ_5	66	66	- 67	172	84	96	96
τ_6	61	62	- 60	- 87			
amine/amide	eq/eq	eq/eq	eq/ax	ax/eq	eq/eq	eq/ax	ax/eq

[a]This angle is a fixed part of cyclohexane ring in U compounds (Table 4A) and a flexible torsion angle in P compounds (Table 4B).

[b]n=0 compounds do not have a methylene bridge between amine and phenyl ring and hence have one less torsion angle (Tables 4A and 4B).

when the methylene bridge between the aminoamide portion of the molecule and the aromatic portion of the molecule is removed (Table 4A, n = 0 analogs). Energy conformational studies of two of these analogs, U47,700 and U47,109, result in a similar low-energy conformer (Table 5) to that of the corresponding n = 1 analogs (U50,488H and U51,754). The low energy conformer of the κ-selective (U50,488H) and the μ-selective (U47,700) are shown in Figure 5A, with overlapping cyclohexyloaminoamide regions. As seen in this figure, the main differences between these analogs are the relative positions of the aminoamide and aromatic portions of the ring. This difference in relative positions must be important in determining relative μ/κ affinities. Judging from the sensitivity of affinity at both receptors, the phenyl group and its substituents make a crucial contact with both the μ- and κ-receptors.

FIGURE 1. Overlap of Low Energy Conformers

A. U50,488H (κ, n=1) and μ-47,700 (μ, n=0); cyclohexane aminoamide overlap

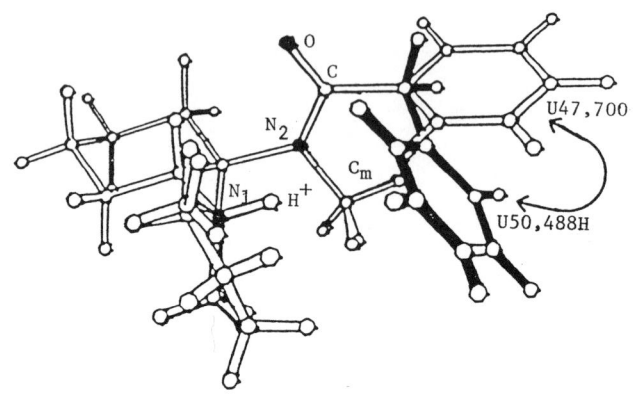

B. U50,488 (eq,eq) and P13 (eq/ax) (RMS = 0.37, $N_1C_1C_2N_2C$); $r_{H^+H^+} = 0.14$)

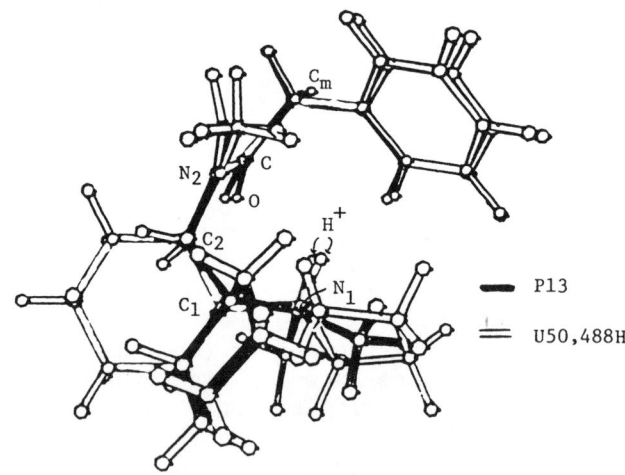

Recently, we have begun extensive synthesis and characterization (SAR) of a new class of achiral 1,1,arylaminoamides (named P compounds by us) shown in Table 4B. We have selected these compounds for study because they are closely related to the Upjohn compounds but are much more easily synthesized and do not have any optically active centers.

Our P compounds have the same four structural moieties as
the U compounds, i.e., the cyclohexane, amine, amide, and
aryl groups; and the amine and amide nitrogens are separated
by two carbon atoms. However, the amine and amide groups
are linked by a rotatable torsion angle τ_2 in the P compounds, rather than one that is part of the cyclohexane ring
as in the U compounds. Moreover, since both substituents
are on the same cyclohexane ring carbon atom, there are only
two rather than four possible relative orientations of the
amine and amide substituents: eq/ax and ax/eq.

Table 4B shows the results of preliminary receptor
binding studies of six of these P analogs. These results
show the similarity of the SAR between the U and P compounds. Like the U compounds, insertion of a methylene
bridge between the aminoamide and aromatic moeities (P1
versus P2) enhances κ selectivity. This selectivity is
further enhanced by changing the amine N-substituent from
a dimethyl group to a piperidine ring (P2 versus P2a).
A third similarity encountered is the ability of a 3,4
dichloro to increase affinity at both μ- and κ-receptors in
both families (P13a versus P13b).

The results of energy-conformational studies of the n=1
P analogs, P2 and P13 in Table 5 provide a possible explanation for the similarities between U and P compounds. These
results indicate that there are many more low-energy conformers than for the corresponding n=1 U analog (U51,754). In
particular, while the eq/eq amino/amide arrangement in the
U compounds was found to be the definitive low energy form,
eq/ax and ax/eq isomers of the 1,1,aminoamide compounds have
equal energies. The torsion angles for the lowest energy
forms of each of these isomers for P13 are given in Table
5. As seen in this table, the conformation for the eq/ax,
but not the equienergy ax/eq form, is very similar to that
of the U compounds. In particular, the value for the rotatable angle τ_2 in P13 matches that of the fixed angle τ_2 in
the U compounds. This similarity is shown in Figure 1B
where the lowest energy eq/eq form of U50,488 is superimposed on the lowest energy eq/ax form of P13. In this figure,
the N1(amine)-C1-C2-N2(amide)-C(=O)-CH2 atoms of each
compound are overlapped with a root mean square of 0.37 Å.
The two amine protons are 0.14 Å apart and point in the same
direction. While there are significant differences in the
position of the remainder of the cyclohexane ring, the
similarity in overlap of these two families of compounds

indicates that they can bind in a similar fashion to the κ-receptor. Thus P-type analogs which have high κ-affinity and selectivity should be possible to obtain.

Insights gained from the studies thus far are guiding further choices for synthesis of this new class of 1,1,aryl-aminoamides. For example, we plan to synthesize analogs of the P compounds that will preferentially enhance the stability of the eq/ax isomers which most resemble the U compound. In particular, addition of bulky alkyl groups to the 4-position of the cyclohexane ring, cis- to the axial amide, should stabilize this form and might also enhance κ affinity and selectivity. Further changes in the P analogs, similar to those which favor κ affinity and selectivity in the U compounds, by localized stereoelectronic modulations rather than gross conformational changes, are being made which should further enhance κ-affinity and κ-selectivity in the P compounds.

Among the compounds selected for synthesis could be κ-selective ones with varying degrees of efficacy which will be useful for further studies of the characteristics and function of the κ-receptor and in receptor isolation. Also nonselective ligands might be found with an unusual spectrum of **in-vivo** activities which could lead to clinically useful analgesics.

REFERENCES

Akiyama K, Gee KW, Mosberg HI, Hruby VJ, Yamamora HI (1985). Characterization of [^{3}H][2-0-penicillamine, 5-0-penicillamine]-enkephalin binding to δ opiate receptors in the rat brain and neuroblastome-glioma hybrid cell line (NG108-15). Proc Natl Acad Sci USA 82:2543-2547.
Blanchard SG, Lee PHK, Pugh WW, Hong JS, Chang K-J (1987). Characterization of the binding of a morphine μ-receptor-specific ligand: Tyr-Pro-NMePhe-D-Pro-NH2, [^{3}H]-PL17. Molec Pharmacol 31:326-333.
Chang K-J, Blanchard SG, Cuatrecasas P (1984). Benzomorphan sites are ligand recognition sites of putative ε-receptors. Molec Pharmacol 26:484-488.
Chang K-J, Cuatrecasas P (1979). Multiple opiate receptors: enkephalins and morphine bind to receptors of different specificity. J Biol Chem 254:2610-2618.

Chang K-J, Hazum E, Cruatrecasas P (1981). Novel opiate binding sites selective for benzomorphan drugs. Proc Natl Acad Sci USA 78:4141-4145.

Chang K-J, Killian A, Hazum E, Cuatrecasas P, Chang J-K (1981). Morphiceptin (NH_2-Tyr-Phe-Pro-$CONH_2$): A potent and specific agonist for morphine (μ) receptors. Science Wash DC 212:75-77.

Cheng AC, Uyeno ET, Polgar W, Toll LR, Lawson JA, DeGraw JI, Loew GH, Camerman A, Camerman N (1986). N-substituent Modulation of Opiate Agonist/Antagonist Activity in Resolved 3-Methyl-3-(m-hydroxyphenyl)piperidines. J Med Chem 29:531-537.

Cowan A (1974). Evaluation in nonhuman primates: Evaluation of the physical dependence capacities of the oripavine-thebaine partial agonists in PATAS monkeys. In Braude MC, Harris LS, Smith JP, Villarreal JE (eds): Narcotic Antagonists," New York: Raven Press, vol. 8, New York, pp 427-430.

Cruciani RA, Lutz RA, Munson PJ, Rodbard D (1987). Naloxonazine effects on the interaction of enkephalin analogs with mu-1, mu and delta opioid binding sites in rat brain membranes. J Pharm Exp Ther 242:15-20.

David M, Moisand C, Meunier J-C, Morgat J-L, Gacel G, Rocque BP (1982). [^{3}H]Tyr-D-Ser-Gly-Phe-Leu-Thr: A specific probe for the δ opiate receptor subtype in brain membranes. Eur J Pharmacol 78:385-387.

Dewar MJS, Thiel W (1977). Ground states of molecules 39. MNDO results for molecules containing hydrogen, carbon, nitrogen, and oxygen. J Am Chem Soc 99:4899.

Fournie-Zalvski MC, Gacel G, Maigret B, Premilat S, Roques BP (1981) Structural requirements for specific recognition of μ or δ opiate receptors. Mol Pharmacol 20:484-491.

Garzon J, Sanchez-Blazquez P, Lee NM (1984). [3H]Ethylketocyclazocine binding to mouse brain membranes: evidence for a kappa opioid receptor. J Pharmacol Exp Ther 231:33-37.

Gillan MGC, Kosterlitz HW (1982). Spectrum of the μ-, δ- and κ-binding sites in homogenates of rat brain. Br J Pharmacol 77:461-469.

Goldblum A (1987). Improvement of the hydrogen bonding correction to MNDO for SCF calculations of aspartic proteinase reactions. In Hadzi D, Jerman-Glazic B (eds): "QSAR in Drug Design and Toxicology," Amsterdam: Elsevier, p. 207

Gmerek DE, Dykstra LA, Woods AH (1987). Kappa opioids in rhesus monkeys III. Dependence associated with chronic administration. J Pharm Exp Ther 262:428-436.

Harris LS (1974). Narcotic antagonists. Structure activity relationiships. In Braude MC, Harris LS, May EL, Smith JP, Villarreal JE (eds): "Narcotic Antagonists," New York: Raven Press, Vol. 8, pp 13-20.

James IF, Fischli W, Goldstein A (1984). Opioid receptor selectivity of dynorphin gene products. J Pharm Exp Ther 228:88-93.

Keys C, Payne P, Amsterdam P, Toll L, Loew G (1988). Conformational determinants of high affinity δ receptor binding of opioid peptides. Mol Phar (in press).

Kosterlitz HW, Leslie FM (1978). Comparison of the receptor binding characteristics of opiate agonists interacting with μ- or κ-receptors. Br J Pharmacol 64:607-614

Kosterlitz HW, Patterson SJ (1981). Tyr-D-Ala-Gly-MePhe-NH(CH$_2$)$_2$-OH is a selective ligand for the μ-opiate binding site. Br J Pharmacol 73:2998.

Lahti RA, Mickelson MM, McCall JM, Von Voightlander PF (1985). [^{3}H]U69,593 a highly selective ligand for the opioid κ-receptor. Eur J Pharmacol 109:281-284.

Leysen JE, Gommeren W, Niemegeers JE (1983). [^{3}H]Sufentanil, A superior ligand for (μ)-opiate receptors: Binding properties and regional distribution in rat brain and spinal cord. Eur J. Pharmacol 87:209-225.

Loew GH, Weinstein H, Berkowitz DS (1976). Theoretical study of the solvent effect on ionization and partition behavior in related opiate narcotics hydromorphine and oxymorphine. In "Opiates and Endogenous Opioids," Amsterdam: Elsevier/North-Holland BioMed Press.

Loew GH, Berkowitz DS (1979). Intramolecular hydrogen bonding and conformational studies of bridged thebaine and oripavine opiate narcotic agonists and antagonists. J Med Chem 22:603-307

Loew GH, Berkowitz DS, DeGraw JI, Johnson, HL, Lawson JA, Uyeno, ET (1976). New N-substituent variations in the morphine series. In "Opiates and Endogenous Opioids," Amsterdam: Elsevier/North-Holland BioMed Press.

Loew G, Hashimoto G, Williamson L. Burt S, Anderson W (1982). Conformational-energy studies of tetrapeptide opiates: Candidate active and inactive conformations. Mol Pharm 22:667-677.

Loew GH, Keys C, Luke B, Polgar W, Toll L (1986). Structure-activity studies of morphiceptin analogs:

Receptor binding and molecular determinants of μ-affinity and selectivity. Mol Pharmacol 29:546-553.

Loew GH, Lawson J. Toll L, Uyeno E, Frenking G, Polar W (1987). Structure-activity studies of a series of 4-(m-OH-phenyl)-piperidines. NIDA Research Monog 75:49-52.

Lutz RA, Cruciani RA, Munson PJ, Rodbard D (1985). Mu1: A very high affinity subtype of enkephalin binding sites in rat brain. Life Sci 36:2233-2238.

Martin WR (1967). Opioid antagonists. Pharmacol Rev 19:463-521.

Martin WR, Eades CG, Thompson JA, Huppler RE, Gilbert PE (1976). The effects of morphine- and nalorphine-like drugs in the nondependent and morphine-dependent chronic spinal dog. J Pharmacol Exp Ther 197:517-532.

Mickelson MM, Lahti RA (1984). [^{3}H]SKF 10047 receptor binding studies: Attempts to define the opioid sigma receptor. Neuropeptides 5:149-152.

Mosberg HI, Hurst R, Hruby VJ, Gee K, Yamamura HI, Galligan JJ, Burks TF (1983). Bis-penicillamine enkephalins possess highly improved specificity toward δ opioid receptors. Proc Natl Acad Sci USA 80:5871-5874.

Munson PJ, Rodbard D (1980). Ligand: A versatile approach for characterization of ligand binding systems. Anal Biochem 107:220-239.

Oie T, Maggiora GM, Christoffersen RE, Duchamp DJ (1981). Development of a flexible intra- and intermolecular empirical potential function for large molecular systems. Int J Quant Chem QBS 8:1.

Pasternak GW, Snowman AS, Snyder SH (1975). Selective enhancement of opiate agonist binding by divalent cations. Mol Pharmacol 11:735-744.

Pasternak GW, Snyder SH (1975). Identification of novel high affinity opiate receptor binding in rat brain. Nature 253:563-565.

Su, T-P (1985). Further demonstration of kappa opioid binding sites in the brain: evidence for heterogeneity. J Pharmacol Exp Ther 232:144-148.

Toll L, Keys C, Polgar W, Loew G (1984). The use of computer analysis in describing multiple opiate receptors. Neuropeptides 5:205-208.

Toll L, Keys C, Spangler D, Loew G (1984). Computer-assisted determination of benzodiazepine receptor hetereogeneity. Eur J Pharmacol 99:203-209.

VonVoightlander PF, Lahti RA, Ludens JH (1983). A selective and structurally novel non-mu (kappa) opioid agonist. J Pharm Exp Ther 224:7-12.

Wolozin BL, Pasternak GW (1981). Classification of multiple morphine and enkephalin binding sites in the central nervous system. Proc Natl Acad Sci USA 78:6181-6185.

THE ROLE OF HYDROGEN-BONDS IN DRUG BINDING

Rebecca C. Wade and Peter J. Goodford,

University of Oxford,
Laboratory of Molecular Biophysics,
The Rex Richards Building,
South Parks Road, Oxford OX1 3QU.

ABSTRACT

Hydrogen-bonds play a crucial role in determining the specificity of ligand binding. Their important contribution is explicitly incorporated into a computational method, called GRID, which has been designed to detect energetically favourable ligand binding sites on a chosen target molecule of known structure. An empirical energy function consisting of a Lennard-Jones, an electrostatic and a hydrogen-bonding term is employed. The latter term is found to be necessary because spherically symmetric atom-centred forces alone may not adequately reproduce the geometry of two interacting molecules. The hydrogen-bonding term is dependent on the length and orientation of the hydrogen-bond. Its functional form also varies according to the chemical nature of both the hydrogen-bond donor and acceptor atoms, and has been modelled to fit experimental observations of crystal structures. The mobility of the hydrogen-bonding hydrogens is considered analytically in calculating the hydrogen-bond energy. The hydrogen-bonding energy functions will be described and their application will be demonstrated on molecules of pharmacological interest where hydrogen-bonds influence the binding of ligands.

INTRODUCTION

Hydrogen-bonds make a particularly significant contribution to the intermolecular interactions of biological molecules. The hydrogen-bond can be defined as an intermediate range interaction between a positively charged hydrogen atom and a strongly electronegative atom, which allows these atoms to approach closer to each other than the sum of their Van der Waals radii. Originally thought to be an essentially electrostatic force (Pauling, 1949), the importance of additional contributions to the hydrogen-bond has been shown by, for example, Dreyfus and Pullman, 1970; these extra components are the attractive charge transfer and dispersion forces, and the repulsive exchange forces. Thus, due to its complexity, the hydrogen-bond has been modelled in a wide variety of ways in empirical energy functions. These functions can be divided into two classes: those that do include an explicit hydrogen-bond energy term (Brooks et al., 1983; Weiner et al., 1984,1986; Momany et al., 1975; Vedani and Dunitz, 1985), and those that do not (Hagler et al., 1974; Snir et al., 1978; Hermans et al., 1984). In the latter functions, it is often necessary to represent individual hydrogen atoms and lone pair electrons explicitly. However, this can be computationally time consuming and requires the estimation of coordinates in large molecules as hydrogen atoms are rarely observed experimentally and lone pair electrons are never seen experimentally. The hydrogen-bond term in the former functions varies, although usually it is an exponential (Lippencott and Schroeder, 1955) or 12-10 (Weiner et al., 1984, Momany et al., 1975) distance dependent function. In the GRID method (Goodford, 1985), an explicit hydrogen-bond function was used, and its further development and application are described in this paper.

THE GRID METHOD

This is a computational procedure which detects energetically favourable ligand binding sites on a chosen molecular target of known structure. The target molecule may be a protein, a polysaccharide, a nucleic acid or a molecule of low molecular weight. The interaction energy of a probe group with the target molecule is calculated at evenly spaced points throughout a chosen region of interest in or near the target molecule. The probe group is an 'extended atom' such as an amine nitrogen atom, a carbonyl oxygen atom or an hydroxyl group. The calculated energies are probe-specific and can be converted into energy contours around the target molecule and displayed using computer graphics.

An empirical pairwise energy potential is employed in program GRID. It is the sum of three terms: a Lennard-Jones term, E_{lj}; an electrostatic term, E_{el}; and a hydrogen-bonding term, E_{hb}. The total energy, E, at any particular grid point, is the sum of the interaction energies of the probe group with each atom of the target molecule.

$$E = \Sigma E_{lj} + \Sigma E_{el} + \Sigma E_{hb}$$

where

$$E_{lj} = A/r^{12} + B/r^6$$
$$E_{el} = q_p q_t / K\alpha \{1/r + [(\alpha-\beta)/(\alpha+\beta)]/(r^2 + 4s_p s_t)^{\frac{1}{2}}\}$$
$$E_{hb} = Er \times Et \times Ep$$

where
A and B are parameters defining the types of interacting atoms;
r is the separation of the probe group and a hydrogen-bonding atom in the target in Å;
K is a combination of geometrical factors and natural constants;
q_p and q_t are the electrostatic charge of the probe group and the target atom respectively;
s_p and s_t are the depths of the probe and the target atom, respectively, in the target molecule (Goodford, 1985)
α and β are the dielectric constants of the solvent and the target molecule respectively.
Er, Et and Ep are functions, respectively, of the length, r, of the hydrogen-bond; the angle, t, made by the hydrogen-bond at the target atom; and the angle p, made by the hydrogen-bond at the probe group.

THE HYDROGEN-BOND FUNCTION IN PROGRAM GRID

This has been modelled to fit the many experimental observations, which were mostly made by x-ray and neutron diffraction, of a variety of hydrogen-bonding molecules, including amino-acids, nucleotides, saccharides, and proteins.

The distance dependence of the hydrogen-bond is given by the function Er:

$$Er = C/r^8 + D/r^6$$

where C and D describe the hydrogen-bonding characteristics of the particular atoms making the hydrogen-bond. This function was chosen after testing a 12-10 function, which resulted in a hydrogen-bond length that was too narrowly defined when compared to the experimental observations of Artymiuk and Blake(1981), and a 6-4 function which was so broad that it was sometimes possible to insert another atom between the hydrogen-bonding atoms. A cutoff energy of -0.25 kcal/mol is applied to prevent the calculation of many very weak hydrogen-bonds.

The angular dependence of the hydrogen-bond is modelled as the product of two terms, Ep and Et. If the probe has the capacity to make only one hydrogen-bond, Ep=1.0 as the probe is assumed to orient itself to form a linear hydrogen-bond to the target molecule. If more than one hydrogen-bond can be formed by the probe group, Ep takes values between 0 and 1. Et depends on the chemical nature of the target atom. It has been derived for each individual type of atom by fitting to the variation of the probability of hydrogen-bond formation with angle t, which was calculated from the experimentally observed distribution of hydrogen-bonds at the target atom considered. These functions favour the formation of linear hydrogen-bonds in the direction of a lone pair or hydrogen atom of the target atom. The decrease in bond energy as the hydrogen-bond deviates from linearity varies according to the type of target atom, and is chosen to fit the structural observations. For instance, hydrogen-bonds to sp^2 hybridized oxygen atoms are modelled to be more constrained to the lone pair directions than those to sp^3 hybridized oxygen atoms.

HYDROGEN-BONDING IN THE CATECHOLAMINES

The catecholamines are pharmacologically important molecules which have been studied in this work in order to demonstrate the properties of the energy function used in the GRID method.

Figure 1 shows the crystal structure of L-noradrenaline (Carlstrom and Bergin, 1967) with energy contours at -3.1 kcal/mol for the interaction with a nitrogen NH probe which is able to donate one hydrogen bond. This type of nitrogen group occurs in the backbone amide group of proteins. The phenolic oxygens of the catecholamine molecule are assumed to have a preferred trigonal conformation due to the back donation of one lone pair to the aromatic ring, and are therefore assumed to be able to donate one hydrogen bond and accept one hydrogen bond. The energy contours close to the two phenolic oxygens lie in the plane of the aromatic ring at ±60° to the C-O bonds. Contoured regions occur on both sides of each C-O bond because program GRID allows the hydrogen atom and the lone pair of each phenolic hydroxyl group to exchange positions within the plane of the aromatic ring in order to make the most energetically favourable interaction with the incoming probe group. Thus, the target molecule is able to adapt to changes in its local environment caused by the approach of the probe group. The presence of both phenolic oxygens is necessary for full beta-agonist activity (Caron et al., 1978), so these energy contours may show regions where interaction with the β-adrenergic receptor occurs.

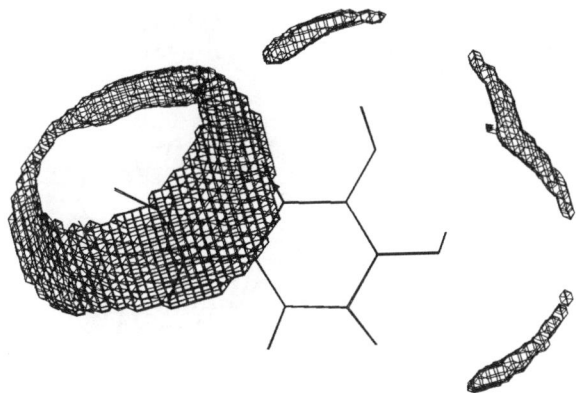

Figure 1. L-noradrenaline with energy contours at -3.1 kcal/mol for an interacting nitrogen NH probe.

The alkyl-hydroxyl group of L-noradrenaline is sp^3 hybridized with two lone pairs and one hydrogen atom. Program GRID considers the rotation of this group at the tetrahedral angle around the C-O bond, and this gives rise to the circular contour in Figure 1. which shows where a linear hydrogen-bond from the NH probe is favoured. The alkyl-hydroxyl group is critically important in the binding and activity of noradrenaline, and these properties are very dependent on its absolute stereochemistry. Figure 2 shows an energy map for the less active D-isomer contoured at the same energy level. The L- and D-isomers are arranged with the catechol and nitrogen moieties in equivalent conformations in Figures 1 and 2. The displaced circular contour in Figure 2 is not complete due to steric interactions with the ammonium nitrogen group which is closer to the alkyl-hydroxyl oxygen in this conformation of the D-isomer than it is in the L-isomer in Figure 1. Figure 3 shows a selectivity map of the difference in interaction energy of the probe with the two isomers. This shows two regions where a nitrogen NH probe can bind to the L-isomer more strongly than to the D-isomer. The gap between the two regions shows where the rings of contours around the alkyl-hydroxyl group shown in Figures 1 and 2 overlap and mutually cancel each other out in the selectivity map. The small 'bite' on the lower left-hand side of the ring contours shows a region where, in the L-isomer a hydrogen-bond is made to the alkyl-hydroxyl group, but in the D-isomer in this conformation, a hydrogen-bond cannot be made to the alkyl-hydroxyl group, and is instead made to the adjacent phenolic oxygen, thus diminishing the effect on the binding energy of moving the alkyl-hydroxyl group.

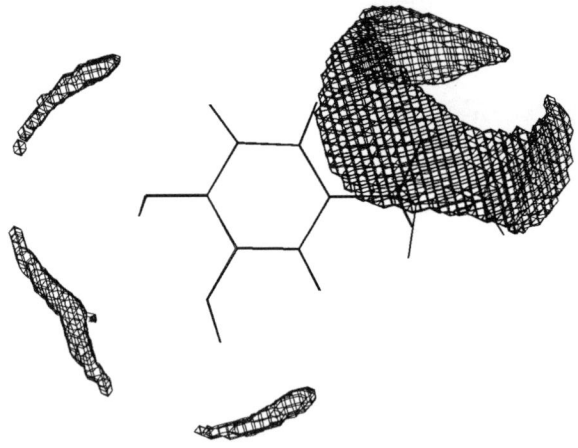

Figure 2. D-noradrenaline with energy contours at -3.1 kcal/mol for a nitrogen NH probe. The molecule is shown in an equivalent conformation to that of L-noradrenaline in Figure 1.

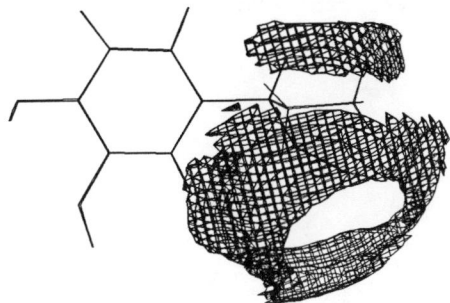

Figure 3. A selectivity map contoured at -2.3 kcal/mol showing regions where the binding energy is at least 2.3 kcal/mol more favourable for a nitrogen NH probe interacting with L-noradrenaline than with D-noradrenaline in the equivalent conformation. See text.

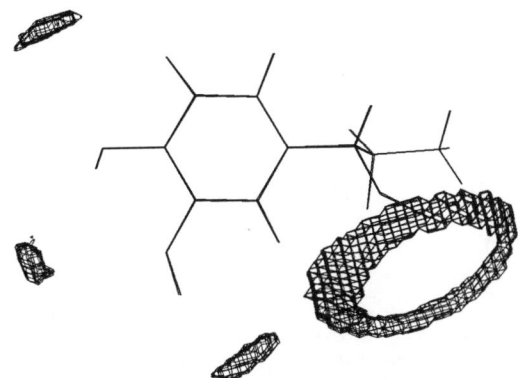

Figure 4. Energy contours at -3.1 kcal/mol for the interaction of L-noradrenaline with a nitrogen probe which is able to accept one hydrogen-bond only.

In Figure 4, the energy contours at the same energy level as in Figure 1 show the interaction, with L-noradrenaline, of a nitrogen probe which has one lone pair, and is therefore able to accept one hydrogen-bond. The contoured regions are smaller than in Figure 1, because noradrenaline is donating a hydrogen bond and its hydrogen atoms are able to exert greater directional constraints on the geometry of the hydrogen-bonds than its lone pair electrons can. In addition, the contours near the phenolic oxygens are of a different shape to those in Figure 1. This is because the angular term Et has a different functional form when these oxygen atoms donate compared to when they accept an hydrogen-bond.

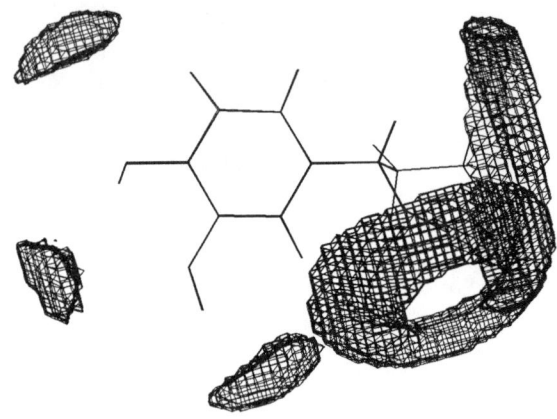

Figure 5. Energy contours at -2.3 kcal/mol showing the interaction of L-noradrenaline with a nitrogen probe which is able to accept one hydrogen-bond only.

At the less negative contour level of -2.3 kcal/mol, shown in Figure 5 for the same nitrogen acceptor probe group, an additional ring of contours is seen near the nitrogen atom of noradrenaline. This nitrogen is assumed to be ionised at physiological pH (Ganellin, 1977) and therefore able to donate three hydrogen-bonds. Its hydrogens are allowed to rotate at the tetrahedral angle to the C-N bond giving rise to the extra ring of contours. These contours are not seen in Figure 4 at -3.1 kcal/mol because a hydrogen-bond to the nitrogen of noradrenaline is weaker than one to an oxygen atom.

In Figure 6, the interaction of a nitrogen probe, which can accept one hydrogen-bond, with adrenaline (Carlstrom, 1973) is seen at the same energy contour level as in Figure 5. Here, the N-methyl group prevents rotation of the two hydrogens attached to the nitrogen atom in adrenaline. Consequently, the contours form two distinct regions in the direction of the hydrogen atoms. One of the contoured regions is adjoining the circular contours around the alkyl-hydroxyl oxygen group.

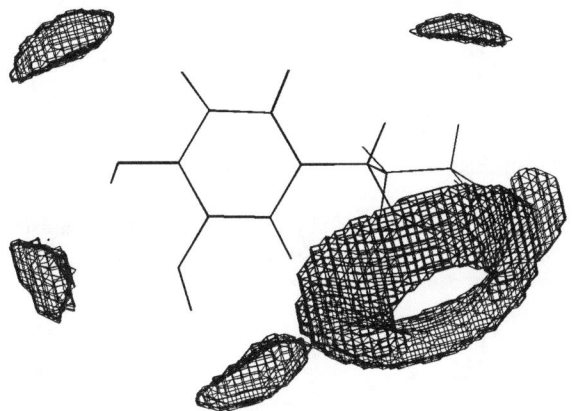

Figure 6. Adrenaline with energy contours at -2.3 kcal/mol for the interaction with a nitrogen probe which is able to accept one hydrogen-bond only.

These energy maps show that atoms with different chemical properties have distinct hydrogen-bonding geometries and strengths which may cause molecules, such as the catecholamines, to show a high degree of specificity in their intermolecular interactions.

HYDROGEN-BONDING IN PROTEINS

The Histidine Residue

Receptors of drug molecules are often proteins and some specific hydrogen-bonding functions have been devised for protein atoms. One protein residue of particular interest is histidine. At physiological pH, it may exist in one of two possible tautomers. In program GRID, both imidazole nitrogens are assumed to be able to donate or accept a hydrogen-bond. The nitrogen to which a hydrogen atom is actually attached is determined by the coordinates and hydrogen-bonding capability of the probe group. The movement of a hydrogen atom between the two ring nitrogens results in a charge redistribution in the imidazole ring and this is taken into account by adjustment of the electrostatic component of the total interaction energy. Thus, program GRID considers the polarization of histidine in the target molecule caused by the presence of an interacting probe group.

At non-physiological pH, the hydrogen-bonding capacity of histidine may be different. At low pH, the imidazole ring will be predominately protonated and able only to donate hydrogen-bonds. At high pH, it will generally be able only to accept hydrogen-bonds.

Cytochrome P450cam

It is important that the energy function of program GRID should be able to reproduce experimental observations. One system which demonstrates the agreement attainable is cytochrome P450cam. It is a monooxygenase haem protein whose crystal structure has been solved to 1.63Å (Poulos et al., 1987).

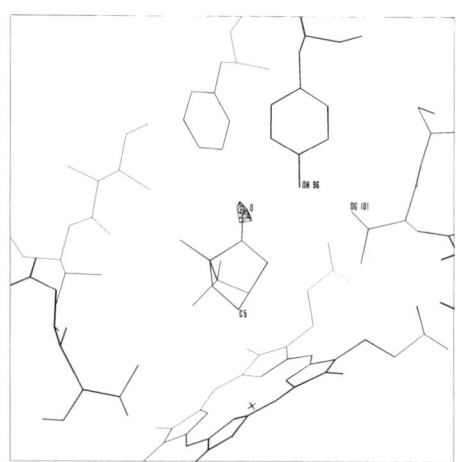

Figure 7. The active site of cytochrome P450cam. The energy contours are at -2.3 kcal/mol for a carbonyl oxygen probe interacting with the protein.

The active site of cytochrome P450cam is shown in Figure 7, with the natural substrate, camphor, situated above the haem ring system. The contours shown are for a carbonyl oxygen probe and were calculated with the camphor substrate removed from the active site. The contours show where the strongest hydrogen-bond is made to tyr 96 and correspond closely with the observed position of the camphor oxygen atom. The interaction of the camphor with tyr 96 is thought (Poulos et al.,1985) to orient the camphor molecule in the correct position for hydroxylation which is a stereo and regiospecific interaction in which only 5-exo-hydroxylation occurs. No energy contour is seen on the other side of the phenolic C-O bond of tyr 96 because an intramolecular hydrogen-bond is made between tyr 96 and the hydroxyl group of thr 101. Therefore, program GRID has been able to clearly locate the unique position of the camphor carbonyl oxygen atom in the active site of cytochrome P450 and to show that the binding energy at this position is much stronger than at any other point in the active site. Program GRID is also able to identify binding sites and indicate their relative selectivity for different ligands in other target molecules with more complex hydrogen-bonding capabilities.

CONCLUSION

A set of hydrogen-bond functions has been devised as part of an empirical energy function used to determine energetically favourable ligand-binding sites on a chosen target molecule. It has been modelled on experimental data on hydrogen-bond formation in crystals. The hydrogen-bond function is dependent on the chemical nature of the hydrogen-bonding atoms and may take into account the motion of hydrogen-bonding hydrogen atoms and lone pairs. It allows these hydrogens and lone pairs to position themselves in the target molecule according to the chemical identity and proximity of the interacting ligand. In both large and small target molecules, the total energy function, incorporating the new hydrogen-bond terms, has been shown to be able to reproduce experimental observations. It should thus be of value in predicting ligand binding and in computer-aided drug design.

REFERENCES

Artymiuk PJ, Blake CCF (1981) Refinement of Human Lysozyme at 1.5Å Resolution Analysis of Non-bonded and Hydrogen-bond Interactions. J Mol Biol 152:737-762.
Brooks BR, Bruccoleri RE, Olafson BD, States DJ, Swaminathan S, Karplus M (1983) CHARMM: A program for macromolecular energy minimization and dynamics calculations. J Comp Chem 4:187-217.
Carlstrom D, Bergin R (1967) The Structure of the Catecholamines. I. The Crystal Structure of Noradrenaline Hydrochloride. Acta Cryst 23:313-319.
Carlstrom D (1973) The Structure of the Catecholamines. IV. The Crystal Structure of (-)-Adrenaline Hydrogen (+)-Tartrate. Acta Cryst B29:161-167.

Caron MC, Mukherjee C, Lefkowitz RJ (1978). β-Adrenergic receptors: SAR determined by direct binding studies. In Smythies JR, Bradley RJ (eds): "Receptors in pharmacology," New York: M. Dekker, pp97-121.

Dreyfus M, Pullman A (1970). A Non-Empirical Study of the Hydrogen Bond between Peptide Units. Theor Chim Acta 19:20-37.

Ganellin CR (1977) Relative concentrations of zwitterions and uncharged species in catecholamines and the effect of N-substituents. J Med Chem 20:579.

Goodford PJ (1985). A computational procedure for determining energetically favourable binding sites on biologically important macromolecules. J Med Chem 28:849-857.

Hagler AT, Huler E, Lifson S (1974) Energy functions for peptides and proteins.I. Derivation of a Consistent Force Field Including the Hydrogen Bonds from Amide Crystals. J Am Chem Soc 96:5319-5327.

Hermans J, Berendsen HJC, van Gunsteren WF, Postma JPM (1984) A Consistent Empirical Potential for Water-Protein Interactions. Biopolymers 23:1513-1518.

Lippencott ER, Schroeder R (1955) A One-dimensional model of the hydrogen-bond. J Chem Phys 23:1099.

Momany F, McGuire R, Burgess A, Scheraga H (1975) Energy Parameters in Polypeptides VII. J Phys Chem 79:2361.

Pauling L (1949) J Chim Phys 46:435.

Poulos TL, Finzel BC, Gunsalus IC, Wagner GC, Kraut J (1985) The 2.6Å Crystal Structure of Pseudomonas putida Cytochrome P-450. J Biol Chem 260:16122-16130.

Poulos TL, Finzel BC, Howard AJ (1987). High-resolution Crystal Structure of Cytochrome P450cam. J Mol Biol 195:687-700.

Snir J, Nemenoff RA, Scheraga HA (1978) Revised empirical potential for conformational, intermolecular and solvation studies. J Phys Chem 82:2497.

Vedani A, Dunitz JD (1985) Lone-Pair Directionality in Hydrogen Bond Potential Functions of Molecular Mechanics Calculations: The Inhibition of Human Carbonic Anhydrase II by Sulfonamides. J Am Chem Soc 107:7653-7658.

Weiner SJ, Kollman PA, Case DA, Singh UC, Ghio C, Alagona G, Profeta S, Weiner P (1984) A new force field for molecular mechanical simulation of nucleic acids and proteins. J Am Chem Soc 106:765-784.

Weiner SJ, Kollman PA, Nguygen DT, Case DA (1986) An All Atom Force Field for Simulations of Proteins and Nucleic Acids. J Comp Chem 7:230-252.

HYDROGEN BONDING IN PROTEIN LIGAND INTERACTIONS:
A THEORETICAL DIMENSION OF ASPARTIC PROTEINASE
CRYSTALLOGRAPHY.

Amiram Goldblum

Department of Pharmaceutical Chemistry, School of Pharmacy, Hebrew University of Jerusalem, Jerusalem, ISRAEL 91120

INTRODUCTION

Much of the current interest in aspartic proteinases is due to a major effort to produce efficient inhibitors of renin which could become clinically useful antihypertensive drugs. Renin, an aspartic proteinase, is known to be involved in the renin-angiotensin cascade, which may lead to essential hypertension.

Many common features were found among enzymes of this family:
1) An aspartic acid pair in the active site (pepsin nos. 32 and 215).
2) High acidity of aspartates: first $pKa \approx 1.0$, second $pKa \approx 4.5$, the lower attributed to ASP-32.
3) Optimal activity of enzymes at acidic pH (renin - in neutral).
4) Conservation of residues in the active site: D32-T33-G34-S35 ; D215-T216-G217-T218; (human renin has A218).
5) Family inhibitor - pepstatin (less active with renin).
6) Homology of primary structures.

This information was obtained mostly from kinetic studies and primary structure determination (Fruton, 1976; Tang, 1979; Foltman, 1981). Much more is known today about the crystal structure of aspartic proteinases , most of them from bacterial origin. The crystal structures of *endothia* pepsin (Pearl and Blundell, 1984), *rhizopus* pepsin (Suguna et al., 1987) and penicillopepsin (James and Sielecki, 1983) have been studied with high resolution and detailed atomic coordinates were deposited in the Brookhaven data bank, including water oxygens' positions for the first two structures mentioned. Much more information appeared in the litterature about the crystallography of some aspartic proteinases and their complexes with reversible and irreversible inhibitors. Based on such studies, it is generally accepted that the crystal structures of pepsins may serve as valuable "starting points" for modeling renin and its interactions (Blundell et al., 1987).

Such studies conclude that the active sites of the aspartic proteinases are rigid , mostly due to the intricate internal hydrogen bonding, and that this rigidity is not destructed by inhibitors , even not by irreversible ones which react chemically with the aspartates. Only a small part of the enzyme, the "flap" moves a little upon the introduction of an inhibitor to the active site. Its role is assumed to be connected to both H-bonding and water extrusion upon inhibitor binding , but the role is unknown with respect to substrates. Tyrosine-75 from the "flap" may have a role in proton transfer during catalysis.

Despite the impressive amount of crystallographic data and kinetic studies of aspartic proteinases, the details of the mechanism of reaction are not clear, yet. It is yet unknown how exactly do the aspartic residues and water molecules involve in catalysis, whether the catalysis is driven by direct nucleophilic attack of an aspartate or is it of the general-acid-general base type, is the active site neutralized by a counter cation , what is the source of the pH profile for the enzymes' reactions, etc. One of the barriers for understanding more about the mechanism is the lack of information about hydrogen positions, mostly those which are involved in H-bonding. To learn more about their positions we must then use , at the moment, theoretical methods, of which the most reliable are quantum mechanical studies. The method should give reasonable descriptions of both the structure and the energetics of hydrogen bonds, since much of the hydrogen bond character is due to the electrostatic nature of the bond and to its sensitivity to surrounding charges and to polarization. The best method would be, then, to subject the whole enzyme to *ab-initio* computations with extended basis sets. But, as this is impracticle, we had to resort to semiempirical quantum mechanical methods, as well as to reduced representations of the protein which will still guard the important characteristics required for such a study to remain meaningful. However- semiempirical methods could not adequately describe the energies and structures of hydrogen bonds, even as small as the H-bond in a water dimer. Of these methods, MNDO (Dewar and Thiel, 1977, modified neglect of differential overlap) is known to reasonbly reproduce reaction mechanisms through enthalpies and entropies of reaction. But MNDO over-estimates bond lengths and underestimates stabilities of hydrogen bonds. This is due to an excessive core-core repulsion term between the proton and the proton acceptor in an H-bond (X—H----Y):

(1) $E_{HY} = Z_H Z_Y (S^H S^H, S^Y S^Y)[1 + f(R_{HY})]$

(2) $f(R_{HY}) = R_{HY} \exp(-\alpha_Y R_{HY}) + \exp(-\alpha_H R_{HY})$

where α_H and α_Y are empirically fitted parameters. Following a recent suggestion (Burstein and Isaev, 1984) we modified MNDO to enable calculations of single and multiple hydrogen bonds for steady states (Goldblum, 1987) as well as for proton transfer reactions (Goldblum, 1988b).

For a proton transfer from X to Y:

$$X—H- - -Y \longrightarrow X- - -H—Y$$
$$(1) \quad\quad\quad\quad (2)$$

we have, in the steady state approach:

(3) $f(1) = R_{XH} \exp(-\alpha_X R_{XH}) + \exp(-\alpha_H R_{XH}) + \exp(-2.0 R^2_{YH})$
and
(4) $f(2) = \exp(-2.0 R^2_{XH}) + R_{YH} \exp(-\alpha_Y R_{YH}) + \exp(-\alpha_H R_{YH})$

For such a transfer, a decision must be made as to which of the electronegative atoms is bonded to the proton, and which is H-bonded to it. For gas phase complexes the solution was to attach eq.3 for $R_{XH} < 1.3 \text{Å}$ and eq.4 for $R_{YH} < 1.3 \text{Å}$. However, for "real" situations in proteins the X---Y distance may be larger, and then a discontinuity should be expected for $R_{XY} > 2.6 \text{Å}$, if the transfer of a proton is along the X---Y axis. To eliminate this problem we suggested to "mix" bonding and H-bonding according to the relative movement of the proton along the transfer path:

(5) $f(R_{XH})^W = (1-\lambda)*f(R_{XH}) + \lambda*\exp(-2.0 R^2_{XH})$

(6) $f(R_{YH})^W = \lambda*f(R_{YH}) + (1-\lambda)*\exp(-2.0 R^2_{YH})$

where $\lambda = (R_{XH}-R^0)/(R^1-R^0)$ gives the "weights" of the proper proton positions, with R^0 being the X—H initial distance, R^1 their final distance following the transfer, and R_{XH} the actual distance. It is a valid approach only for transfers of the "double-well" type. The correction was applied to several systems which enabled comparisons to experiments or were studied by *ab-initio* with extended basis sets.

For the water dimer, MNDO/H predicted correctly that the linear (C_s) dimer is the most stable, with a total free energy of 3.17 kcal-mol^{-1}, compared to the experimental value of 3.3 kcal-mol^{-1} (Goldblum, 1987). Many other interactions which may occur in biochemical reactions have been surveyed, and thus the necessary "building blocks" for larger calculations were obtained, as well as an appreciation of the possible limitations.

RESULTS

The first problem we have studied was the differential acidity of the two aspartates: Is one of them much more acidic than the other, as suggested by experiments? Our largest model for these studies included all the eight residues which hold the aspartic pair together (Figure 1):

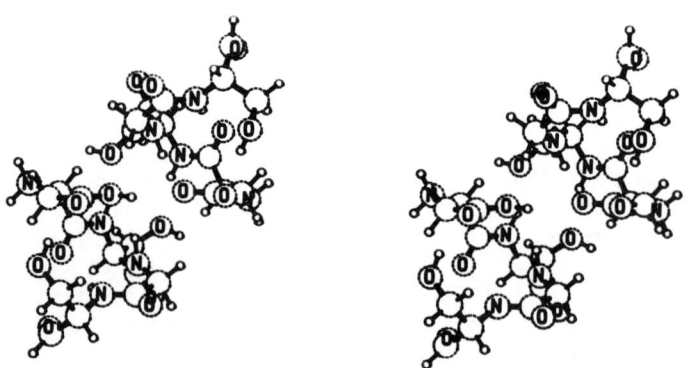

Figure 1. Stereo plot of the model for the active site of apsartic proteinase, from *endothia* pepsin coordinates with optimization of all hydrogen positions.

In order to study the effects of protein residues on the acidity (Goldblum, 1988a), the structure was initially reduced to just the two aspartic side chains, in their crystal positions from X-rays. Heavy atoms (carbons, oxygens, etc.) were left in their crystal positions, and hydrogen positions were geometry optimized for the two possible alternative ionizations — ASP32$^-$----H-ASP215, and ASP32-H---ASP215$^-$. The results for the three pepsins which were reported with higher resolution demonstrated that in all three enzymes a preference to be ionized by 2-3 kcals was found for the aspartic acid no. 32, and the calculated entropies were nearly similiar for the two alternatives in each case.

The next step was to study the influence of water molecules on the relative acidity of the aspartates. Three water molecules, the maximum which could be accomodated in close proximity to the aspartic pair, were introduced to interact with the anionic couple— one at a time, with full optimization of their positions. The results show a very small influence on the relative acidity, and thus we could conclude that water does not support a change in the basic different preference of this pair to ionize.

This preference was studied by energy partitioning of the interactions among the acids in the two alternatives (Table 1). Single contributions give the values for each of the species in its optimized position while disregarding the presence of its neighbour. The pair stabilizations (in parenthesis) were calculated by the difference between the energy of the pair and the single contributions to it. Similar energies of the ionized acids are found in the two alternatives (-110.6 kcal-mol^{-1}) but protonated ASP-32 is more stable than protonated ASP-215 by 4.4 kcals, due to conformational differences. On the other hand, the pair stabilization from ionized 215 and protonated 32 contributes 6.8 kcals less than the interaction of ASP 32$^-$---H-ASP215.

TABLE 1. Single and pair contributions to the energy of interaction among the aspartic pair in two alternative ionizations, based on penicillopepsin coordinates.

	ASP 32$^-$----ASP 215	ASP 32----ASP 215$^-$
1. Single :		
ASP 32	-110.62	-91.48
ASP 215	-87.12	-110.68
2. Pair:	-210.59 (-12.85)	-208.15 (-6.00)

The difference in the geometry and charges of the alternative H-bonding schemes for the two ionizations are responsible for the excessive stabilization of ASP32⁻----ASP215 with respect to the other ionization: The oxygen-oxygen distance is constant in both, but H----O⁻ is shorter (2.022Å), the O---H----O⁻ angle is larger (145.7°) and the O---H----O⁻ charges are larger: -0.406, +0.391 and -0.689, in that order. For ASP32---ASP215⁻ the respective values are 2.407Å, 110.1°, with charges of -0.363, +0.355 and -0.661 electrons. This combination of geometric and coulombic factors is thus the source of a basic preference for one ionization option of the two. Any calculation which is not based on a quantum mechanical description must attempt to mimic such a result with proper potential functions for hydrogen bonding.

To learn further about the active site in aspartic proteinases, the basic model of two side-chains was enlarged to include the influence of surrounding residues. This study was done with each of the neighbours separately and in combination with other residues. Finally we studied the relative ionization options in the "full" active site, which included eight residues. Again, the positions of heavy atoms were left in their crystal coordinates, while all hydrogen positions were fully optimized. It was found that the presence of other residues **changes** the basic tendency of ASP 32 to be ionized, and reverses it in favour of ASP 215, so that its enthalpy of ionization is more stable by about 0.5 kcal-mol^{-1}. This finding is contrary to most of the ideas expressed in the litterature about the acidity preference in the active site of aspartic proteinases. However, a similar idea was recently raised in favour of equal sharing of the proton by the two aspartic moeities (Pearl and Blundell, 1984). Such sharing requires a low energy transition state for proton transfer, and we are currently engaged in an effort to study this mechanism. If it is correct, than the acidity in the active site of aspartic proteinases is a property of the full active site, and not of a single acid. The active site may thus react as if it were an organic diacid with two carboxylates in close proximity. The pKa's of many such organic diacids are known and should be compared to the acidity in the active site. To calculate those acidities for comparison one should include the effect of water. Disregarding this important contribution, we assumed "gas phase" ionizations according to:

(7) $AH_2 \longrightarrow H^+ + AH^{-1}$, for which

$$\Delta H_r = \Delta H_f(H^+) + \Delta H_f(AH^{-1}) - \Delta H_f(AH_2)$$

and (8) $AH^{-1} \longrightarrow H^+ + AH^{-2}$, for which

$$\Delta H_r = \Delta H_f(H^+) + \Delta H_f(A^{-2}) - \Delta H_f(AH^{-1})$$

Some results are given in Table 2. No proton acceptor was implied in the calculations, and the reaction enthalpies are thus high due to the enthalpy of the proton (365.7 kcal-mol^{-1}, Ford and Scribner, 1983). In this table, lower deprotonation enthalpies correlate well with lower pKas, for each of the two ionization steps. For the largest model of the active site the second ionization was not yet computed, but is expected to be in line with the values for smaller models. It is found , by the calculation, to be the most acidic of the diacids in this table.

TABLE 2. Enthalpies (kcal-mol^{-1}) and pKa values for ionizations of some diacids.

Name	ΔH_1 [a]	ΔH_2 [b]	pK$_1$	pK$_2$
diacid				
cyclobutane-1,2-COOH	329.17	441.70	1.12	7.63
maleic	329.19	446.96	1.83	6.07
fumaric	344.61	421.91	3.03	4.44
o-hydroxybenzoic	339.61	460.74	2.97	13.40
models for active site				
aspartic side chains	338.57	439.13	≈1.0	≈4.7
partial model[c]	315.66	399.93	"	"
partial model + H$_2$O	314.80	395.04	"	"
largest model	312.10	d	"	"

[a] calculated by eq.7. [b] by eq.8. [c] includes reduced representation for peptide bonds and for side chains. [d] not calculated.

With the large model for the active site, we have recently undertaken a study of its interaction with a model of the family inhibitor, pepstatin. The central statine residue in this pentapeptide has a 3S configuration on the carbon carrying the alcohol function, at P_1-P_1' (Figure 3).

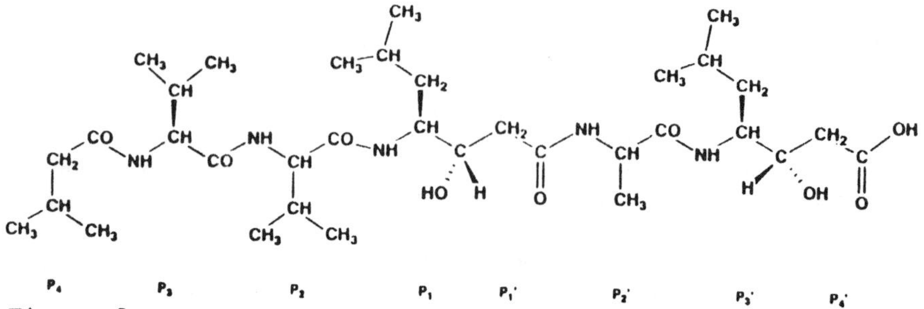

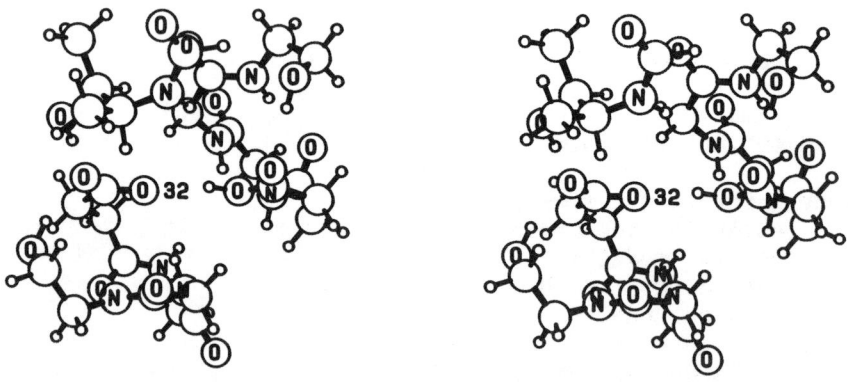

Figure 3. Pepstatin, the natural inhibitor of aspartic proteinases.

Reversal of this configuration leads to a large reduction (2000 fold) in binding affinity to pepsin (Rich, 1985). Thus it was a reasonable test for the methodology, to find why such a difference occurs. To follow the affinities of the two configurations we used a model of the central statine residue only, in both configurations. The optimized position of (3S)-statine with the active site is shown in Figure 4.

Figure 4. Interaction of (3S)-statine with the active site of *endothia* pepsin. Two H-bonds are formed by OH and NH of the inhibitor to the active site (stereo plot).

Starting from this final position, the configuration was reversed to (3R)-statine, and the strong H-bonding interaction could not be maintained further due to collision between the statine methyl group and the active site (Figure 5). The final position of (3R)-statine is further away from the close interaction found for the (3S) configuration.

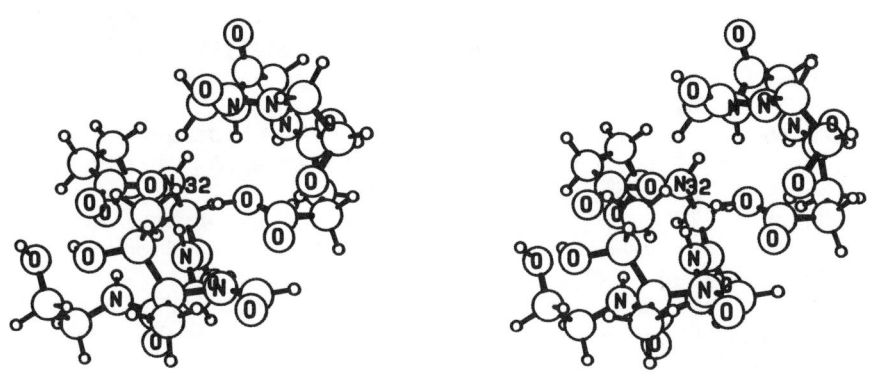

Figure 5. Repulsion between (3R)-statine and the active site of *endothia* pepsin (stereo plot).

We may thus conclude that a combination of atomic positions from protein crystallography with a reliable quantum mechanical method which can properly describe hydrogen bonds proves to be useful for probing mechanistic problems of protein activity

REFERENCES

Blundell TL, Cooper J, Foundling SI, Jones DM, Atrash B and Szelke M (1987). On the rational design of renin inhibitors: X-ray studies of aspartic proteinases complexed with transition state analogues. Biochemistry 26: 5585-5590.
Burstein KYa and Isaev AN (1984). Theoret Chim Acta 64: 397-405.

Dewar MJS and Thiel W (1977). Ground states of molecules. 38. The MNDO approximations and parameters. J Am Chem Soc 99: 4899-4906.

Foltman b (1981). Gastric proteinases- structure, function, evolution and mechanism of action. Essays in Biochemistry 17: 52-84.

Ford GP and Scribner JD (1983). MNDO calculations of proton and methyl and ethyl cation affinities of neutral carbon, nitrogen, and oxygen bases. J Comput Chem 4: 594-604.

Fruton JS (1976). The mechanism of the catalytic action of pepsin and related acid proteinases. Adv Enzymol Relat Areas Mol Biol 44: 1-36.

Goldblum A (1987). Improvement of the hydrogen bonding correction to MNDO for calculations of biochemical interest. J Comput Chem 8: 835-849.

Goldblum A (1988a). Theoretical calculations on the acidity of the active site in aspartic proteinases. Biochemistry 27: 1653-1658.

Goldblum A (1988b). Calculation of proton transfers in hydrogen bonding interactions with semiempirical MNDO/H. J Molec Struct, in press.

James MNG and Sielecki AR (1983). Structure and refinement of penicillopepsin at 1.8A resolution. J Molec Biol 163: 299-361.

Pearl LH and Blundell TL (1984). The active site of aspartic proteinases. FEBS Lett 174: 96-101.

Rich D (1985). Pepstatin derived inhibitors of aspartic proteinases. A close look at an apparent transition-state analogue inhibitor. J Med Chem 28: 263-273.

Suguna K, Bott RR, Padlan EA, Subramanian E, Sheriff S, Cohen GH and Davies DR (1987). Structure and refinement of the aspartic proteinase from *rhizopus chinensis*. J Mol Biol 196: 877-900.

Tang J (1979). Evolution in the structure and function of carboxyl proteases. Molec Cell Biochem 26: 93-109.

COMPUTATIONAL STUDIES OF LIGAND/RECEPTOR INTERACTIONS

Sid Topiol and Michael Sabio

Department of Medicinal Chemistry, Berlex Laboratories, Inc., 110 East Hanover Avenue, Cedar Knolls, NJ 07927.

ABSTRACT

An analysis of common features of many proteins provided the basis for a general model for the origin of receptors (Topiol, 1987). In this model, receptors are derived from fully operational parent systems. The parent system is rendered inactive by "deletion" of some critical component, thereby converting it to a receptor for the deleted entity. Such a model could explain the use of common molecular machinery by different biological systems (e.g., receptors and enzymes), the origin of receptor subtypes, the use of common effector systems by different receptors, the sequence homologies between varied proteins, the relationship between endogenous ligands and biological "building blocks", and the selectivity and compatibility between natural receptors and endogenous ligands. Some examples of the use of this model to analyze a number of different biochemical systems and processes will be given. Preliminary insights from these studies as a guide to developing molecular models for the action of cyclic nucleotide second messengers will be discussed.

I. INTRODUCTION

Modern drug design can be viewed as moving into a new era. In the past, with little knowledge of the sequences and structures of receptor proteins and the associated mechanisms of ligand/receptor interactions, drug design was heavily dependent on the use of structure activity relationships. Recently, however, there has been a rapid increase in our knowledge of protein structure and action. It is clear that this will allow for more direct routes toward drug design. On the other hand, many protein sequences and many more protein structures will not be known for some time (DeLisi, 1988). Thus we will now be faced with the challenge of gleaning as much information as possible from systems we know in greater detail towards the understanding of "similar or related" systems.

One example of an important feature of many biochemical systems is the use of common entities and processes. This tendency of different biological systems to use similar entities and processes is continually reaffirmed by new evidence. A model has recently been proposed (Topiol, 1987) which could be used to provide possible explanations for this overwhelming tendency for the use of common molecular machinery in a diversity of receptor systems. In this scheme, beginning with this observation of the use of common molecular machinery, a model for the origin of receptor systems was developed in which ostensibly different but related receptor systems can be derived quite naturally from such common machinery and thus are in fact related to each other. It has also been shown how this scheme could relate two previously proposed, detailed models for drug action at 5-hydroxytryptamine and histamine H_2-receptors systems (Topiol, 1987; Osman et al., 1987; Weinstein et al., 1986). This model also explains the apparent coincidence between the components of proteins (e.g., amino acids) and endogenous ligands (e.g., neurotransmitters) as well as a number of other issues regarding receptor systems. This model is reviewed below along with some of the proposed implications. In particular, we outline how this model has been extended to relate catalytic and regulatory binding sites in the development of a unifying description of the role of the cyclic nucleotides adenosine-3'5'-cyclic monophosphate (cAMP) and guanosine-3'5'-cyclic monophosphate (cGMP) as agonists and substrates of various proteins (Topiol and Sabio, 1988).

II. THE DELETION MODEL

We begin by considering a complete functioning system, S, made up of a collection of subunits, α_i, which together perform some dynamic process, i.e., S is a dynamic system.

We now wish to design a means of inactivating and
reactivating this system, i.e., a control mechanism. The
simplest approach is to delete some element, α_i, necessary
for its functioning. The receptor subsystem $S(i)$ is now
defined as that part of the system derived from S by the
removal of the element α_i. Because α_i is required for the
complete functioning system, the derived subsystem $S(i)$ no
longer undergoes the dynamic process characteristic of S,
i.e., $S(i)$ is inactive. We now observe that $S(i)$ serves as
a receptor system for α_i such that binding of α_i
reactivates the system by converting $S(i)$ back to the fully
functional system S. It is immediately obvious how such
deletions provide an explanation of the relationship of
endogenous ligands, such as 5-HT and histamine, to the
components of biological systems (the amino acids
tryptophan and histidine, respectively, in this example).
The similarity between these naturally occurring ligands
and the biological building blocks is clearly not
fortuitous in this model. (It should be understood that
the term deletion is used here in a general sense. It
should not be confused with the more common usage of the
term deletion whose connotation is the removal of an amino
acid. For instance, removal of the indole group of
tryptophan to form glycine, which corresponds to a
substitution of one amino acid for another, is considered a
deletion herein.)

Consider now possible mechanisms of ligand/receptor
binding and activation (Fischer, 1894; Koshland, 1958,
1959, 1963; Burgen et al., 1975). The receptor site,
ligand, and some coupling mechanism constitute a switching
system often described as a lock-and-key system (see Figure
1). When the ligand binds to the receptor site it
activates the system via this coupling mechanism. The
coupling mechanism need not be a separate physical entity
from the receptor site. Several questions regarding the
origin and efficiency in the design of the switching system
exist. If we consider these elements in the arbitrary
order ligand, receptor site, and coupling mechanism, the
first question is how to select or design the ligand.
Next, given the ligand, it is necessary to design an
efficient and selective complement to it, i.e., the
receptor site. Finally, it is also necessary to design a
coupling mechanism such that when the ligand is bound to
the receptor site, the system S can be activated. All of
these problems are resolved by the deletion model for the
development of receptors. A ligand, as a separate entity,
(i.e., not otherwise a part of the system S) need not be

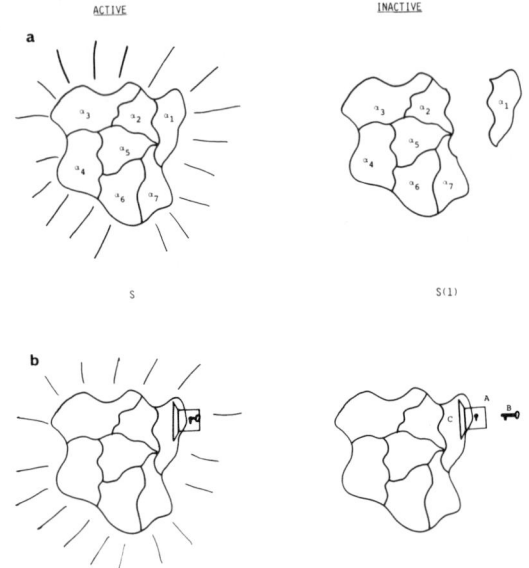

Figure 1. Schematic representations of active and inactive forms of receptor systems for: a) the deletion model for an α_i receptor; b) the classical lock-and-key model.

selected or designed. Similarly, the ability of the ligand α_i to reactivate S(i) is no longer an issue since it does not need to couple to a system but rather is an integral part of the system S. No new entities need to be invoked. The compatability of the ligand with the receptor system S(i) is ensured.

We consider below a number of different manifestations of this model using (a) a common parent system to generate different receptor systems; (b) different parent systems to give rise to receptor systems for the same ligand; or (c) a given parent system to be subdivided in a number of different ways.

a. A Common Parent System for Different Ligands

Beginning with the initial basic system S, two (or more) different receptor subsystems S(i) or S(j) can be created through removal of α_i or α_j, where α_i and α_j could be any of the essential components of S. Thus S(i) and S(j) are different receptor subsystems derived from the same parent system S. This explains the conservation of the same functioning elements in different biological systems S, which are common (or closely related) parent systems to different receptor subsystems. Moreover, when we consider that the two bound ligand/receptor pairs, $[\alpha_i, S(i)]$ and $[\alpha_j, S(j)]$, lead to the same parent system S, it becomes clear how different receptor systems can couple to common effector systems. An example might be a set of different hormone receptors all of which may be derived from a common parent system and form part of the adenylate cyclase system. In each case, the binding of the ligand to its receptor generates the same parent system and therefore precipitates the same molecular events. Some interesting examples of the presence of such common elements in different biological systems have been reported recently (Bourne, 1986).

It is also interesting that the deleted entity, α_i, need not correspond to a physical entity; in the rhodopsin system the incident light and the energy provided by it could be considered α_i. In zymogen activation, e.g., conversion of chymotrypsinogen to chymotrypsin, α_i might correspond to the proper positioning of key functional groups.

b. Different Parent Systems for a Common Ligand

It is possible for two parent systems to give rise to receptor systems for the same ligand. In general we may assume the existence of more than one parent system (denoted SA, SB, ...,SX,...). Each parent system gives rise to its own set of receptor systems. The sets of receptor systems derived from each of these parent systems SX need not be mutually exclusive. Thus two or more parent systems could give rise to a receptor system for the same ligand, α_i, i.e. SA(i) and SB(i). If the environment of the deleted entity is similar in the two parent systems then we might expect the biochemical properties to be similar. For instance, histamine H_2-receptors in the brain and in the gastrointestinal tract may be related in this way. Perhaps more interestingly on the other hand, the two receptor systems derived from SA and SB could have significant differences in the local environment of the deleted unit α_i. This could form the basis for receptor subtypes where each subtype is derived from a different parent system. The histamine H_1- and H_2-receptors could for example be related in this way. As the same building blocks are used for different biochemical systems, it is not difficult to imagine deletion of the same entities from different systems. Furthermore, for receptor proteins, there is an inherent maximum of twenty possibilities for single amino acid 'deletions', of which it is intuitively obvious that certain deletions are more likely to be effective as deletions (e.g., histidine, serine, threonine, aspartate, etc.). Thus, the deletion of common entities to form receptors from different parent systems is easily envisioned.

c. Division of a Parent System into Multiple Components

The notion of deleting components of a common parent system could also be extended to multiple deletions. For instance, deletion of two entities from a common parent system would lead to the three components $S(i,j)$, α_i and α_j. The assignment of these labels to the three components is arbitrary as the system is not operative until all three pieces are bound together. These three pieces could, for instance, correspond to a ligand, receptor, and effector. Furthermore, the order in which the three components recombine is also not likely to be critical (see, e.g., the following example).

Recent studies on Bacillus amyloliquefaciens subtilisin (Carter and Wells, 1987) could serve as an interesting example of how a system corresponding to two deletions could be designed. Based on the x-ray structure of subtilisin they modeled the binding of a variant of the peptide substrate L-Phe-L-Ala-L-Ala-L-Tyr-L-Gly-L-Phe (FAAYGF, with similar notation used below; representing residues P4 to P2' respectively) in which the Ala in the P2 position was replaced by His. It was observed that the His in the P2 position could overlap with His_{64} of the catalytic triad of subtilisin: Ser_{221}, His_{64} and Asp_{32}. His_{64} was then converted to an Ala by site directed mutagenesis (see Figure 2b and 2c). This mutant would thus be expected to be a binary receptor system requiring both a substrate and the histidine group for activity. Indeed

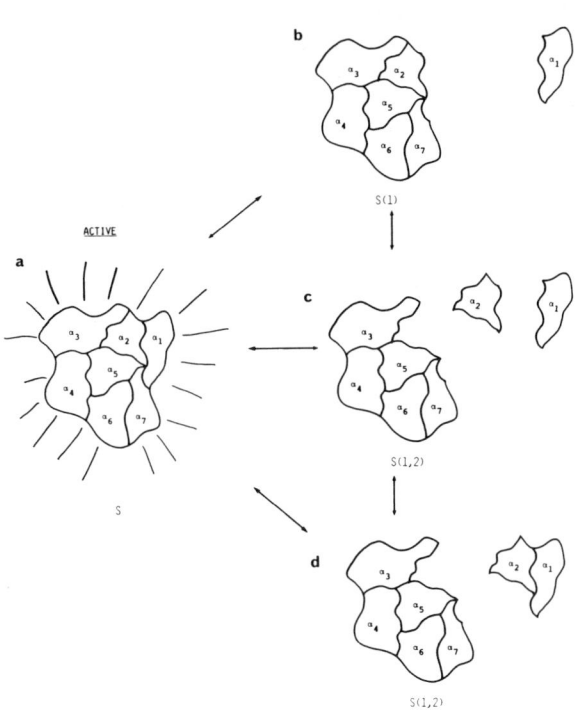

Figure 2. Schematic representation of the deletion model for generating binary receptors: a) the fully functional parent system; b) the receptor S(1) and its ligand α_1; c) the receptor S(1,2) and its two separate ligands α_1 and α_2; d) the receptor S(1,2) and its two adjoined ligands α_1 and α_2.

this mutant showed a reduction in the catalytic efficiency by a factor of a million for the substrate N-succinyl-L-Phe-L-Ala-L-Ala-L-Phe-p-nitroanilide (sFAAF-pNA) alone, as compared with the wild form of subtilisin. At this stage the two ligands were precombined to form sFAHF-pNA (Figure 2d). This replacement of the Ala in the P2 position of the substrate by a His (see Figure 2d) resulted in a 400 fold increase in the rate of catalysis. Interestingly, only small changes in K_m were observed (see below). In their analysis of this system of "substrate-assisted catalysis" Carter and Wells have in fact suggested that at some earlier evolutionary stage the His group may have been a part of the natural substrates such as FAAYGF (Carter and Wells, 1987). In terms of the deletion model one could speculate that not only were these two combined, but all three (i.e., including the subtilisin) were part of one parent system.

III. A UNIFIED SCHEME FOR cAMP AND cGMP REGULATORY AND CATALYTIC BINDING SITES

The deletion model and associated recognition of the use of common molecular machinery by different biochemical systems has recently been used to develop a model relating regulatory (e.g., cAMP dependent protein kinase) and catalytic (e.g., cAMP phosphodiesterase) binding sites for cAMP and cGMP (Topiol and Sabio, 1988). The model is based on the x-ray structure of catabolite gene activator protein in Escherichia coli, (CAP; McKay and Steitz, 1981; Weber and Steitz, 1987) complexed with cAMP. cAMP allosterically activates this protein, i.e., acts as an agonist. Significant sequence homologies have been identified between CAP and the regulatory domains of types I and II, cAMP-dependent protein kinases (cA-PK; Weber et al., 1982). Our own computational chemical studies have confirmed the essential features of the interactions derived from the x-ray structures (Topiol and Sabio, 1988). Recognizing the significance of this sequence homology as an example of the use of common structural features for the design of cAMP regulatory sites, Weber et al. have developed a model for the structure of the regulatory binding sites of cA-PK (Weber et al., 1987).

It has also been shown (Takio et al., 1984) that there are significant homologies between the regulatory domains of cA-PK and cGMP-dependent protein kinase (cG-PK). Indeed, given the similarity in the structures of cAMP and cGMP, as well as the tendency for cAMP to bind to cGMP regulatory (and catalytic) sites, this seems almost predictable. In order to understand the "cross-binding" of cGMP and cAMP we have done computational chemical studies of the binding of cGMP to the cAMP binding site (Topiol and Sabio, 1988). These studies clearly identify a possible binding model.

The catalytic domains of cAMP and cGMP phosphodiesterases have also been found to have significant homologies (Charbonneau et al., 1986). Recently, it has been claimed that a seven amino acid sequence in the cAMP binding region of the **drosophila** dunce$^+$ gene product (a phosphodiesterase) also exists in the cA-PK and cG-PK binding regions (Chen et al., 1986). Upon merging the sequences of the regulatory and catalytic regions of all of these proteins a dramatic feature appears. In the structural region to which the cyclic phosphates of the cyclic nucleotides bind, a highly conserved Arg and Ala of the regulatory sites are replaced by highly conserved Ser's (or Thr's) in the catalytic binding sites (Topiol and Sabio, 1988). It has thus been suggested that the hydroxyls of these Ser's/Thr's are primary components of the hydrolytic machinery of the phophodiesterases as they are in other enzymes (e.g., serine proteases or DNAse I). By analogy to the work of Carter and Wells on subtilisin, we could suggest that nature has used the machinery of the phosphodiesterase enzymes to create regulatory receptor sites by 'deletion' of key catalytic groups. Moreover, this analysis of the preserved structural and mechanistic components of these various cyclic nucleotide binding proteins has provided a basis for the development of related models for these systems.

IV. CONCLUSIONS

Using the simple deletion model for the origin of receptor systems it is possible to explain a number of different general features known for such systems. The use of this deletion model to analyze related systems could provide insights from systems for which more detailed data (e.g., x-ray structures) exists, to the extrapolation of models for systems for which there is only limited information available (e.g., sequence data). These more 'educated guesses' at the models for ligand/receptor interactions should provide for more efficient guides into the design of new drugs for these systems.

REFERENCES

Bourne HR (1986). One Molecular Machine Can Transduce Diverse Signals. Nature 321:814-816.
Burgen ASV, Roberts GCK, Feeny J (1975). Binding of Flexible Molecules. Nature 253:753-755.
Carter P, Wells JA, (1987). Engineering Enzyme Specifity by "Substrate Assisted Catalysis". Science 237:394-399.

Charbonneau H, Beier N, Walsh KA, Beavo JA (1986). Identification of a Conserved Domain among Cyclic Nucleotide Phosphodiesterases from Diverse Species. Proc Natl Acad Sci USA 83:9308–9312.

Chen, C-N Denome S, Davis RL (1986). Molecular Analysis of cDNA clones and the Corresponding Genomic Coding of the **Drosophila** dunce[+] Gene, the Structural Gene for cAMP Phosphodiesterase. Proc Natl Acad Sci USA 83:9313–9317.

DeLisi C (1988). Computers in Molecular Biology: Current Applications and Emerging Trends. Science 240:47–52.

Fischer E (1894). Einfluss der Configuration auf die Wirkung der Enzyme. Ber Deut Chem Ges 27:2985–2993.

Koshland DE Jr (1958). Application of a Theory of Enzyme Specifity to Protein Synthesis. Proc Natl Acad Sci USA 44:98–104.

Koshland DE Jr (1959). Mechanisms of Transfer Enzymes. In Boyer PD, Lardy H, Myrback, K (eds): "The Enzymes," New York: Academic Press, vol. 1, pp 305–345.

Koshland DE Jr (1963). The Role of Flexibility in Enzyme Action. Cold Spring Harbor Symp Quant Biol 28:473–482.

McKay, DB, Steitz, TA, (1981). Structure of Catabolite Gene Activator Protein at 2.9 A Resolution Suggests Binding to Left-Handed B-DNA. Nature 290:744–749.

Osman R, Topiol S, Rubenstein L, Weinstein H (1987). A Molecular Model for Activation of a 5-Hydroxytryptamine Receptor. Mol Pharmacol 32:699–705.

Takio K, Wade RD, Smith SB, Krebs EG, Walsh KA, Titani K (1984). Guanosine Cyclic 3′,5′-Phosphate Dependent Protein Kinase, a Chimeric Protein Homologous with Two Separate Protein Families. Biochem 23:4207–4218.

Topiol S (1987). The Deletion Model for the Origin of Receptors. Trends Bioch Sci 12:419–421.

Topiol S, Sabio M (1988; unpublished).

Weber IT, Takio K, Titani K, Steitz TA (1982). The cAMP-Binding Domains of the Regulatory Subunit of cAMP-Dependent Protein Kinase and the Catabolite Gene Activator Protein are Homologous. Proc Natl Acad Sci USA 79:7679–7683.

Weber IT, Steitz TA, (1987). Structure of a Complex of Catabolite Gene Activator Protein and Cyclic AMP Refined at 2.5 A Resolution. J Mol Biol 198:311–326.

Weber, IT, Steitz, TA, Bubis, J, Taylor, SS (1987). Predicted Structures of cAMP Binding Domains of Type I and II Regulatory Subunits of cAMP-Dependent Protein Kinase. Biochem 26:343–351.

Weinstein H, Mazurek AP, Osman R, Topiol S (1986). Theoretical Studies on the Activation Mechanism of the Histamine H_2-Receptor: The Proton Transfer between Histamine and a Receptor Model. Mol Pharmacol 29:28–33.

DESIGN AND SYNTHESIS OF ANTIMUSCARINICS BASED ON PHYSICAL AND MECHANICAL ATTRIBUTES

Richard K. Gordon, Ruthann M. Smejkal,
Eli Breuer, and Peter K. Chiang
Department of Applied Bichemistry,
Walter Reed Army Institute of Research,
Washington, DC 20307-5100.

Exposure to organophosphate agents results in acute toxicity due to the accumulation of acetylcholine. Atropine, an antimuscarinic which displays high affinity specifically for the muscarinic receptor, has been used in the prevention and treatment of these nerve agent casualties. Several functional groups are usually required in a molecule to achieve potent antimuscarinic properties: a protonated nitrogen atom is found near one end of the molecule (the cationic site); the center of the compound contains an electronegative ester group (the anionic site); and a relatively bulky hydrophobic and lipophilic portion is located at the opposite end of the molecule. The effect of altering these critical portions of the molecule provides information for the synthesis of more potent antimuscarinics and maximizes their protective effects.

Many efforts to design drugs for specific actions have been attempted based on predictions of quantitative structure-activity relationships (QSAR), that is, expressing aspects of structure in quantitative terms with respect to a standard compound and relating these values to corresponding changes in activity. Historically, chemists have tried to correlate differences in such physicochemical parameters as lipophilicity, dipole moment, bond angles and distances, steric effects, ionization states or electron withdrawing/donating ability with changes in biological or chemical activity. A molecular mechanics approach, such as using energy minimization models, x-ray crystallography, and nuclear magnetic resonance spectroscopy to study the interaction of a compound with solute molecules, has also been utilized. The objective of the present study was to use the principles of QSAR to quantitate the relationships between the biological activity and physical/mechanical properties of a series of atropine analogs in order to design more potent antimuscarinics and define the topology of the antagonist binding site of the muscarinic receptor based on their physical attributes.

One of the more stringent means of comparing several antagonists with their efficacy is to determine the distances between significant pairs of atoms occuring in each molecule (1). We used distance geometry analysis to describe the preferred linear distance between the carbonyl oxygen and the protonated nitrogen required for optimal antagonist interaction with the muscarinic receptor. To evaluate as clearly as possible only the effect of the bond distance on the potency of these muscarinic antagonists, the compounds contained as many features consistent with each other as possible (Figure 1). The analogs consisted of a constant hydrophobic ester

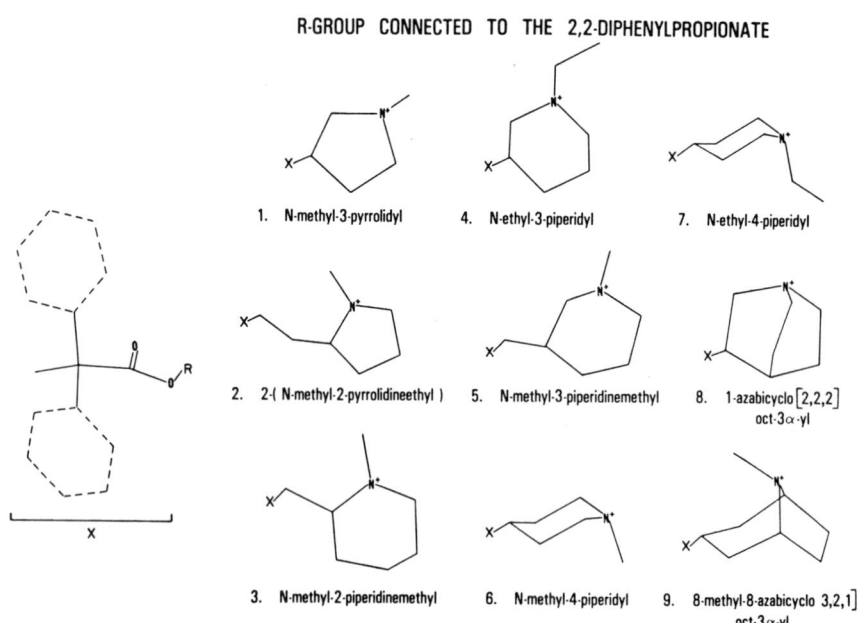

Figure 1
Structure of the 2,2-diphenylpropionate compounds.

moiety, 2,2-diphenylpropionate, and an amino alcohol portion of the molecule, containing a protonated nitrogen in the assay (pH 7.4). The modified portion of the compounds surrounding the nitrogen was altered so as not to introduce a wide variability of chemical structures, and the rings of the synthesized analogs containing the NH_4^+ were always saturated and cyclic. Topological and geometric values of the compounds were computer-generated using a classical molecular mechanics approach. The shortest calculated bond distance was 4.4 Å, and the longest bond distance was about 5.1 Å.

The effect of altering these critical portions of the molecule on the ability of the analogs to function as antimuscarinic agents was assessed by three biological assays. The analogs were tested for their ability 1) to antagonize acetylcholine-induced contraction of guinea pig ileum, 2) to block carbachol-stimulated release of α-amylase from pancreatic acini, and 3) to inhibit the binding of [^{3}H]N-methyl-scopolamine to the muscarinic receptors of N4TG1 neuroblastoma cells (2,3,4).

A linear relationship was established for bond distances between the protonated nitrogen and the carbonyl oxygen (N^+ and C=O) of the tested compounds and the log of the three biological assays of antimuscarinic potency: (a) the K_i values determined from the inhibition of [^{3}H]NMS binding to the muscarinic receptors of the N4TG1 neuroblastoma cells (Figure 2), (b) the K_i values from the inhibition of α-amylase secretion (Figure 3), and (c) the K_B values from the inhibition assay of ileum contraction (Figure 4). Linear regression analysis was performed as described (2). The equations defining these three lines, the regression coefficients (r) and the significance (Student's t-test) were, respectively:

$y = 2.81x - 5.65$; $r = 0.86$; $p < 0.005$;

$y = 2.83x - 6.85$; $r = 0.89$; $p < 0.002$;

and $y = 2.02x - 1.84$; $r = 0.76$; $p < 0.02$.

The longest calculated bond distance in this series of compounds was about 5.1 Å, but a maximum inhibitory potency may not have been reached with these compounds.

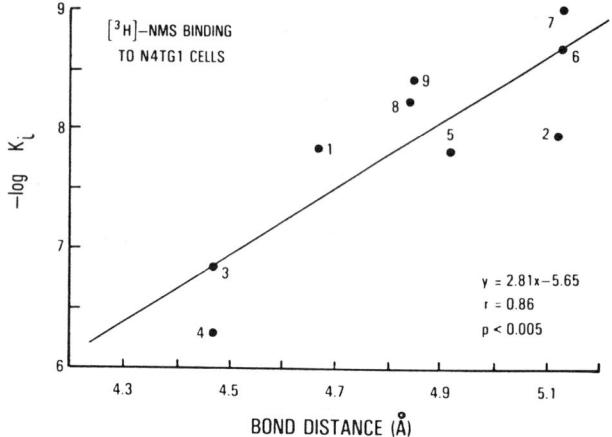

Figure 2
The bond distance between the carbonyl oxygen and protonated nitrogen of each analog was linearly related to the log of the K_i values obtained from the [^{3}H]NMS binding inhibition assay in N4TG1 neuroblastoma cells, performed as described previously (3).

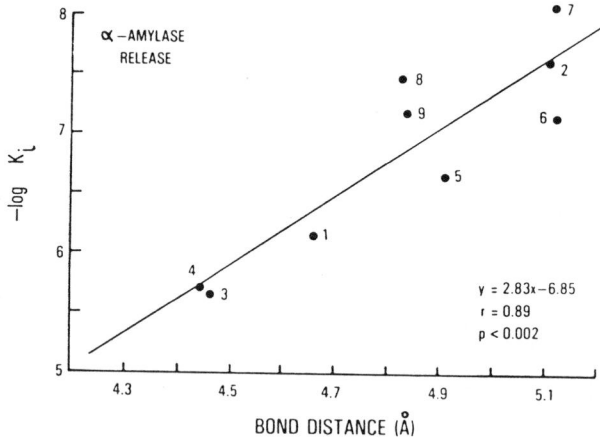

Figure 3
The bond distance between the carbonyl oxygen and protonated nitrogen of each analog was linearly related to the log of the K_i values obtained from the inhibition of α-amylase release in pancreatic acinar cells, prepared as previously described (2).

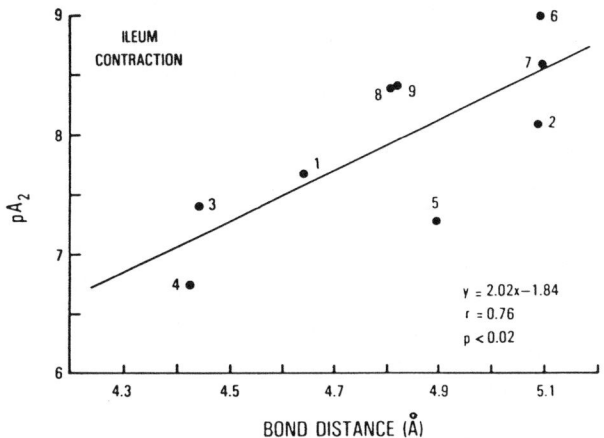

Figure 4
The bond distance between the carbonyl oxygen and protonated nitrogen of each analog was linearly related to the pA_2 values obtained from the inhibition of guinea pig ileum contraction, as described previously (4)

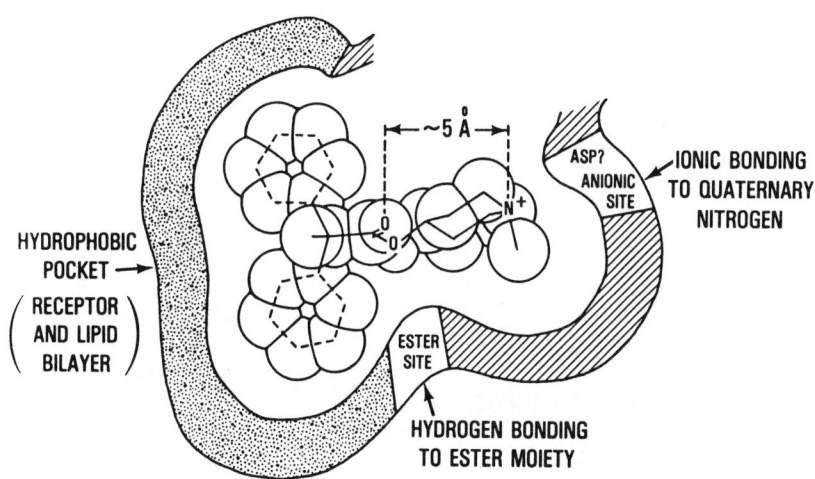

Figure 5
Modeled structure of the muscarinic receptor.

These results demonstrate that the potencies of muscarinic antagonists can be predicted on the basis of distance geometry. In addition, these results also provide information on the size constraints established with antagonists between their cationic site and ester site and their corresponding binding sites of the muscarinic receptor. A possible structure of the receptor site is shown in Figure 5.

The bond distance values calculated by the XIRIS molecular modeling system are similar to the values obtained by Takemura (5). Using a series of relatively planar muscarinic agonists, rather than antagonists, he reported an oxygen site and protonated nitrogen site distance of 4 to 5 Å. However, our results disagree with the statement by Broutsyna et al (6) that the distance between the nitrogen and oxygen atoms could not be an important geometric feature of an antagonist as it is for an agonist of the muscarinic receptor.

In conclusion, this investigation provides information on the ligand binding site of the muscarinic receptor and a basis for the design of more potent antidotes of organophosphate poisoning. The results demonstrate that potent antagonists can be synthesized and predicted on the basis of distance geometry and readily tested by diverse biological assays.

REFERENCES

1. Pauling, P. and Datta, N. Proc. Natl. Acad. Sci. (USA) 77:708-712 (1980).
2. Gordon, R.K. and Chiang, P.K. J. Pharmacol. Exper. Ther. 236:85-89 (1986).
3. Ahmad, A., Gordon, R.K., and Chiang, P.K. FEBS Lett. 214:285-290 (1987).
4. Witkin, J. M., Gordon, R. K., and Chiang, P. K. J. Pharmacol. Exper. Ther. 242:796-803 (1987).
5. Takemura, S. J. Pharm. Dyn. 7:436-444 (1984).
6. Broutsyna, N.B., Khromov-Broisov, N.V., Losev, N.A., Zhorov, B.S., and Govyrin, V.A. Farmakologiya I Toksikologiya 43:219-230 (1981).

SECTION V. ROUND TABLE DISCUSSION

Round Table Discussion

Panel members of the round table discussion were:

Prof. Harold A. Scheraga — Cornell University
                            Chairman

Dr. Jonathan Greer   — Abbott Pharmaceuticals
Prof. Barry Honig   — Columbia University
Prof. Edward M. Kosower   — Tel-Aviv University
Prof. Michael Levitt   — Stanford University
Prof. Jon Lindstrom   — Salk Institute
Prof, Garland R. Marshall — Washington University
Prof. Israel Silamn   — Weizmann Institute of
                            Science

Scheraga:
The panel will make a few comments in response questions that I will read in a moment and the discussion will be open to all the participant of the conference. It is my understanding that this conversation is being tape recorded and will be published, and properly edited, so will you please say your name before you talk. Now a few questions have been provided to us for discussion here, so let me read them to you. First one is: What can the present state of the art of computation and modeling contribute to ligand-receptor recognition, drug design, and the mechanism of biological function, for example: ligand activation of receptors, such things as opening channels and trans-membrane signals. Then we would like to discuss some specific examples. Competitive inhibitors of enzymes and competitive antagonists of receptors, and finally, what can we expect in the future. First of all, are bigger computers and related graphics the answer, or are we so enamored with computers that we are not bothering to think clearly, I suppose this is a rhetorical question, and or to realize that new ideas are required. Finally, to be a little more specific, it would be interesting to have some thoughts on the role of solvent and ionic effects. The panel is going to take these in random order, and I will entertain whoever who wants to start off.

Lindstrom:
A few comments that have occurred to me, of a semi heretical nature. It seems to me that model building should be primarily done to provide testable alternatives to aid experimental determination of receptor structure. I don't think that modeling techniques are sufficient with the available data to make detailed models of AChR ligand binding sites or channels. What is needed is not more computers, but more data. I think that modeling can sometimes be actually destructive because it can take a simple idea and dress it up in computer-generated, color, 3D pictures which give the illusion that the problem is solved, but are really pseudo data that may mislead some into believing that the problem is actually solved, and inhibit some experimental studies that can solve the problem. On the other hand trying to model a well-defined protein structure as a method of testing modeling techniques, does seem like a good idea. Modeling systems ahead of testabi-

lity or not in conjunction with experiments seem to me to be a waste. However, pragmatic modeling of drug efficacy, when large drug families are known, seems to me to be a good idea. It may or may not yield a literal representation of the receptor site, but if done properly, should almost certainly lead to the design of more effective drugs. The most useful kind of receptor modeling I can think of involves binding and functional assays using human receptors. This is now becoming possible using cloned cell lines and transfected cell lines. For example, all the drugs used to effect human muscle nicotinic receptors have not been tested yet on human AChR receptors but on AChR receptors from other species, because until recently no human cell line expressing human AChR receptors was available; it was assumed that human receptors will behave exactly like those of the species that have been tested, and that, in fact, isn't true.

Kosower:
Since I have been involved in trying to model receptors without doing any experiments, I feel that I should make some sort of comment on this approach. Its been historically true in molecular sciences and chemical sciences, that when you have a structure, the most famous case being the benzene molecule, that when you have a structure, it is easier to think about a problem, it is easier to design experiments, to either show the model to be valid or to go on to something else. Thus, I think that in spite of the fact that we have, in many cases, no clear idea of exactly how receptors are built, I think it is terribly important that we think about how the receptor is built and especially how it could work. It is not sufficient as we know, to get a amino acid sequence and let it go at that and hope that some day someone will do an x-ray structure and then we will understand what we are doing. I think that it is important, in the most conservative way possible, to build up a structure, try to understand how it could work and go as far as possible. One must keep in mind that you are building a model and that during the model building process, one may think of experiments with which one could test ideas and go further.

Honig:
I basically agree with Dr. Lindstrom's comments. It is clear that you have to make a distinction between globular proteins with a well defined crystal structure and membrane proteins. With regards to globular proteins, I think that there is a tendency these days for people to do calculations and get numbers which appear to be very accurate; numbers that are within a Kcal/mole or two from experiment. When that happens, I think it may be misleading, because in reality it may well never be possible to predict binding energies, certainly not conformational energies, to within experimental accuracy. What we should be trying to do is to gain enough insights to help think about designing and interpreting experiments and if we can make some sense out of what is going on, then we are useful as modellers since that's the way modellers have always been useful. Now with regard to membrane proteins, where there are very few examples, homology model building is very difficult. I think in some cases you can make a real claim that it is harmful to predict membrane protein structures, since many people will actually believe a bad model and design experiments accordingly. In general, anyone who actually believes his model, is probably making a serious mistake. On the other hand, if one makes a rational set of assumptions, and states that "these are the assumptions and this is what I have come up with", its probably a useful exercise.

Greer:
I would not go so far as to say that one should never believe ones models. I think what is critical, is that one should approach this with tremendous trepidation. When one does do modeling, what becomes very obvious is the insecure feeling that one has and how that insecure feeling vanishes as experimental data appear on the scene to either support the model or require its change. The way we like to look upon a model, is that it is not a solution, it is just our current working hypothesis. Many of our problems, in fact are just the situation that we are talking about, where we are looking at a ligand or a series of ligands, a small or large number of analogs that have been made, where very little is known about the receptor. We have heard a number of talks that I found

encouraging from Garland Marshall and Gilda Loew and others that are teaching us to work with these analogs when we do not know anything about the receptor. But the more experience that I find in dealing with these analogs, one finds very strange phenomena, phenomena that are difficult to explain, such as partial agonists, the relationship between binding affinities and agonist versus antagonist behavior, is sometimes very anomalous. One feels the crying need, more and more, for obtaining a true crystal structure, for a good receptor ligand interaction. Certainly my thinking on this subject, I find, is colored by our much deeper knowledge of substrate or inhibitor enzyme interactions, and there may be aspects of that that are not really extrapolatable to the receptor interaction. I think that there is good evidence that there are other things going on in the receptor and we really have no experimental referents for them, and that makes it a very difficult situation to model. So, I would like to see a tremendous effort to try and get crystal structures for some of these receptors. We are certainly thinking about it and others are thinking about them and I think that is going to revolutionize our thinking about this whole problem as soon as some of these structures become available. And then, I think, the modeling will be tremendously more powerful. The smaller our extrapolation, the better off we are when we are doing this modeling; and the more solid the experimental basis upon which it sits, the more likely it is to be correct and therefore the more likely it is to be useful.

Scheraga:
Well, it is certainly admirable to want to have more crystal structures, but something has to go before that. What's the present status of isolating receptors?

Greer:

Well, I think we have heard one, if not more talks this week, where the amounts of these proteins, the number of receptors that are being cloned and expressed in significant quantities is increasing. It is increasing much faster than our ability to handle them in terms of obtaining crystal structures, but certainly there are probably going to be a half a dozen to a dozen of these pro-

teins available almost immediately and the challenge is
going to be to sit down and go through the arduous years
of figuring out how to crystalize them. That is a
thankless, not a very fundable task from a grant prospec-
tive, but it is a critical one. I think availability of
the protein, is not going to be the problem, I think that
it is going to be picking the one to work on and finding
the conditions to get it crystallizable.

Levitt:
I want to make some general comments about modeling, which
should perhaps have preceeded what has been said here. In
many ways, the value of molecular modeling is like "beauty
in the eyes of the beholder". It depends on what the
experimentor wants to know, and success often depends on
luck. For example, the fact that a small peptide from the
acetylcholine receptor protein, should actually bind a
toxin is certainly a piece of luck. Who would have
believed that a small peptide could act like a whole enzyme? In many ways the beautiful work that Jonathan Greer
presented here, had elements of luck in it. For example,
his crystal structure happens to be disordered in a region
of a helix rather than a more complicated series of turns.
With modeling, it is useful just to try and see.
I think that it is also very important that modeling be
done by the person who knows the system that is being
modeled. It is a very educational experience, broadening
ones horizon. Often just picking up plastic models and
playing with them is as useful as using giant computers.
Another general point, is that predictive modeling value,
for example, designing a better enzyme, doesn't have to be
a hundred percent right. If you can suggest twenty
mutants and only two of them work, that would still be a
major contribution. Provided one can reduce the odds from
one in a thousand to one in ten, one has made enormous
progress. If one felt that what he said did not have to
be a 100% right but only 10% right, it would give one more
confidence.
Finally, I want to make a point about the value of
building models of structures that are already known. It
is very very difficult to predict models of proteins that
are about to be solved as it might take up to five years
before the model is proved or disproved. By that time
everyone has forgotten about the model, and it is often

very difficult to know "what did I do right"? Now that there are computers, one can do modeling in a completely automatic way. This has the great advantage of enabling one to model things that are known. For example, as a homology modeling exercise, a computer program could take the chymotrypsin coordinates and build the coordinates of trypsin. If this could be done to say, half an Angstom accuracy, one would be enormously encouraged. The advantages of automatic modeling is that you can limit the amount of information given to the computer program and repeat the exercise as many times as necessary to get a statistically reliable result.

Scheraga:
Before I give up the microphone, I would like to reiterate and emphasize something that Michael just said. I think that, in order to give the field of modeling credibility, one has to establish the validity of the methods and the procedures. I think therefore, that we have to do a lot of work on problems where we already know the answers, not that we are going to pretend that we are predicting them, but to test the procedures and validate them. Unless one can do that, I think that the field is not going to have any credibility.

Silman:
I wanted to insert a note of scepticism which may be tough on both modelers and crystallographers. Jonathan Greer implied that receptors having been cloned and expressed, the task would be to crystallize them as a step to working out their mechanism of action. In the case of the nicotinic acetylcholine receptor, at least, we may soon have crystals and, subsequently, a three-dimensional structure. A receptor, however, is not an enzyme, but a transmembrane protein embedded and functioning in a lipid bilayer. In my opinion, we do not really have any concept of how a channel in the bilayer is induced by interaction of a ligand with its receptor, and I don't think that the mechanism will become apparent from the crystal structure. I think it's going to be much more complex than that: One is going to have to take into account lipid composition, lipid-protein interactions, surface pressure and charge, etc. I think a good analogy would be the case of oxidative phosphorylation in mitochondria. Twentyfive years

ago, people dealt with this phenomenon in terms of a linear array of chemical reactions catalyzed by a series of enzymes which were suitably arranged on the inner mitochondrial membrane. Along came Mitchell, introduced vectorial coupling of phosphorylation to proton and other ion gradients, and changed everyone's concepts completely. I think that for agonist-induced channel opening via receptors, it will not be enough to solve three-dimensional structures and to carry out reconstitution experiments. Somebody will have to introduce completely new concepts. Obviously, I have no idea what these will be, but I feel sure that they are bound to come eventually.

Greer:
I just want to respond, I am delighted that there will be a transcript of this, but I think that the record will show that I certainly did not imply that crystallizing receptors is going to be easy, I think it is going to be difficult and I think that it is going to take a considerable effort. On the other hand, the basic tenent of biochemisty and molecular biology has been that you can learn a lot if not everything about the system, by taking it to its components and studying the nature of its components. I believe that it will be true to a great extent with receptors and that we are not going to learn everything about this, there is no doubt about that. But I, for one, would like to see some detailed structural information about the way that many of these agonists or antagonists interact with the receptor and what the structural implications and consequences are, experimentally, of those interactions. There will, of course be more that needs to be learned. I agree with you, but let us start somewhere.

Silman:
Now that was precisely my point. In the case of oxidative phosphorylation, taking the system apart and reconstituting it was a very valuable process, up to a point, and cloning and expression of mitochondrial genes was also worth doing. What Mitchell did, however, although in a sense much more macroscopic, brought in totally new concepts and completely changed our ideas about how the process works.

Marshall:
I think that there are a couple of things that are at least worth pointing out. One danger is being too intimidated by all the things that we do not know and we leave out, and I think that in almost everything that was discussed, we left out a lot in all the models. If you are going to model membranes and in particular, the systems that you are dealing with are dependent on having voltage across the membrane, where you are talking about 60 to 100 millivolts across 30-40 angstroms, that is a significant force that is impinging on the system, which I do not think you can ingnore. I think that is going to dramatically impact what your model is going to look like and what your helix is going to look like, how it is going to orient and everything else. I do not think that we currently have a methodology to deal with that discontinuous and complicated system with regard to electrostatics. I have not seen anybody demonstrate that they can handle that problem. Certainly molecular dynamics on bilayers have been a problem so far. But I do not think that we should not try, that is the only way we are going to find out where these problems lie. I think the real problem is, is doing what Prof. Scheraga suggests and that is finding good model systems to model. Whether the data is really all that good, where we really know the answers and where we really can compare the results. We have seen so many examples this week where we showed comparisons with experimental data and to my eye, the comparisons do not fit the data at all. It is the old issue, is the glass half full or half empty? I see the glass as being half empty and there is a big hole in the bottom and you know, somehow we have got to plug the hole and fill the glass and fit the data that exists. I think that we have to start getting serious, because we are misleading ourselves if we cannot fit the data. I think modeling is part of an iterative loop. You model and you make predictions and then go to the lab and make compounds and do the experiments and come back and change the model. People ask me: "Why do you never talk about your failures, you know, you never talk about the models that were wrong". A model is not right or wrong. A model is how you keep track in three dimensions of what the data is that you have. It evolves with time. So I guess for me, it is part of an intellectual process, it is not a distinct thing, just a tool that goes with experimental science.

Levitt:
This is an "anecdotal" technical point, Garland Marshall mentioned, the importance of the membrane potential across a protein. It turns out that if you convert millivolts per angstrom to kilocalories per electron per angstrom, the force that is generated by the membrane potential on a net electronic charge is close to .05 kilocalories per angstrom. Those people who do molecular mechanics minimization know that this is a very small net force, one which is as close to zero as one can get.

Marshall:
The summation of many small, aligned components makes significant effects.

Kosower:
I would like to give one example. After I worked on the acetylcholine receptor channel, Numa published the sequence of the sodium channel, so I looked in it for a similar amphiphilic segment, couldn't find it and got very worried. I found, however an amphiphatic segment, which has become known as S4, so I put that into the bilayer. Since there were four of them, that put 20 plus charges into the bilayer. You have to neutralize them. A comment that I was going to make this morning, but didn't get a chance to, was that, in the middle of the sodium channel, there is a long string of negative charges. Some examples were mentioned before by Joel Sussman. I put the string of negative charges into the membrane and lo and behold the whole became neutral plus one could see more or less, a working model of some kind of channel, a model channel, but still a channel. Now it turned out later that the S4 migrated in everyone else's model into the bilayer, and after that it was discovered that S4 is characteristic of channels so far, and is used as an identifying element. What I am trying to say is that modeling at a certain level can lead you to ideas that you didn't have before and that can be helpful in analyzing structures which one gets. It is not the end, it is on the way, it is part of the iterative process, in doing what we are trying to do, which is to understand how these molecules are built and after that, how they work.

Scheraga:
Well, Lets throw this open to the floor, -- either comments to the general audience or directed to the panel.

Rein:
I would like to take a somewhat more optimistic point of view regarding the role of modeling. What I see as the the main advantage of modeling is that it offers a formalism for hypothesis testing on one hand and it permits a more systematic interpretation of experimental data on the other hand. It helps to test constistency of assertions which can be made from experiments. I see that we can take examples from this meeting for both these cases. Lets take the hypothesis of the receptor ligand interaction proposed by Prof. Kosower. Obviously there are points in this hypothesis which can be tested by modeling. If acetylcholine can open the channels, it means that the subunits which form the rosette interact with acetylcholine in a fashion that can bring it to channel opening. Thus, the interaction which is consistent with channel opening, must be both structually and energetically compatible with such a rearrangement. This is not an easy task to perform, but in principle this is possible. What we need for this is, mutual understanding between the experimentalist and the theoretician. The theoreticians have to be familiar with the experimental facts and with the system and the experimentalist with the proposed model. Such a mutual education is in fact the primary object of this meeting. The second example is concerned also with the acetylcholine receptor system. Work like that of Dr. Gershoni of the Weizmann Institute of Science in which the sequence of residues of the receptor interacting with antagonists such as snake toxins are identified and provide the information required for model building. Modeling technique, like molecular mechanics or molecular dynamics yield models of the relevant structures and interaction and tests whether the experiments are consistant with the model. These two examples illustrates essentially the interplay of experiment and theory and underscore the fact that modeling is an iterative process going back and forth between experiments and theory to tests the consistency and go deeper into the understanding of the molecular nature of the underlying phenomena.

Honig:
We seemed to be involved in a discussion as to whether we should be funded anymore, I think most of us here feel that modeling is a good thing to do so what we are really discussing are the differences between good modeling and bad modeling. I just wanted to point that out. I don't think that any one here feels that modeling can't be successful if it is done properly. In conjunction with experiments, its a useful enterprise. I think that we all agree with that and I thus suggest that we move on to another topic.

Lindstrom:
I just want to make a small point, not relevant to the general concept of modeling, but specific to an issue you raised, about proposals for the localization of acetylcholine binding sites and how someday that might be testable. I showed you data of an experiment using toxin-labeled acetylcholine receptor in a 2-D crystal-line array that localized the toxin binding sites. As you can see from those experiments, these sites are quite widely separated, much more than the length of an acetylcholine molecule; so in fact, you can say that experimental data exists incompatible with the idea that you have an acetylcholine molecule at the interface between subunits forcing open a channel. This is a minor point irrelevant to the general issue of modeling.

Rein:
The model which I used was also contrary to this. I brought it out as a possible example.

Jardetzky:
Since so much has been said about modeling, it might be useful to define the term. There is purely geometric modeling that is based on chemical or very weak physical evidence, there is modeling based on biological evidence, there is modeling based on physical measurements and there is modeling on the basis of more or less strong physical theories, such as molecular mechanics and molecular dynamics. The virtue of these different kinds of modeling is very different. A model based on strong physical evidence is not a problem. A model based on weak or indirect evidence or on theoretical considerations should not be

reported as if it were a fact. I think that the critical point is worth underscoring: the most important measure of the value of modeling in the absence of real convincing evidence, is its predictive ability. If you can predict an experimental result, then it is certainly a worthwhile exercise.

Guy:
Now that most of us agree that modeling in conjunction with experiments is useful, I would like to examine the question as to whether computational calculations in conjunctions with computer graphics is useful or is a dangerous exercise. I think that most of us agree that when we either develop a technique, we would like that technique to be completely objective, for it to be reproducable and for it to be valid, for it to be rigorous and examine all the possibilities before we come out with the model. In practice that's often difficult to achieve. We have tried to do that with the helix backing problem for membrane proteins and in our hands the problem blew up in our faces, and became computationally too complex to do. Because of all the different environments and solvents involved, we weren't even confident that the energies we were calculating would be valid. So we used an alternative approach of using computer graphics, applying surfaces, docking helices and manipulating them manually, and then doing energy calculations. I think that with the advent of coming new generations of machines - nicknamed by Bob Langridge - crayola - with graphic capabilities and high computational capabilities, we are going to see a lot more people using the computer graphics and the energy calculations and mixing those together. The difficulty with that is that you loose all that criteria of objectivity and reproducability with the computer graphics. So, I would like some discussion on whether in fact it is a dangerous development that people might be so attracted by the computer graphics and believe that because some energy calculations have been done that the rigor is there when in fact it is missing.

Scheraga:
Well let me respond to that. I agree with you, when you are going to do an energy calculation you have to have a systematic way of exploring the conformational space. It has to be systematic and rigorous, whether you do it with or without graphics. Graphics could be a crutch to help you evaluate where you are and evaluate your results, but I think that there has been too much reliance on graphics to provide the initial conformations with which to start a calculation.

Honig:
The other side of that, is that in my particular department, there are a number of molecular biologists who have this and that cloned and have sequences, and they want to start thinking about them. So they come to my lab, or to someone else's lab for ideas, and in general we don't have good ideas and we don't have the time to devote to thinking about their problems. I think that an important development in the next few years is going to be the transfer of the tools of modeling to people who aren't professional modelers. If proper software can be developed, people can sit and play by themselves, as long as they have educated themselves properly. We keep talking about things being dangerous but they are not really dangerous. If they are good scientists, and they read the literature, they will know what's right or wrong. Thus, I believe it to be important to provide tools, not only for the professionals but for the non-professionals as well.

Greer:
I think the idea of having programs that work on their own and are completely objective is certainly what one is striving for; but I feel that certainly in the short term or perhaps in the forseeable future this is an ideal. The great thing that graphics has done is to provide the information for us in a way that we can apply our own scientific knowledge and intuition and input this into the problem. The advent of significant computing capability closely coupled to the graphics, I think will extend the ability that we have to input our intuition into these problems to try and get to the answers faster than by a brute force approach. But if we do that, it does place certain very critical obligations upon us. In particular,

it places upon us the obligation to describe exactly what we have done -- because that is the only way that our fellow scientists can then attempt to reproduce it and learn from it. If we do that, then I feel that system is probably going to get us to the answers that we want, which is usable models or structures that can be useful for predictions. Useful in an applied sense and useful in pure science sense. Faster than brute force approaches. I think that everybody agrees in principle that brute force would be the best way to do it; but, in practice, what graphics has done is given us the ability to put in our intuition. I am wondering whether in fact, close coupling of large computing and real time graphics is going to do to the field of molecular modeling what real time rotation graphics did to the field 10 to 15 years ago. On the other hand, I would like to respond to Barry. I am very nervous about taking people who do not have expertise in the area of structural biology and giving them what are superficially user friendly programs that operate on these sophisticated machines and allow them to think that they can be theoretical chemists. I do not really think that is wise. I think that you need experts, as in every field, and that with the guidance of the experts and with close interaction with the experts, these people can do a lot by themselves, be they molecular biologists or synthetic chemists or whatever. But it needs to be watched very carefully because many of these programs are very easy to use and you get out exactly what you put in.

Honig:
We agree, certainly about the need for experts. On the other hand, good molecular biologists will learn to use the programs properly, they'll speak to the right people and they will learn to use them. They are perfectly capable people.

Greer:
If they think that it is important they will, if it is not important to them, they will not.

Levitt:
Looking at molecular modeling, there are two different ways of going about it: Modeling by analogy and ab initio modeling.

Modeling by analogy will be become increasingly important as more structural and sequence information is available. If this information is combined in a powerful computer graphics workstation, one will be able to "play games" with unknown protein structures. This will change our attitude to molecular modeling and could be a major breakthrough.

Energy calculations, which are <u>ab</u> <u>initio</u> in the sense of starting from basic forces and going on to structures, will be important for evaluating different models, but they will not be the major way we will solve protein conformations over the next few years.

Sussman:
I agree with Mike to a certain extent, but I think what comes out best from model building today, and I see this for the near future, is modeling by sort of analogy. I think that it should be stressed, however, that if you just look at the data bank of sequences and then compare them with 3-D structures, what we are <u>not</u> going to find are the <u>surprises</u>. To mention just a few such <u>surprise</u> structures that have recently been determined: 1) the troponin C by Osnat Herzberg which is an incredibly unusual structure, that no one was able to predict. It has a dumbbell at the top, a long α-helix, really long, and then another dumbbell at the bottom and 2) the Z-DNA stucture by Rich and Wang, which is so different than anyone imagined, that even Alex Rich would have had difficulty trying to publish it based solely on theoretical arguments. In fact when Pohl and Jovin published their high salt CD spectra of DNA, several years ago, and noticed the drastic change versus the CD spectra at low salt, the referees would not allow them to speculate as to what they really thought was going on, in the now well understood B --> Z transition. Thus, what I am getting at, is if we want to make a big jump into something new, that this is just not possible with the current generation of model building techniques, and I do not see any easy way around it.

Rein:
I just wanted to say that the concern which was expressed about the danger with some of the user friendly programs when they get in unexperienced hands, is not justified, these programs are not so much user friendly, just to give some reference to it. My experience with new graduate students with either a physics or biophysics background is that at least 2 - 3 months work is required before they start to be able to do anything with Charm or Gromos or Mogly, so we are still quite a long way from programs that are so user friendly that everyone can use them.

Greer:
Let me just respond to that. What I really was referring to is that a number of commercial companies are now working very hard on modifying those programs that we have just named with user friendly interfaces. They are the same programs underneath, but now they have these interfaces on top to make them user friendly. So all the problems that your students are having, learning about what is going on in the guts of them for two or three months or more, under your careful guidance, are just being papered over very carefully, and that is what worries me. I am not worried about the students in your lab.

Youkharibache:
I do not want to discuss that subject too much. But, the people who are developing the software are scientists, and the market for companies to use this software exists, and it has to be addressed by scientists. So Polygen is developing this software with scientists and it is what it is today and it is going to get better and more and more user friendly and verified, checked by scientists, that is all I want to say.

Greer:
I am certainly not advocating that it should not become user friendly, let me stress that right away, because I think that the professionals need the programs to be user friendly, and the friendlier it is, the easier it is for us to use it. I agree with that entirely. It is just that when that happens, it then opens the way for others to think that they can use the programs and that is the problem. Certainly user friendly programs are entirely desirable, as far as I am concerned.

Youkharibache:
In addition to the last point, what I want to say is that we also train the people, scientists train the people to use them exactly like in the lab that you would train your new students. So this will be solved.

Stern:
I think what Jonathan is trying to say is that it is not just a question of a program being user friendly. Anyone using these programs, knows that they are not universal and it is naive to think that you can take a program that is instated to solve your particular problem and therefore it will solve a particular problem that you have applied it to without finding some bug that has never come up before or having to develop a change in the potential function, or something. I mean, it very often requires interaction on the part of the scientist and someone with experience to get the program to work properly. It is not just a question of being able to activate it simply.

Scheraga:
I think that Peter has made a good point here, which I would like to expand on. If you just turn any of these programs loose on any starting conformation, all you are going to get is the nearest local minimum, and you've got to know what you are doing. Maybe we ought to switch now to another subject.

Ramachandran:
I just want to say a few words regarding receptor availability and also sound an optimistic note about experimental data coming in rather rapidly. It is now really possible to express these membrane receptors in mammalian cells at very high levels, as I discussed in my talk yesterday. The human insulin receptors which I didn't talk about, we have a cell line expressing a million sites per cell, which means I am able to isolate 1 to 2 milligrams in a few weeks. The other thing that I want to mention is, that among the three classes of receptors I described, the ion channels may be the most difficult ones. The growth factor receptors, on the other hand, are very hydrophilic except for the single transmembrane domain, and in fact there are efforts under

way already to express the extra cellular domain involved in binding the ligand. I am confident that crystal structures will become available in the not-too-distant future. The fact that CD_4, the T-cell surface glycoprotein is able to neutralize HIV in vitro, shows that the extra cellular domain that is expressed without the transmembrane anchor, retains all the properties in terms, at least, of binding to the ligand. I think that many people are beginning to do this and one of our objectives was to obtain significant quantities of recombinant receptors to do spectroscopic and other types of measurements in the presence and absence of the ligand.

Haas:
I think that modeling will be with us forever. It is not only until the day that we have crystal structure, because then the story just begins, because the crystal structure is just a corner in the story and and we know that when we bind a ligand to a receptor, the receptor does something, it changes, so it transmits information to the inside of the cell. The active site definitely doesn't have the same conformation when it is bound or not bound, so dynamics is important, we don't see it in the crystal structure. The model is also helpful in incorporating data obtained by all methods. If it's spectroscopic, mutagenesis, crystal structure which is the beginning point, but then the model helps us in integrating and incorporating all the data that was obtained by other methods.

Chava Brender:
As an outsider coming from the Monte-Carlo simulation of polylelectrolytes, we are looking after the changes of conformations of polylelectrolytes under some certain conditions of the surroundings, like pH or temperature or even changing the charge on the polylelectrolyte, I can feel that the modeling does have a very important aim, maybe of more value than we think, in the way that maybe it also influences the experimentalist where to search for some phenomena, and also other modelists can use the consequences of their friends from the theoretical part. We have here many models and many methods and I do feel that the results of those models, even if they are not correct can be used as assumptions or as a beginning or a starting point to other modeling, in the future, so maybe again as an outsider, this is my feeling.

Karfunkel:
I am concerned about one point when the audience speaks about modeling. I had the impression that the audience considers modeling as a pure application. For me modeling has very high importance in the invention of algorithms. It is often much more difficult to invent an algorithm than to use it. Everyone uses it for instance, the Connolly surfaces, but it is not trivial to invent such an algorithm or to produce it, or everyone uses distance geometry, but nobody was concerned, I have just heard from Crippen how difficult it was and how by chance he reached such a result, and I believe that at least an important part in the future development of modeling will be the inventions of new algorithms, and not pushing buttons.

Scheraga:
That is part of what I meant when I introduced this question of new ideas.

Tapia:
I would like to comment on the problem of interfacing the world of theory and the world of real experiment. I will present our laboratory as one example. We have strong groups in protein and virus x-ray crystallography. A group in theoretical chemistry was implanted. Our role is the study of enzyme catalysis mechanism, dynamical properties of biomacromolecules, macromolecular recognition mechanisms. The interaction with the experimentalist is made via the particular problems under study. We get the molecular biology know-how, while in return, they get our know-how interactively. In absence of theroeticians, the application of computer programs is usually done as if they were black boxes. This is not a good practice. New problems claim for more sophisticated theoretical tools. The creation of such tools is the business of professional theoretical chemists.
Theoretical chemists isolated from the molecular biology world usually produce poor applications. Molecular biologists trying to develop new computational tools or modifying those in use, commonly produce poor theory. To avoid this problem, I think it is worth trying to create groups of professional theoretical chemists inside molecular biology departments. The proba-bility of creating good science, not publication noise, will be certainly increased.

Kosower:
I would just like to make one brief comment. There are many levels of modeling that are coming out, but a lot of people have the idea that the model is like a photograph of a result after you have gotten it. However, one kind of model is built before you know the final answer and then you think about how to get further information on it. Quite a number of times people have said to me and I have heard it said to others, well your model is interesting, but we'd like it better if you proved it.

Marshall:
One area that is worth mentioning is the idea of transformations. One of the questions that we were given was, are bigger computers going to do it for us. I think that my response is, more is not necessarily better. I think that we can generate enough confusion already with the computers that we have. You know I am like everyone else, I would love to have a factor of ten or a factor of a thousand and I am sure that I am going to eventually get it. Better graphics can help you. The problem is that I can display too much already. What I need to be able to do is to figure out what is the important thing that I need to see. How do I transform my data into some meaningful representation. How do I deal with multi-dimensional data. How do we take things that are 10 or 15 parameters, i.e., torsional angle dependent things, and extract that so that I can understand what it means. I think that there are real opportunities for dealing with those issues, and that that is where the real challenge is. The other point is a problem which underlines particularly the last statement. We are on the verge of making a non-science out of something that should be a science. We have a great disparity in the available facilities to which we have access. How many of the latest and greatest results that you have seen published have been repeated by anybody. There are lots of impediments to doing that. I do not have a Cray of my own; I do not have this guy's program; it is going to take me six man months to implement this algorithm. There are all these impediments for us to be scientists. The idea, I thought when I grew up and was young and naive, was that if someone else could do the experiment then I could go into my lab and do it and get the same result. I think that we

have got to get really serious about making sure that that
is still true. Because if I cannot get the same results
with my parameters as someone else, then we ought to know
about that collectively as a community and figure out what
is wrong. There has been a letter around recently among
some of the worst offenders of this, among the
crystallographic community, saying we should share our
coordinates, i.e., we really should not be able to publish
unless we publish our coordinates. If you read the names
of the people that have signed the letter, they are all
there, who would not give you their coordinates. I happen
to be editor of a small journal called The Journal of
Computer-Aided Molecular Design, and we are very close to
saying that you cannot publish a model in our journal,
unless you provide machine-readable coordinates. The
reason for that is, because everybody who is interested in
that problem, not only has a machine that they could put
those coordinates in, and look at and really understand
what you were saying, but some of them have got data that
you do not know about that they could test your model.

Levitt:
I think that Garland in his second part of his statement
has answered the question raised in the first part: he
would really like to have more computer time. One of the
most important things to realize about simulations and
energy calculations is that they have a lot in common with
experiments in biology. It is remarkable how biologists
manage to get hard answers from complicated systems like
cells. They do this by changing very few variables and by
repeating experiments many times. We need many more hours
of computer time so that we can repeat calculations and do
controlled experiments. For example, one could repeat a
calculation with someone else's force fields to ensure
that results were similar. With more computer power
there is going to be more willingness to spend some of the
time improving algorithms, repeating calculations under
slightly different conditions and repeating runs to get
better statistics. This is going to be very important for
the whole field.

Honig:
I want to reiterate what was said and also make a comment on a comment that came from a gentleman in the audience. Part of the field of modeling is building models, on protein or receptors for example, but part is the development of new algorithms and new methodologies. It is very clear that in the last few years there has really been major progress. I remember five, ten years ago, every time a theoretician gave a talk, someone would ask: "What about the solvent"? and the theoreticians would answer: "We know that water is important but we don't know what to do with it". Now in many labs around the world, the solvent is being accounted for with one method or another and as we go on, we are also understanding principles much better than I think we did a few years ago. I think fast computers are very important for that. Solvation is clearly an important problem. I see it is also one of the topics we are supposed to discuss on this panel. You can only deal with large numbers of solvent molecules with a lot of computer time, so I think that we all need computer time, if used judiciously.

Scheraga:
I think that there is an optimistic note that Barry introduced and that I would like to reiterate so that hopefully we'll leave the meeting in that spirit. When this whole field got started, we began very simply and solved simple problems, problems that couldn't have been solved without this methodology, always checking with experiment. As we escalated to more complex problems, we always checked with experiments. This provided confidence to go on to the next step. Unfortunately, at the present time, in the next step we are running into the problem that we require more massive computer power but, in many laboratories, there is at least a sound rational approach to problems, always checking with experiments. So I think that there is credibility when these things are done properly.

Kohn:
I would like to refer to some of those remarks. You may be familiar with the so called Pygmalion effect. Getting enamored with your models. The caveat story which occurred about 20 years ago is the story of polywater. If you remember, there was experimental evidence about a

material that was called polywater, a polymeric type of water. 400 Papers were published on it before it turned out that it did not exist. Even Allen, a theoretical chemist built a very beautiful theory to show why this polywater should exist and what was its structure. So there you are, getting enamored with a model may be a draw back in that field.

Scheraga:
Although as far as that example is concerned, there were enough skeptics to raise the question about the validity of that calculation.

Rein:

I would like to come back for a minute to this question of more computer time. The emphasis of this meeting was not very much on quantum chemistry, but we have to realize that if we are interested in chemical mechanisms or enzymatic catalysis, that eventually we have to use quantum mechanical methods and in this case, let's say, even for a system like 20 atoms, to explore the energy hypersurface along the reaction coordinates, tens to hundred of hours of Cray time may be required. So it is quite obvious that if we want to follow covalent chemistry and catalysis which is also an essential part of the whole problem, we would need much more computational power. The second remark is concerned with Garland's concern of publishing x-ray data from modeling and diffraction data. I would like to remark that there is a smart little program, I guess written by one of the crystallographers, that from a stereopair can actually reproduce the coordinates. There is argument pro- and contra as to how ethical this procedure is, but it is a way to check coordinates.

Marshall:
There is a lot of people that don't publish stereo diagrams any more because of that too.

Greer:
Or they publish them and they look like brillo pads and it is impossible to trace the chain from them.

Scheraga:
I don't think that you are seriously suggesting that that is a substitute for requiring the publication of coordinates.

Honig:
I just want to make a brief comment about the comment about polywater. Some of you may not believe it, but I am actually aware of experimentalist who have made mistakes.

Topiol:
I just want to make a couple of comments. First of all in terms inspired by thinking about quantum chemistry, when we are looking at analogies between proteins aside from looking at just structural analogies, we should start using more elaborate methods to find tuned details out and carry mechanistic analogies from one system to another. The other thing again, going back to quantum chemistry, we are talking about how we test our models and how much we can rely on our models by comparing to people who have done calculations on the same system. Well the other thing that we can do is make models of our models, in other words, make smaller pieces of the system that we are studying and callibrate them with more accurate methods and I think that in itself will help a lot, about reliability and credability.

Scheraga:
Well, I don't know if we have provided satisfactory answers to all the questions that were raised here, but I think that it is clear that there is certainly a strong interest and ferment that is bringing together various groups to discuss these modeling problems. Many drug and other companies are getting involved in setting up groups that bring together modelers and molecular biologists and so on. There seems to be an assessment that these things are worthwhile, and I think that we can only hope that we can do these things properly and rigorously. With that issue, if no one else has any more comments, I will call this meeting to a close. Before we all leave, I think we ought to give the last words to our organizers.

Rein:
I guess, all of you would like to spend the last hour here next to the pool and not listen to further talks, so I will be very brief. I would like to thank our excellent speakers who certainly greatly contributed to the quality of this meeting. I would also like to thank the local Organizers and particularly Dr. Golombek for choosing this beautiful place for the meeting which made the atmosphere so pleasant here. Thank you.

Golombek:
I would like to thank the speakers from abroad, who gave excellent talks and who made such an effort to come to Israel. I would like to thank the speakers from Israel, who gave such very good and intriguing talks. I would like to thank all the participants who came to this conference and contributed to the discussions. I would like to thank Professor Rein and Professor Israel Hertman, who is not here this afternoon, for initiating the idea of this conference and for their efforts in its organization. I would like to see you all again in one of our next conferences. Thank you.

Index

Accessible surface area, 23, 33–34
3-Acetoxyquinuclidine, 192, 193, 195, 196
Acetylcholine
  phase transitions, 199
  see also specific acetylcholine receptors
Acetylcholine acylhydrolase, 347
Acetylcholine chloride, 195, 196, 198, 199, 200
Acetylcholine receptor, muscarinic, 327–337
  antimuscarinic design and synthesis, 465–470
  functional diversity, 331–337
  partial molal volumes of agents, 192–194
  structural diversity, 328–331
Acetylcholine receptor, nicotinic, 245–259, 289–280, 479, 482
  cholinergic binding site, 341–345
  muscle and nerve
    muscle, 252–253
    neuromedulloblastoma cell line, 252–253
    neurons, 253–259
    transmembrane orientation, 247–252
  structural and dynamic model, 279–291
    antagonists, non-competitive, 286–287
    bilayer portions, 282–284
    desensitization and resensitization, 289–290
    dynamics, 285–286
    exobilayer portions, 284–285
    single-group rotation theory, 280–282
    tests of model, 287–289
Acetylcholine receptor-ligand interactions, 39–52
  amino acid sequences, 49–50
  binding sites, 51–52
  crystal structures, 46, 47, 48
  geometric parameters, 49
  hydrogen bonding and ionic contact distances, 42, 43
  proton acceptor, 45–46
  proton donor, 46
  residues, 42, 43–44, 136–142, 185–196
Acetylcholine receptor protein, 478
Acetylcholinesterase
  antimuscarinic design and synthesis, 465–470

  sequence similarities of, 347–356
  structural and immunochemical properties, 305–314
Actinidin, 405–407
Adaptive importance sampling Monte Carlo, 13–14
Adenine, DN pairing, 138
Adenosine-5′-monophosphate, 211, 212
Adenylate kinase, modules, 35
Adenyl cyclase, acetylcholine receptor binding, 334
Adrenalin, hydrogen bonds in drug binding, 436–441
β-Adrenergic receptor
  and acetylcholine receptor, muscarinic, 329, 330
  hydrogen bonds in drug binding, 436–441
AF-14, 195, 196, 198
AFDX116, 332
Affinity alkylation, acetylcholine receptors, 284
Agonists
  acetylcholine receptor, nicotinic, 289
  cholinergic, action of, 197–200
AIDS virus, 115, 116, 117, 119, 121, 122–123, 124
Alcohol dehydrogenase, 35
Algorithms, 492
Alkyl groups
  acetylcholine receptors, nicotinic, 284
  cholinergic binding site, 341–345
  organophosphoryl conjugates of chymotrypsin, 293
Alpha-carbon coordinates, ferredoxin, 174–175, 181
Alpha helix, 8
  GABA receptor. *See* GABA receptor
  pardaxin, 234, 235, 236
  poly(L-alanine), 7, 11–12
  ribosomal CTF, 56–57
Alpha subunit
  cholinergic binding site, 341–345
  see also specific receptors

499

Amino acid sequences
  acetylcholine receptor, muscarinic, 327, 328, 329, 331
  acetylcholine receptor, nicotinic, 255–256, 280, 281, 282
    exobilayer, 284–285
    ion channel element, 283
  acetylcholinesterase
    monoclonal antibody studies, 307, 308–309, 310, 313
    vs. related proteins, 347–356
  complement components, 386
Amino acid side chains, nicotinic acetylcholine receptors, 281, 282
Aminoacyl-tRNA, ribosome binding, 137–138
Amphi, 234
Anaphylatoxins, 386–394
ANCHOR, 152–153
Anchorage domains, membrane, 125
Anfinsen experiment, 157–158
Angular rotation, ferredoxin, 178
Annealing, 14
Anomalous scattering, ferredoxin, 176, 178
Antagonists
  acetylcholine receptor
    muscarinic, 327, 332
    nicotinic, 286–287, 289
  receptor interactions, 483
Antibodies
  acetylcholine receptors, 246, 247, 248, 249, 250
  MPCP-603, 22
  see also Monoclonal antibodies
Anticholinesterases. See Organophosphates
Antigenic sites
  normal-mode dynamics, 87–93
  see also Monoclonal antibodies
Antimuscarinics, design and synthesis, 465–470
Anti-parallel β-sheets, 7
Anti-syn isomerization, RNA, 138
Aphanothece sacrum, 172
Apo retinol binding protein (RPB), 56, 59–60
APPEND, 153
Arecoline, 195, 196, 198
Arginine codons, 137
Aspartic acid
  acetylcholine binding, 51, 52
  acetylcholine receptor, muscarinic, 328, 330

acetylcholinesterase, 306, 313, 352, 353
  deletion model, 460
Aspartic proteinase
  crystallography, 445–453
  modules, 35
Aspergillus nidulans, 30, 32
Atomic charges, and base pairing, 133
ATP binding region, modules, 35
Atrial tissue, acetylcholine receptors. *See specific acetylcholine receptors*
Atropine, 194, 195, 196, 197, 332, 334, 465

Backbone torsion angles, 234
Bacteriophage f1 gIII, 125
Barbiturates, 204
Base pairing, 131–139
Basis functions, 375
Bcl-2 protoonocogene, 112, 113, 126
B-DNA, 488
Benactyzine, 195, 196, 198
3-Benziloxyquinuclidine (QNB), 192, 193, 195, 196
Benzomorphans, 411, 419, 420
Benztropine, 195, 196, 198
Beta-bend, 7
Beta sheet prediction, build-up procedure, 7
Bethanechol chloride, 199, 200
BHRF1, 112, 113, 126
Bidentate bond, metal binding, 98
Binary receptors, deletion model, 461
Binding energies
  cluster, 96–97, 105
  predictability of, 476
  quantitative structure-activity relationships, 268, 272–274
  self-consistent-field, 97
Binding site
  acetylcholine receptors, 247, 281, 282, 289–290
  cholinergic, 341–345
Biophase, differential solubilities in, 191
Bloch equations, 146
BNLF1 protein, 117, 118
Boat conformation, pardaxin, 237, 238
Bond angle, ferredoxin, 176
Bond distance
  antimuscarinic design, 467, 468, 469, 470
  ferredoxin, 176
  PROTEAN data, 151–152

Bonds, polymorphism, 210–212
Born polarization energy, 100, 101–102
Born radii, metals, 104
Boundary layer water, electrostatic interaction, 65–73
Bovine pancreatic trypsin inhibitor (BPTI), 6, 158–169
Brain
    acetylcholine receptors, 253–254, 255–256, 256; *see also specific acetylcholine receptors*
    acetylcholinesterase, 310
Bromoacetylcholine, 284
Brownian fluctuations
    high-amplitude, 158
    and fluorescence decay, 159–160
Build-up procedure, 5–7
Bungarotoxin
    acetylcholine receptor subunits, 246, 247, 249
    cholinergic binding site, 343, 344, 345
3-Bu-Q, 195, 196, 198
3-Butyroxyquinuclidine, 195, 196, 198
Butyrylcholinesterases, 310–311, 347–356

Calcium, binding sites, 98–104
Canonical base moieties, base pairing, 131–139
Carbachol, 199, 200, 332, 334, 335
Carbachol chloride, 199, 200
Carbonyl group, 43
Carboxyesterases
    acetylcholinesterase sequence comparisons, 349–356
    see also Acetylcholinesterase
Carboxyl residues, ferredoxin, 182
Carboxypeptidase A, 34
Carboxy terminal fragments. *See* C-termini
Cardiac muscle, acetylcholine receptors. *See specific acetylcholine receptors*
Catabolite gene activator protein, 462
Catalytic sites
    acetylcholinesterase, 310, 313, 356
    modules and, 34–35
Catecholamines
    acetylcholine receptor, muscarinic, 330, 332
    hydrogen bonds in drug binding, 436–441
Cation binding sites, 95
Cation channels
    acetylcholine receptors, 251, 258
    see also Ion channels

Cayley-Menger determinant, 14–15
cDNA, 477
    acetylcholine receptors, 255–256
    acetylcholinesterase, vs. related proteins, 347–356
    cholinergic binding site, 344
Cell fusion, proteins involved in, 119–123
Central nervous system, acetylcholine receptors. *See specific acetylcholine receptors*
Centripetal profile, 30, 31
Charge
    electrostatic interaction, 65–73
    ferredoxin, 182
    see also Electrostatic interactions
Charge-solvent term, 72
Charge triad, acetylcholinesterase, 306
CHARM program, 235, 489
Chinese hamster ovary (CHO) cells, 331–337
Chloramphenicol, 206–209
Chloroplast carboxy terminal fragment, 56–58
Chlorpromazine, 286
Cholinergic receptors
    acetylcholine binding, 39
    binding site, 341–345
    partial molal volumes or agonists and antagonists, 192–194
    see also specific acetylcholinesterase receptors
Cholinesterases
    organophosphoryl conjugates, 294
    see also Acetylcholinesterase
Chromomycin, 217, 225–226
Chymotrypsin, 293–303, 313, 479
Chymotrypsinogen, 459
Circular dichroism (CD) spectra
    organophosphoryl conjugates of chymotrypsin, 295–300, 302
    pardaxin, 238, 239
Circularly polarized luminescence spectra, 298, 299, 300, 302
Cluster binding energies, 96–97, 105
Cluster reaction energetics, 98–100
CNDO/2 net atomic charges, 133
Cobratoxin (CTX), 39–52
Coding region
    acetylcholine receptor, muscarinic, 327
    see also Exons
Codon-anticodon interactions, base pairing, 131–139

Coherent Instance Generator, PROTEAN, 153–154
Cohesive energy density, 191, 192
Colicin E1, 20–22
Colloidal gold electron microscopy, 248
Combinatorial algorithms, 78, 80, 82
Compact conformation, 8
Compact effective potentials, 96
Complement pathway, 385–395
  C3A crystal, 386–387
  C3A model structure, 387–390
  C3A N-terminus, 390–391
  C5A model vs. NMR, 392–394
Computational demands
  multiple-minima problem. *See* Multiple-minima problem
  PROTEAN, 150
Computational studies, ligand-receptor interactions, 455–463
  cAMP and cGMP regulatory and catalytic binding site, 462–463
  deletion model, 456–462
    common parent system for different ligands, 459
    different parent system for common ligands, 459–460
    division of parent system into multiple compounds, 460–462
Computer graphics systems, 362–364, 485–487
Configurational polymorphism, 204
Configuration interaction studies, 373–382
Conformational change
  multiple-minima problem. *See* Multiple-minima problem
  organophosphoryl conjugates of chymotrypsin, 293–294
  pardaxin, 237, 238
Conformational energy calculations, 70–71
Conformational polymorphism, 204
Connoly surfaces, 492
Conservation
  acetylcholine receptors, 248
  see also Sequence homologies
Constraints, PROTEAN, 151
Continuum treatment, solvent, 72
Coordinates, 496, 497
CORELS, ferredoxin, 174
Coupling

acetylcholine receptor, 331–337
lock-and-key system, 457, 458
monoclonal antibody probes, 317–325
vectorial, 480
CPM-morphine, 419
Crayola, 485
CRAY-XMP, 24
CRO repressor, 35
Crystallography
  aspartic proteinase, 445–453
  homology model building, 19–25
  protein-ligand hydrogen bonding, 445–453
Crystal structure
  complement component, 386–387
  and modeling systems, 474, 475, 476, 477
  quantitative structure-activity relationships, 267
C-termini
  pardaxin, 234, 235, 236
  ribosomal, 56–58
CYBER-205, 22
Cyclic AMP
  ligand-receptor interactions, 456, 462–463
  receptor-effector coupling and, 319
Cyclic GMP, 456, 462–463
Cysteines
  acetylcholine receptors, 251, 256, 287
    binding site, 39–40, 258
    subunits, 246
  acetylcholinesterase, 309–310, 313
  cholinergic binding site, 341, 342
  ferredoxin, 178, 179
  HIV envelope, 121, 123
Cytochrome C, 167
Cytochrome P450cam, 442–443
Cytoplasmic surface
  acetylcholine receptors, 257
  see also Transmembrane region
Cytosine, spectral features, 376–382

DADL, 417, 418
DAGO, 415, 417, 418
Data base methods, 76
Dealkylation, organophosphoryl conjugates of chymotrypsin, 293
7-Deazanebularin, base pairing, 131–139
Dechromose-A chromomycin, 217
Decoys, molecular, 345

Index / 503

Deletion model, ligand-receptor interactions, 456–462
Delphi program, 69, 234
Density functions, 146
Deprotonation enthalpy, 451
Desensitization, nicotinic acetylcholine receptor, 289–290
Dielectric constant
  conformational energy calculation, 70–71
  electrostatic interactions, 65–73
Differential solubility, biophase, 191
Dihedral angles, electrostatic-optimization procedure, 7
Dihydrofolate reductase marker, 184, 331–332
Diisofluorophosphate (DFP), 306, 310, 312, 347–356
Dimaprit, histamine interactions, 361–369
Dimensionality, increase of, 14–15
Dimer, CTF, ribosomal, 57–58
Dimethylacetylene, 211
3-DiPh-Ac-Q, 195, 196, 198
2,2-Diphenylpropionate, 466
DISCOVER, 387
Distance geometry analysis
  antimuscarinic design and synthesis, 466–470
  intramolecular distance distributions in BPTI, 160–163, 166
  PROTEAN data, 151–152
Distance restraints, 13
Disulfide bridges
  acetylcholine receptors, 251
    acetylcholine binding, 39–40
    nicotinic, 287
    subunits, 246
  acetylcholinesterase, 309, 313
  Anfinsen experiment, 157–158
  cholinergic binding site, 341, 342
  intramolecular, 7
Disulfide intermediates, BPTI, 167
DNA
  B and Z structures, 488
  base pairing, 131–139
  chromomycin binding, 225
  see also cDNA
DPDPE, 412
DPLPE, 412
Drosophila acetylcholinesterase, 307, 349–356
Drug molecules

acetylcholine receptor, nicotinic
  agonists and antagonists, 289
  antagonists, 286–287
  hydrogen bonds in, 433–443
  modeling systems, merits of, 475, 477
  partial molal volume and, 189–200
  polymorphism in, 203–214
  Voroni receptor site models, 267–276
DSLET, 412, 415, 417, 418
DTLET, 412
Duplex-drug contacts, chromomycin, 224–225

Eel, acetylcholinesterase, 310
Effector coupling. *See* Coupling; Receptor-ligand interaction
  acetylcholine receptor, 331–337
  monoclonal antibody probes, 317–325
EKC, 417, 418
Electrical potential, 70
Electron-density map, ferredoxin, 174, 175
Electronegative substituents, and base pairing, 132, 133
Electron transfer, ferredoxin, 179
Electrostatically driven Monte Carlo, 10–12
Electrostatic interactions, 65–73, 481
  acetylcholine receptors, 258
  applications, 70–72
  and base pairing, 132, 133
  ferredoxin, 182
  histamine and dimaprit interactions, 362, 364, 366, 367, 368
  hydrogen bonding in protein ligand interactions, 449–450
  5-hydroxytryptamine receptor, 402, 406, 407, 408
  microscopic and macroscopic models, 66–69
  morphine analogs, 416
  mutations, 71
  numerical and computational methods, 69
  optimization of, 7–9
  potential around proteins, 71
  solvation contributions to protein stability and ion binding, 71–72
  Voroni models, 273
Electrostatic-optimization procedure, energy minimization, 7–9
ELISA
  acetylcholine receptors, 249
  acetylcholinesterase, 308

Empirical potantial function, 133, 134
Endothia pepsin, 445-453
Energetics, 485-487, 488
  hydrogen bonding in protein ligand interactions, 449
  5-hydroxytryptamine receptor, 402-403, 406
  total electrostatic energy, 71-72
  Voroni models, 272-274, 275, 276
Energy conformational studies, opiate analogs, 416, 427
Energy minimization, 8, 55
  conformational energy calculation, 70-71
  dimensionality increase, 14-15
Energy minimum, pattern recognition techniques, 15-16
Energy transfer, non-radiative, 158-160
Enkephalin
  DAGO, 417
  high-dimensional global minimum, 15
  random starting conformations, 9
Enthalpy, 23
  of deprotonation, 451
  in drug-receptor interaction, 194-197
  of hydration, metal cations, 100-101
  of pairing, base pairs, 132
Entropy, 23
  cholinergic-receptor-drug binding, 195-197
  conformational, 4
  Voroni models, 275-276
Envelope proteins, viral
  HIV, 119, 121, 122-123, 124
  hydrophobic domains, 126
Epistemology of modeling, 474-497
Epitopes
  acetylcholine receptors, 249, 250
  lysozyme, 91-92
  myoglobin, 92
  see also Monoclonal antibodies
Epstein-Barr virus proteins, 112, 113, 117, 118, 125, 126
Equations of motion, 89
Equilibrium properties, polymers, 138-139
Erythrocytes
  acetylcholine, 308
  acetylcholine receptor, muscarinic, 328
  acetylcholinesterase, 310
Escherichia coli
  carboxy terminal fragment ribosome, 56-58

RNA polymerase, 136-137
Ester oxygen, 43
Ethylketocyclazocine, 411
Etorphine, 419, 420
Evolution
  acetylcholine receptors
    nicotinic, 289
  receptor subunits, 246
  Halobacterium marismortui, 171-184
  tertiary structure, conservation of, 30, 32
  see also Sequence homologies
Excitation energy transfer, non-radiative, 158-160
Exobilayer regions, nicotinc acetylcholine receptor, 284-285
Exons
  acetylcholine receptor, muscarinic, 327
  hemoglobins, 28-29
Experimental data
  and model building, 474-480, 485, 492
  PROTEAN data, 151-152
  Voroni models, 273, 275, 276
Extension profile, 30, 31

FASTRUN(9), 24
Fc receptor, receptor-effector coupling and, 318-325
Ferredoxin, 171-184
  adaptation to high salt, 181-184
  conformation, 177-179
  vs. *Spirulina platensis* ferredoxin, 179-181
  structure determination, 173-177
  surface charge distribution, 181
Finite difference Poisson-Boltzmann equation, 68, 69, 70, 71
First coordination shell cluster binding energy, 101-102
Flavodoxin, 82, 83
Florey-Huggins correction, 191
Fluorescence
  ferredoxin, 179
  organophosphoryl conjugates of chymotrypsin, 295-296, 298
Fluorescence spectroscopy, time resolved, 157-169
  BPTI
    intramolecular distance distribution, 160-163
    reduced, folding conditions, 165-167
    reduced unfolded, 163-165

current experiments, 167–169
nonradiative energy transfer, 158–160
Force-fit computation methods, 133
Formamide ligands, 96, 98–104
Formate anion, 96
Formate ligands, metal binding, 98–104
Four helix bundle topologies, 79
Free energy
  binding, quantitative structure-activity relationships, 268, 272–274
  folding, 157–158
  partial molal volumes, 192, 193
Free-Wilson model, 276
FRODO, 55, 174
Fruitfly, 123; see also Drosophila
Function, modules and, 34–35

GABA receptor, 257, 258
  glycine receptors, 239–242
  pardaxin, 234–239
Gag protein, HIV, 115, 117
Gene structure
  acetylcholine receptor
    muscarinic, 327
    subunits, 246
  see also Exons; Introns; Sequence homologies
Genetic engineering
  acetylcholine receptor, nicotinic, 287
  see also cDNA
Geometric representations, PROTEAN, 152–154
Global minimum, 8, 75
  dimensionality increase, 14–15
  enkephalin, 9
  pattern recognition techniques, 15–16
  poly(L-alanine) chain, 7
Globular proteins, 4
Globule state
  modules, 29
  molten, 167, 168
Glutamate dehydrogenase, 184
L-Glutamic acid, 211, 212
Glycine, 457
Glycine receptors, 239–242, 258
Gramicidin S, 3, 4
Growth hormone, human (HGH), 78, 80–82
Graphics systems, 55, 485–487
GRID method, 433–443
Gromos, 489
Growth factor receptors, 490

GTP gamma-S, 332

Halobacterium marismortui, ferredoxin, 171–184
Hartree-Fock self-consistent field (SCF), 373–382
Hartree-Fock-Slater (HFS) procedure, 363
Heart, acetylcholine receptors. *See specific acetylcholine receptors*
Heats of hydration, metal cations, 100–101
Helical structure
  alpha. *See* Alpha helix
  four helix bundle topologies, 79
Hemoglobin, 28, 32, 33, 34, 35
Herpes simplex virus-1, 119, 122, 125, 126
Heuristic control, PROTEAN, 151
Highly constrained proteins, 19–25
Histamine H-2 receptor
  deletion model, 456, 459–460
  histamine and dimaprit interactions, 361–369
    charge, 366, 367, 368
    geometry, 364
    N- and S-fits, 364–369
    structures, 362–363
Histidine
  acetylcholinesterase, 306, 313, 352, 353
  deletion model, 460
  hydrogen bonds in drug binding, 441–442
  subtilisin, 70
Histrionicotoxin, 286
Holo-retinol binding protein, 58–61
Homologous sequences. *See* Sequence homologies
Homology model building, 19–25
HONDO code, 96
Host-guest random copolymers, 7
Human growth hormone (HGH), PLANS, 78, 80–82
Human immunodeficiency virus (HIV)
  envelope protein, 119, 121, 122–123, 124
  gag protein, 115–117
  hydrophobic domains, 126
Humans
  acetylcholine receptors, nicotinic, 280
  acetylcholinesterase, 307, 308, 347–356
Hybrids, nicotinic acetylcholine receptor, 289
Hydrates, 204
Hydration, globular protein, 4
Hydration energy, 72
Hydration shell, metal cations, 100–101
Hydrogen bonds, 42
  AChR-ligand interactions, 51

cobratoxin-AChR complexes, 43
drug binding, GRID method, 433–443
    catecholamines, 436–441
    proteins, 441–443
formamide, 104
morphine analogs, 416
nucleic acids
    base pairing, 131–139
        melting temperature effects, 138–139
        RNA polymerase mediated pairing, 136–137
        wobble type base pairs, 134–136
    protein-ligand interactions, 445–453
Hydrogen exchange, PROTEAN data, 151–152
Hydrolysis, polymorphism and, 208–209
Hydrophobicity, 110
    acetylcholine receptor
        muscarinic, 328, 329
        nicotinic, 282
    acetylcholinesterase, sequence comparisons, 352, 353
    complement components, 388
    GABA receptor. See GABA receptor
Voroni models, 273
5-Hydroxytryptamine receptor
    computation, 401–402
    deletion model, 456
    proton transfer, 402–407
        individual helices and, 405–407
        total protein effect, 402–405
Hyoscine, 192, 194, 195, 196, 197
Hyoscine hydrobromide, 199, 200
Hypervariable loops, 22

Imipramine, 195, 196, 198
Immunoglobulin, 27; see also Monoclonal antibodies
Immunoglobulin E, receptor-effector coupling studies, 318–325
Information theory entropy, Voroni models, 275–276
Input-output characteristics, PROTEAN, 151–152
Insulin receptor, 490
Interleukin-2, 78–80
Intermediate local structures, 158
Internal salt bridge, ferredoxin, 183
Interproton distances, chromomycin, 219–221
Intramolecular distance distributions in BPTI, 160–163, 166

Introns, 27
    large proteins, 29–32
    location, 32–34
    in small domains or subunits, 28–29
In vitro mutagenesis, acetylcholine receptors, 248
Ion building, solvation contributions, 71–72
Ion channels, 479, 480, 482, 490
    acetylcholine receptors, 251, 258
        antagonists and, 286–287
        nicotinic, 282, 283
    colicin E1, 20–22
    inhibitors, 232
Ionic conditions, *Halobacterium marismortui*, 171–184
Ionic interactions
    cobratoxin-AChR complexes, 42, 43
    and dielectric function, 71
Ion-pair interaction, binding site, 105
Ion solvation, electrostatic interaction, 65–73
Iron-sulfur proteins, ferredoxin, 171–184

Junctions, location of, 32

Kinetics, organophosphoryl conjugates of chymotrypsin, 293–294
Kowsower model, 282

α-Lactalbumin, 167
Langevin dipoles, 67
Latent membrane protein (LMP), Epstein-Barr virus (EBV), 117, 118
Least squares fit, Voroni models, 274
Lectins, acetylcholine receptors, 247, 248
Left-handed α-helix, poly(L-alanine), 12
Leghemoglobins, 28
Lennard-Jones type term, 133
Ligand-receptor interactions. See Receptor-ligand interactions
Lipid-protein interactions, 479
Liquid-crystal phase transition, cholinergic agent action, 197–200
Local maxima, 8, 30, 31
Lock-and-key system, 267, 457, 458
Long range interaction, 77
δ-Lysin, 231–232
Lysozyme
    accessibility, 33
    modules, 32, 34–35

Index / 507

normal-mode dynamics, 90–93

Magainin, 231, 232
Magnesium binding sites, 98–104
Magnetization transfer in macromolecules, 146
Main immunogenic region (MIR), acetylcholine receptor subunits, 246
Malate dehydrogenase, 184
4(N-Maleimido) benzyltrimethyl ammonium iodide. *See* MBTA
Masoncogene
 cellular protein, 117, 119, 120
 hydrophobic domains, 126
Mast cells, 318
MBTA
 acetylcholine receptors, 253, 254
 cholinergic binding site, 341–345, 343–344
McN-A-343, 195, 196, 198, 199, 200
Melittin, 111–113, 126, 231–232
Membrane structures, 479–480, 481, 479
 acetylcholine receptors
  orientation of, 247–252
 GABA receptor, 231–242
 ion channels, 20–22
 modeling, validity of, 476
 organization and function, 109–126
  inner membrane, 115–117
  multiple hydrophobic domains, 117–123
  outer membrane, 115
  prepromellitin and mellitin, 111–113
  proteins without signal peptide, 113–115
 see also Hydrophobicity
Metabolism, polymorphism and, 208–209
Metal binding sites, 95–106
 ab initio calculation, 96–98
 model calculation, 98–104
Metazocine, 419
Methacholine chloride, 195, 196, 198, 199, 200
Methotrexate, 332
Metropolis Monte Carlo, 10
Minima
 multiple. *See* Multiple-minima problem
 pattern recognition techniques, 15–16
Minimum energy conformations, 75, 267
MNDO, 363, 364
 hydrogen bonding in protein ligand interactions, 447
 morphine analogs, 416
Modeling systems, epistemology of, 474–497

Modules, 27–35
 functions, 34–35
 large proteins, 29–32
 location of, 32–34
 secondary structures, 34–35
 small domains, 28–29
Mogli program, 235, 489
Molal volume, partial, 192, 193
Molar attraction constant, partial molal volumes, 192, 193
Molecular biology, 486, 487
Molecular decoys, 345
Molecular dynamics, 55–62, 75, 481, 483, 484
 conformational energy calculation, 70–71
 retinol binding protein, 58–61
 ribosomal protein CTF, 56–58
Molecular geometry, polymorphism, 210–212
Molecular mechanics, 483, 484
Molecular orbital wavefunctions, 373–382
MOLMEC, 415
Molten globule state, 167, 168
Monoclonal antibodies
 acetylcholine receptors, 246, 247, 253–254
 acetylcholinesterase, 306, 308–309, 310–313, 312–313
 receptor-effector coupling processes, 317–325
Monte Carlo methods, adaptive importance sampling, 13–14
Monte Carlo-plus-minimization, 10
Morphine, 411, 419, 420
Morphine analogs, 415–416
Mulliken population analysis, 402
Multidimensional data, 493
Multiple-minima problem, 4, 75
 adaptive importance sampling Monte Carlo, 13–14
 build-up procedure, 5–7
 electrostatically driven Monte Carlo, 10–12
 inclusion of distance restraints, 13
 increase of dimensionality, 14–15
 Monte Carlo-plus-minimization, 10
 optimization of electrostatics, 7–9
 pattern recognition-plus-minimization, 15–16
Muscarinic cholinergic receptor
 partial molal volumes of agents, 192–194
 see also Acetylcholine receptor, muscarinic
Muscle, acetylcholine receptors, 252–253, 253–254

subunit stoichiometry, 258
Xenopus, 246
see also specific acetylcholine receptors
Mutations
  acetylcholine receptor, nicotinic, 287
  cholinergic binding site, 344
  and electrical charge, 70
Myasthenia gravis, 252
Myoglobin, PLANS, 78
Myristic acid, 123

Naloxone, 417, 418
Naltrexone, 419
Nerve
  acetylcholine receptors, 253–259
    nicotinic, 280
    subunit stoichiometry, 258
  see also specific acetylcholine receptors
Net atomic charges, 133
Neuromedulloblastoma cell line, 252–253
Neurotoxin binding site, 41
Neurotoxins, 483
  acetylcholine receptor subunits, 246, 247, 249
  cholinergic binding site, 343, 344, 345
Neurotransmitters, deletion model, 456
Newton-Raphson method, 235
N-glycosylation, acetylcholine receptors, 253, 256, 328, 330
Nicotine, acetylcholine receptor affinity, 254
Nicotinic acetylcholine receptor. See Acetylcholine receptor, nicotinic
NOESY spectrum, chromomycin, 223
Nonradiative excitation energy transfer, 158–160
Normal-mode dynamics, prediction of antigenicity, 87–93
Nortriptyline, 195, 196, 198
N-termini
  acetylcholine receptors, 247–248, 255–256
  complement components, 389, 390–391, 392
  pardaxin, 234, 235, 236
Nuclear magnetic resonance (NMR) spectra
  C5A model vs., 392–394
  cholinergic binding site, 344
  ferredoxin, 183
  geometry, 151–154
  input-output characteristics, 151–152
  PROTEAN expert system, 149–151
  theoretical considerations, 145–149
  two-dimensional, chromomycin conformation, 217–226
Nuclear Overhauser Enhancement, 146–147, 148
  chromomycin, 219–221
  PROTEAN data, 151–152
Nucleic acids, base pairing, 131–139
Nucleotide sequences
  acetylcholinesterase vs. related proteins, 347–356
  see also Sequence homologies

Oncogenes
  bcl-2, 112, 113
  hydrophobic domains, 126
  masoncogene, 117, 119, 120
Opiate receptor heterogeneity, 411–428
  analog binding, 417–428
    affinity profiles and relative capacities, 419–420
    in vivo responses, 421–422
    mu and kappa receptors, 421–423
    N-substituents, 420–422
    P compounds, 425–428
    U compounds, 415, 416, 417, 422–425, 426
  experimental binding studies, 414–415
  theoretical studies, 415–417
  types of receptors, 412
OPSYN oligodeoxynucleotide probes, 348
Optimization of electrostatics, 7–9
Organophosphates
  acetylcholinesterase sequence comparisons, 347–356
  antimuscarinics, design of, 465–470
  chymotrypsin conjugates, 293–303
    aged vs. non-aged, 296–302
    kinetics, 293–294
    spectroscopic findings, 299–303
    spectroscopic methods, 295–296
Oripavine, 419, 420
Ovomucoid, 29, 32
Oxidative phosphorylation, 479–480
Oxotremorine, 195, 196, 198, 332, 334
Oxymorphone, 419, 420
Oxymyoglobin, normal-mode dynamics, 90–93

Packing angles, 80
Packing quality, 82, 83
Pairing interactions, base pairing, 131–139

Pancreatic secretory trypsin inhibitor, 29
Parallel binding, 289–290
Pardaxin, 231, 233, 234–239, 232
Partial arrangements, PROTEAN, 150
Partial molal volume, pharmacodynamic activity, 189–200
   agonist liquid-crystal state transition, 197–200
   complex molecules, 192–194
   differential solubilities in biophase, 191
   enthalpy-entropy relationship in drug-receptor interactions, 194–197, 198
   theory, 190–191
Pattern Language for Amino and Nucleic Acid Sequences (PLANS), 77
Pattern recognition plus minimization, 15–16
PBEP-Cht, 299, 300, 302
PBEPF (1-pyrenebutyl ethyl phosphorochloridate), 295, 296, 297
PBP-chymotrypsin (pyrenebutyl hydroxyphosphoryl-Cht), 297, 298, 299, 300, 301, 302
PBPDC (1-pyrenebutyl diphosphorochloridate), 295, 296, 297
Penalty function, 14–15
Penicillopepsin, 445–453
Pepsin, aspartic proteinase crystallography, 445–453
Pepstatin, 452
Peripheral nervous system, acetylcholine receptors. *See specific acetylcholine receptors*
Perpendicular binding, acetylcholine receptor, 289–290
PGLa, 231
Pharmacodynamics, partial molal volume and, 189–200
Phase determination, ferredoxin, 176–177
Phases
   cholinergic agent action, 197–200
   ferredoxin, 174, 176–177
Phencyclidine, 286
Phosphodiesterases, catalytic domains, 462–463
Phosphoinositides, 319, 334
Phosphoinositol linkage, 123
Phosphorylation, oxidative, 479–480
Physical properties, Voroni models, 273
Pilocarpine HCl, 195, 196, 198, 199, 200
Pirenzepine, 327, 334
Plants, hemoglobins, 28
Poisson-Boltzmann (PB) equation, 65, 66–67

Poisson equation, 65
Polarization
   metals, 104
   molecular orbital states, 376
   organophosphoryl conjugates of chymotrypsin, 295–296
Polygen, 489
Poly(Gly-Pro-Pro), 3
Poly-(L-alanine), 7, 11–12
Polylysine, 137
Polymers
   equilibrium properties, 138–139
   nucleic acid binding, 137–138
Polymorphic solvates, 204
Polymorphism, drug, 203–214
   definitions, 204
   examples, 205–209
   methods, 213–214
   variable geometric features, 209–212
Polynucleotides, nucleic acid binding, 137–138
Polyphosphoinositides, receptor-effector coupling and, 319
Poly(tripeptides), 3
Potassium channels, 20–22; *see also* Ion channels
Potential energy
   conformational energy calculation, 70–71
   globular protein, 4
   5-hydroxytryptamine receptor, 402–403, 406
Potential maps, 70
Prealbumin, 59, 60, 61
Prediction of antigenicity, 87–93
Prediction of structure, 75–84
Prepromelittin, 111–113, 126
Problem abstraction, PROTEAN, 151
Problem decomposition, PROTEAN, 150
3-Propionoxyquinuclidine, 195, 196, 198
PROTEAN expert system
   geometry, 152–154
   input-output considerations, 151–152
   structure, 149–151
PROTEIN (program), 174
Protein binding energies, 101–102
Protein kinases, 123, 462
Proteins
   hydrogen bonds
      drug binding, 441–443
      ligand interations, 445–453
   normal-mode analysis, 87–93

stability, solvation contributions, 71
Proteolysis, cholinergic binding site, 343
Proton resonances, chromomycin, 219–221
Proton transfer
   in CTX-AChR complexes, 45–46
   histamine and dimaprit interactions, 362, 368
   5-hydroxytryptamine receptor, 400, 401–407
Proton transfer model. 400, 401–407
Protooncogenes
   bcl-2, 112, 113
   cellular protein, 117, 119, 120
3-Pr-Q, 195, 196, 198
PRUNE function, PROTEAN, 153
Pseudopolymorphism, 204
Pseudo-Watson-Cirick base pairs, 134–136
Purines
   base pairing, 131–139
   configuration interaction studies, 375
Pygmalion effect, 495
3-Pyridine aldoxime methiodide (3-PAM), 301
Pyrimidines
   base pairing, 131–139
   configuration interaction studies, 375

QNB, 197
Quantitative structure-activity relationship (QSAR)
   antimuscarinic design, 465–470
   approaches, 267–268
Quantum chemistry, 496, 497
   histamine and dimaprit, 363, 364
   proton transfer, 402
   spectral comparisons, methodology, 373–382
Quaternary oximes, 293
Quiniclidinyl benzilate, 332

Raft conformations, 232, 233, 237, 238
Real space rotation search, ferredoxin, 178
Receptor-effector coupling
   acetylcholine receptor, 331–337
   monoclonal antibody probes, 317–325
Receptor-ligand interactions
   acetylcholine receptor, 39–52
   computational studies, 455–463
   crystal structure and, 477
Reciprocal space least-squares techniques, 174
Recombinant DNA, 477
   cholinergic binding site, 344
   see also cDNA
Redfield density matrix, 146

Reducing agents, cholinergic binding site, 341
Renin, 445
Renormalization group, 5
Resensitization, nicotinic acetylcholine receptor, 289–290
Resonance assignments, chromomycin, 222–224
Retinol binding protein, molecular dynamics, 58–61
Rhizopus pepsin, 445–453
Ribavarin, 211
Ribosomes
   aminoacyl-tRNA binding, 137–138
   protein, molecular dynamics, 56–58
Right-handed α-helix, poly(L-alanine), 11
RNA, base pairing, 131–139
RNA polymerase mediated pairing, 136–137
ROCKS, 174
Rotating anode diffractometer, 174

Salt, *Halobacterium marismortui*, 171–184
Salt bridge
   ferredoxin, 183
   pardaxin, 237
Scattering, ferredoxin, 176
Scrapie agent, hydrophobic domains, 126
Scrapie-like protein precursor, 114, 115, 116
Secondary structures
   modules, 35
   prediction of, 76–77
   see also specific receptors
Second genetic code, 157
Self-consistent field (SCF), 7–9, 96, 373–382
Sequence homologies, 76
   acetylcholine receptor, nicotinic, 287–289
      exobilayer, 285
      monoclonal antibody studies, 247, 248
   acetylcholinesterase, 307, 308–309, 349–356
Serine
   acetylcholinesterase, 306, 313, 352, 353
   deletion model, 460
Serine protease inhibitors, modules, 29
Serine proteases
   acetylcholinesterase, sequence comparisons, 349–356
   modules, 35
Serotonin, 324, 325
Side chains, nicotinic acetylcholine receptors, 281, 282
Siftware, 489, 490

Single-group rotation theory, 280–282
Single isomorphous derivative, ferredoxin, 178
Single isomorphous replacement, ferredoxin, 174, 176
Single site mutations, 70
Site-specific mutations
 cholinergic binding site, 344
 and electrical charge, 70
SKF 10047, 411, 419
Smooth muscle acetylcholine receptors. *See specific acetylcholine receptors*
Snake toxins, 483; *see also* Neurotoxins
Sodium channels, 20–22, 482
Solubility, polymorphism and, 208–209
Solvates, 204
Solvation
 computational considerations, 495
 electrostatic interaction, 65–73
Solvent polarization terms, 72
Soy bean, 28
Specificity of drug action, 191
Spectral density functions, 146
Spectroscopy
 organophosphoryl conjugates of chymotrypsin, 295–300
 quantum mechanical SCF/CI studies, 373–382
Spin diffusion, 146
Spin-spin coupling constants, PROTEAN data, 151–152
Spirulina platensis, 173, 179–181
STAR-100, 22, 23, 24
Statine, 452–453
Steroids, 204
STO-3G basis set net atomic charges, 133
Stoichiometry
 acetylcholine receptor subunits, 246
 colicin E1, 20
Structure, prediction of, 75–84
Suberoyl bischolines, 279–280
Substrate-assisted catalysis, 462
Subtilisin, 67, 70, 102–104, 313
Sulfapyridine, 205–206
Sulfhydryl groups, acetylcholine binding, 39–40
Sulfonamides, 204
Superoxide dismutase, 70
Surface area, accessible, 23, 33–34
Surface charge, ferredoxin, 182

Tertiary structure, conservation of, 30, 32

Tetracyanoplatinate, 183
Tetracycline resistance protein, 119, 120, 126
Tetrahedral coordination, ferredoxin, 178
Tetramer, pardaxin, 238, 239
Theory, prediction of structure, 75–84
Thermodynamics
 cholinergic-receptor-drug binding, 195–197
 see also Enthalpy
Thioureum, 363
Three-dimensional global energy minimum, 15
Threonine, deletion model, 460
Thy-1 membrane glycoprotein precursor, 114, 115, 116, 126
Thy-1 protein, 123
Thyroxine binding prealbumin, 59, 60, 61
Time resolved fluorescence spectroscopy. *See* Fluorescence spectroscopy, time resolved
Torpedo californica
 acetylcholine receptors, 41, 245–259, 280
 acetylcholinesterase, 306, 308, 309, 310, 311, 349–356
Torpedo marmorata, 280
Torsional space, 89
Torsion angles
 pardaxin, 234
 polymorphism, 210–212
Total electrostatic energy, 71
Toxins. *See* Neurotoxins
Transfer efficiencies, BPTI, 166
Transformations, modeling, 493
Transmembrane potential, colicin E1, 20–22
Transmembrane region
 acetylcholine receptors, 247–252
  muscarinic, 328, 329, 330–331
  nicotinic, 282–284
 theoretical problems, 479–480, 481, 490
Trimethylammonium group, 43, 45
Trinucleoside diphosphates, nucleic acid binding, 137–138
Triose phosphate isomerase, modules, 30–32, 35
tRNA binding, ribosomes, 137–138
Troponin C, 488
Trypsin, 479
Tryptophan
 ferredoxin, 179
 retinol binding protein, 60
Tubular crystalline arrays, acetylcholine receptors, 247

Turn prediction, 77
Tyrosine kinase, HIV envelope, 121
Tyrosines, HIV gp160 envelope protein, 123, 124

U69,593, 415, 418, 423

Valinomycin, 21
Van der Waals forces, PROTEAN data, 151–152
VAX, 23
Vectorial coupling, 480
Viral proteins
    Epstein-Barr virus, 112, 113, 117, 118, 125, 126
    herpes simplex virus-1, 119, 121, 122, 125, 126
    human immunodeficiency virus-1 (HIV-1), 126
        envelope protein, 119, 121, 122–123, 124
        gag protein, 115–117
    hydrophobic domains, 126
    membrane-bound, 110, 112, 113, 115
Virazole, 211
Vitamin A, retinol binding protein, 58–61
Vitamin B12, 205
Voroni receptor site models, 267–276
    binding constants, experimental, 267
    free-energy of binding, 272–274
    least squares fitting of data, 274
    linearized representation, 270–272

Water
    cluster binding energies, 97
    electrostatic interaction, 65–73
    hydration
        energy of, 72
        globular protein, 4
        metal cations, 100–101
    immobilization of, 23
    subtilisin, 103
Water-ice transition, cholinergic agonists, 197–200
Watson-Crick pair, base pairing, 131–139
Wheat germ agglutinin, 247
Wobble type base pairs, 134–136

Xenopus, acetylcholine receptors, 246, 251
XIRIS, 470
XPF, 232
X-ray diffraction
    gramicidin S, 3, 4
    structures of parent molecules, 55
    see also Crystallography

YOKE, PROTEAN, 153, 154

Z-DNA, 488
Zymogen activation, 459